Challenges and New Opportunities for Tourism in Inland Territories:

Ecocultural Resources and Sustainable Initiatives

Gonçalo Poeta Fernandes
CITUR, Polytechnic Institute of Guarda, Portugal

IGI Global
PUBLISHER of TIMELY KNOWLEDGE

A volume in the Advances in Hospitality, Tourism, and the Services Industry (AHTSI) Book Series

Published in the United States of America by
IGI Global
Business Science Reference (an imprint of IGI Global)
701 E. Chocolate Avenue
Hershey PA, USA 17033
Tel: 717-533-8845
Fax: 717-533-8661
E-mail: cust@igi-global.com
Web site: http://www.igi-global.com

Library of Congress Cataloging-in-Publication Data

Names: Fernandes, Gonçalo, 1969- editor.
Title: Challenges and new opportunities for tourism in inland territories : ecocultural resources and sustainable initiatives / Gonçalo Fernandes.
Description: Hershey, PA : Business Science Reference, 2022. | Includes bibliographical references and index. | Summary: "This book identifies and disseminates entrepreneurial strategies in tourism and leisure ecocultural resources that can be the determining factors of development for the mobilization of economic and socio-cultural initiatives, promoting tourism and generating conditions of sustainability"-- Provided by publisher.
Identifiers: LCCN 2021027093 (print) | LCCN 2021027094 (ebook) | ISBN 9781799873396 (hardcover) | ISBN 9781799873402 (paperback) | ISBN 9781799873419 (ebook)
Subjects: LCSH: Sustainable tourism--Case studies. | Ecotourism--Case studies.
Classification: LCC G156.5.S87 C527 2022 (print) | LCC G156.5.S87 (ebook) | DDC 910.68/4--dc23
LC record available at https://lccn.loc.gov/2021027093
LC ebook record available at https://lccn.loc.gov/2021027094

This book is published in the IGI Global book series Advances in Hospitality, Tourism, and the Services Industry (AHTSI) (ISSN: 2475-6547; eISSN: 2475-6555)

British Cataloguing in Publication Data
A Cataloguing in Publication record for this book is available from the British Library.

For electronic access to this publication, please contact: eresources@igi-global.com.

Advances in Hospitality, Tourism, and the Services Industry (AHTSI) Book Series

Maximiliano Korstanje
University of Palermo, Argentina

ISSN:2475-6547
EISSN:2475-6555

Mission

Globally, the hospitality, travel, tourism, and services industries generate a significant percentage of revenue and represent a large portion of the business world. Even in tough economic times, these industries thrive as individuals continue to spend on leisure and recreation activities as well as services.

The Advances in Hospitality, Tourism, and the Services Industry (AHTSI) book series offers diverse publications relating to the management, promotion, and profitability of the leisure, recreation, and services industries. Highlighting current research pertaining to various topics within the realm of hospitality, travel, tourism, and services management, the titles found within the AHTSI book series are pertinent to the research and professional needs of managers, business practitioners, researchers, and upper-level students studying in the field.

Coverage

- Service Training
- Service Management
- Tourism and the Environment
- Casino Management
- International Tourism
- Travel Agency Management
- Cruise Marketing and Sales
- Destination Marketing and Management
- Food and Beverage Management
- Health and Wellness Tourism

IGI Global is currently accepting manuscripts for publication within this series. To submit a proposal for a volume in this series, please contact our Acquisition Editors at Acquisitions@igi-global.com or visit: http://www.igi-global.com/publish/.

The Advances in Hospitality, Tourism, and the Services Industry (AHTSI) Book Series (ISSN 2475-6547) is published by IGI Global, 701 E. Chocolate Avenue, Hershey, PA 17033-1240, USA, www.igi-global.com. This series is composed of titles available for purchase individually; each title is edited to be contextually exclusive from any other title within the series. For pricing and ordering information please visit http://www.igi-global.com/book-series/advances-hospitality-tourism-services-industry/121014. Postmaster: Send all address changes to above address.

Titles in this Series

For a list of additional titles in this series, please visit: www.igi-global.com/book-series

Rebuilding and Restructuring the Tourism Industry Infusion of Happiness and Quality of Life
André Riani Costa Perinotto (Universidade Federal do Delta do Parnaíba, Brazil) Verônica Feder Mayer (Federal Fluminense University, Brazil) and Jakson Renner Rodrigues Soares (Universidade da Coruña, Spain & Universidade Estadual do Ceará, Brazil)
Business Science Reference • copyright 2021 • 330pp • H/C (ISBN: 9781799872399) • US $195.00 (our price)

Handbook of Research on the Impacts and Implications of COVID-19 on the Tourism Industry
Mahmut Demir (Isparta University of Applied Sciences, Turkey) Ali Dalgıç (Isparta University of Applied Sciences, Turkey) and Fatma Doğanay Ergen (Isparta University of Applied Sciences, Turkey)
Business Science Reference • copyright 2021 • 906pp • H/C (ISBN: 9781799882312) • US $415.00 (our price)

Socio-Economic Effects and Recovery Efforts for the Rental Industry Post-COVID-19 Strategies
Maximiliano Korstanje (University of Palermo, Argentina)
Business Science Reference • copyright 2021 • 281pp • H/C (ISBN: 9781799872870) • US $195.00 (our price)

Risk, Crisis, and Disaster Management in Small and Medium-Sized Tourism Enterprises
Diego R. Toubes (University of Vigo, Spain) and Noelia Araújo-Vila (University of Vigo, Spain)
Business Science Reference • copyright 2021 • 308pp • H/C (ISBN: 9781799869962) • US $195.00 (our price)

Resiliency Models and Addressing Future Risks for Family Firms in the Tourism Industry
Anita Zehrer (MCI The Entrepreneurial School, Austria) Gundula Glowka (MCI The Entrepreneurial School, Austria) Katrin Magdalena Schwaiger (MCI The Entrepreneurial School, Austria) and Victoria Ranacher-Lackner (MCI The Entrepreneurial School, Austria)
Business Science Reference • copyright 2021 • 361pp • H/C (ISBN: 9781799873525) • US $195.00 (our price)

Strategic International Restaurant Development From Concept to Production
Angelo A. Camillo (Sonoma State University, USA)
Business Science Reference • copyright 2021 • 497pp • H/C (ISBN: 9781799843429) • US $195.00 (our price)

ICT Tools and Applications for Accessible Tourism
Celeste Eusébio (University of Aveiro, Portugal) Leonor Teixeira (University of Aveiro, Portugal) and Maria João Carneiro (University of Aveiro, Portugal)
Business Science Reference • copyright 2021 • 356pp • H/C (ISBN: 9781799864288) • US $215.00 (our price)

IGI Global
PUBLISHER of TIMELY KNOWLEDGE

701 East Chocolate Avenue, Hershey, PA 17033, USA
Tel: 717-533-8845 x100 • Fax: 717-533-8661
E-Mail: cust@igi-global.com • www.igi-global.com

Editorial Advisory Board

Table of Contents

Detailed Table of Contents

The study examines tourist attractions and creative experiences of Fethiye, a coastal destination in Mugla, Turkey, based on theoretical overlaps in creative tourism. Creative tourism is suggested as an initiative for sustainability of local lifestyle, unique culture, and authenticity of Fethiye in this study. Therefore, the study was maintained with an exploratory and conceptual viewpoint. The results are presented as Fethiye's tourist attractions (13 categories) and creative experiences (six themes). The prominent creative experiences are specified as house wine, squeeze of olive, cheesemaking, mushroom (morel) picking, agriculture cultivation (local fruits and vegetables), Dastar weaving (geotagging product), carpet weaving (carpet of Kaya), wool spinning, local festivals, traditional bazaars, and soap making. Finally, a creative tour case was offered, and two creative tour routes were proposed in the appendix. The study revealed that creative tourism has importance for sustainability of authentic culture, responsible consumption, and economic development of the local community.

Section 2
Sustainability and Challenges for the Inland Tourism Products

Chapter 6
Sustainable Tourism: How Is This Issue Effectively Approached? 91
Fatima Lampreia Carvalho, FE, University of Algarve, Portugal
Silvia Fernandes, University of Algarve, Portugal

This work intends to verify if there is academic research that proposes innovative strategies for sustainable tourism. It analyses 70 valid documents including theses, dissertations, scientific papers, and reports. Main objectives to explore are which themes stand out most; the difference between academic production in Portugal and Brazil; and how the topics of sustainability, environment, governance, planning, and regulation relate to tourism. A software for qualitative analysis is used to enrich the discussion around sustainable tourism discerning preferential governance strategies. Few works use sustainability indexes, and, if used, most of them are descriptive. For real sustainable tourism, its governance has to be based on appropriate indicators. Measures can empower destinations, providing the information needed to decide accurate and creatively. A challenge for the Portuguese academy on tourism and hospitality is to have an ongoing role in implementing key indicators and in their validation and monitoring.

Chapter 7
Slow Tourism in Low-Density Territories: NUTS III – Beiras and Serra da Estrela 106
Mafalda Marques, Estoril Higher Institute for Tourism and Hotel Studies, Portugal
João Reis, Estoril Higher Institute for Tourism and Hotel Studies, Portugal

This research focuses on slow tourism in low-density areas of Beiras and Serra da Estrela, Portugal, and explores how these genuine and authentic spaces can become attractive for tourism development. Compared to visitors who travel to a destination to visit as many attractions as possible in the shortest time, slow tourists refuse to be treated as a number and seek to receive from and give to their destination in the same proportion. Their involvement in a host community, practiced in a personalized, deep, and informed way, contributes to their well-being. The community's inhabitants recognize that their cultural identity is valued, feel motivated to serve as ambassadors, and note that tourist activities can lead to improvements in their quality of life. This research includes the results of interviews, a benchmarking in Italy, and an analysis of a village tourism. The findings reveal the opportunities they hold for slow tourism, with very positive effects for local development.

The essence of slow tourism is the sustainability of the locals. Slow tourism includes tourism activities by visiting slow cities, tasting slow food, and enjoying the moment, avoiding the worry of uploading photos to social media accounts while experiencing all these. Slow tourism is based on the need to experience travel within the framework of sustainability and to prioritize the locals during these travels. What is important in this tourism activity is not how long a destination is spent or how many destinations are visited, but how productive time is spent at the destination and the emotions that emerge afterward. In addition to these, many practices are carried out to ensure the sustainability of local culture, especially in slow cities, within the scope of slow tourism. Practices such as the continuation of local handicrafts, ensuring the continuity of local life under the influence of ancient civilizations, and sustainability of the local architectural style are included in slow city destinations. The essence of all this is to ensure the continuity of the local culture.

This research aims to reflect on the importance of inland surface waters for the development of local communities and their potential to foster tourism and leisure activities, promoting differentiated tourism products that project this inland tourist destination. It seeks to inquire into new business opportunities, while simultaneously reflecting on ways of controlling uses and issues of preservation and environmental sustainability. The study is developed in the Serra da Estrela territory, framing the destination and establishing an approach to the evolution of recreational fishing and its contributions to the strengthening of tourism and leisure, given the exceptional conditions of water mass supply and its environmental value for an activity that has an intense relationship with nature. The purpose is also to establish lines of research to be followed and methodologies to be implemented for a better knowledge of this activity and identify action strategies for the future.

This chapter describes the evolution of research on social responsibility in the tourism sector until November 2017. The authors have compiled a database of 66 articles focused on the field published in academic journals. Bibliometric methods and techniques were used to describe the evolution of scientific activity, countries and active institutions, most productive authors, applied methodologies, and the main

Foreword

The book *Challenges and New Opportunities for Tourism in Inland Territories: Ecocultural Resources and Sustainable Initiatives* organises and develops a highly topical subject in a rigorous and innovative way, pointing out skills that promote fruitful interpretations highlighting the relationships between territory, development, low functional and population densities, sustainability and tourism.

The tripartite organization of the book specifies the strands of research and analysis: a) tourism development, b) sustainability associated with hinterland territories and c) enhancement of culture and heritage, within the tourism offer. Nevertheless, the main focus of the whole work is the hinterland territories. These are approached by exploring, in a qualitative and quantitative, exploratory and interpretative way, their best practices. The diversity presented proves the quality of the selection of chapters.

This takes us to various processes of local development and sustainability, within the scope of tourism. Emphasising the way in which cultural heritages, international recognition of local resources, cross-border cooperation, gastronomy, wine tourism and agricultural activities, the valorisation of quality time associated with slow time, local customs and cultural habits, outdoor sports activities, local personalities, local music and local events, endogenous resources are perceived as contributions to the valorisation of inland territories which, with the pandemic, have gained an unusual prominence.

The relevance of the work gains, therefore, more scope, as it lets us understand how endogenous resources and local communities of low density areas have much to offer. They stimulate diverse forms of tourist animation, creativity and relationships, valuing the experience (Chs. 5 and 7, 8, 11, 12, and 14) and its expression of authenticity (Chs. 4, 5, 8). Simultaneously, emotions (Chs. 8 and 11) and memorable activities (Ch. 5) are explored, transporting us to an interpretation of the tourist experience that instils the possibility of transformation (Chs. 2 and 5).

We are also led to look at the world in a different way, escaping the pressure of daily life and conquering a quality time (Chs. 8 and 12) associated with slow time (Chs. 7, 8, and 14) which is combined with another way of looking at the Other, of understanding and participating in the life of local communities (Chs. 1, 2 5, 7, 8, 9, 10, and 12). Alongside all this, the book pays special attention to sustainability (Chs. 1, 2, 4, 5, 6, 8, 9, 12, and 13), expressing its importance when referring to territorial, cultural, economic, nature, destinations, local flavours, material goods, immaterial goods, heritage, development, employment and, of course, leisure and tourism.

It follows that inland territories and sustainability have a strong correlation, clearly expressed in this work. Sustainability is a complex, systemic and cross-cutting concept that needs to be perceived in a holistic way. For that very reason, sustainability is seen as the aggregating element of an integrated form of territorial management (public and private stakeholders, academia and civil society). Management is also seen here as a subject for discussion (Chs. 1, 2, 3, 9, 10, 11, and 12) and a qualifying reference for the analysis. This integrated governance is filtered by culture (Chs. 1, 3, 5, 8, 11, 14) which, in addition to being part of the development process, manages to be one of its pillars or, going further, to mediate sustainable development or even to be foundational for this development, anchoring on creativity, innovation, authenticity and identity as transformational values of contemporary society.

These readings, analyses and interpretations are a legacy of great value given to us by the book *Challenges and New Opportunities for Tourism in Inland Territories: Ecocultural Resources and Sustainable Initiatives*, throughout its 3 parts and 14 chapters, by the pen of 27 authors from 5 European countries. Its reading allows all those interested in the issues of tourism, sustainability and low density territories a captivating and high quality reading, transporting the reader to different realities.

Assuming tourism as an activity capable of creating synergies for the promotion of local development, the image of places, local culture, the authenticity of practices and cultural heritage (tangible and intangible), this book pays special attention to the environment and good tourism practices. In this way, through the study of Geoparks, mountain areas, archipelagos, coastal destinations and river resources, it manages to identify environmentally friendly modes of intervention. Their importance is highlighted by being able to create economic sustainability for local populations and maintain environmental quality, fundamental to the goal of sustainability that the political and cultural guidelines today propose.

The work is thus presented with an organization that gives it thematic dynamism, paying special attention to the relations between economy, society, environment and culture. It shows that the territory (support, living space and experiences, people, actions and interventions, borders and belongings - these of great symbolic value) has to be understood as an equation in which no aspect can be minimized, serving tourism and the relationship between nature and culture, as aggregators of wills, promoters of development and creators of resilience.

Norberto Santos
Universidade de Coimbra, Portugal

Preface

Tourism practices increasingly assume Complexity in the forms of their realisation, in the means involved, in the required experiments, and in the connections with the territories that support them. Tourism, in its territorial relationship and the transformations generated, promotes the constitution of resources as products, to which services are associated and functionalities are generated for their enjoyment.

Forms of ownership must be increasingly careful and properly planned to ensure their best use and sustainability in the face of resource fragility, environmental impacts, and the ways in wich communities and their ways of life are involved.

The frequent divergence between global needs for nature protection and local needs for socio-economic development has been, in many situations, an obstacle to the management of inland spaces, in particular low-density territories, as it can disqualify their perception and the activities established by their people.

These positions require specific and concerted attention from the several interlocutors, especially management entities, capable of creating conditions for the sustainability and qualification of these spaces, the valorisation of their productions, and a greater capacity to enhance the resources held as products and services of value to society and the economy.

In this context, ecocultural resources are determining factors of development for the mobilisation of economic and socio-cultural initiatives, promoting tourism, and generating conditions of sustainability. They are spaces of opportunity, owning resources and heritage with high preservation, enhancing new perceptions and forms of use, generating territorial cohesion, promoting self-esteem for local communities, and arguments for diverse and differentiated tourist experiences. The involvement of the community is decisive in valuing destiny, understanding local ecocultural realities, and developing the process of preservation and service creation. It is considered a necessary approach of inclusion and protection, aggregating the ecological and cultural binomial as a determinant for a deeper knowledge of territorial realities and their specificities. Thus, sustainability and participation are crucial for the long-term future of inland tourism activities, with local governance assuming an important role in building tourism capacity, mobilising resources, and streamlining entrepreneurial initiatives.

At present, the COVID-19 epidemic, and all the constraints and restrictions imposed as a result, have caused a deep crisis in tourism, the impacts of which are incalculable, requiring new forms of experience and development in tourism. Whereas before we were faced with a culture of unbridled consumption extended to a global scale, which tourism in its multiple forms and contexts represented, we are today facing a profoundly different scenario, where consumption is blocked by the pandemic and the restrictions and fears it raises, leading us to question the models and/or ways in which tourism practices will be carried out in the future.

We can affirm that we consumed products and tourist spaces in a nonconscientious way, given the simple and attractive way everything was proposed to us, to quickly freeze our movements, reconsider all the security concerns, and question which practices and in which spaces we can develop them. We are in a phase in which tourism truly establishes a process of deceleration, of reconstruction of trends and destinations. The frequent divergence between global needs for nature protection and local needs for socio-economic development has been, in many situations, an obstacle to the management of interior spaces, in particular low density, as it can disqualify their perception and the activities established by their people. These positions require specific and concerted attention from the various interlocutors, especially management entities, capable of creating conditions for the sustainability and qualification of these spaces, the valorisation of their productions, and a greater capacity to enhance the resources held as products and services of value to society and the economy. The inland territories arise, based on their natural and cultural resources, as challenges in building new models or ways of living tourism and leisure, strengthening the surroundings and valuing natural spaces, local traditions, and cultures, ways of life, and experiences of territories with authenticity. In this context, ecocultural resources are determining factors of development for the mobilisation of economic and socio-cultural initiatives, promoting tourism, and generating conditions of sustainability in inland territories.

They are spaces of opportunity, closing resources and heritage with high preservation, enhancing new perceptions and forms of use, generating territorial cohesion, promoting self-esteem for local communities, and arguments for diverse and differentiated tourist experiences. The involvement of the community is decisive in valuing destiny, understanding local ecocultural realities, and developing the process of preservation and service creation.

It is considered a necessary approach of inclusion and protection, aggregating the ecological and cultural binomial as a determinant for a deeper knowledge of territorial realities and their specificities. Thus, sustainability and participation are crucial for the long-term future of inland tourism activities, with local governance assuming an important role in building tourism capacity, mobilising resources, and streamlining entrepreneurial initiatives.

Rural areas and inland tourism gain expression and fans in times of pandemic, due to the less densified lodging modalities, the valorisation of nature, the activities promoted in the open air, the meaning of their identities, the affectivity, and the closeness and empathy of the communities that welcome them. Greater knowledge of these trends, territorial innovation, good governance practices, and new projects, among other aspects, are at the focus of this publication.

This book seeks to promote an approach of value for the low-density territories and those that, due to their territorial specificities, present new challenges of development in the face of social and economic devitalisation trends, and that, where their eco-cultural resources, assume new positioning and interests. In this sense we attempt to:

- Frame and describe tourism in inland territories (Concepts and Practices)
- Promote ecocultural resources as tourism activities
- Foster tourism as a factor for local development
- Identify forms and models of governance
- Disseminate strategies and good practices of inland tourism
- Dynamise activities and tourist products based on natural and cultural heritage
- Identify and disseminate entrepreneurial strategies in tourism and leisure

The 14 chapters that organise this book are structured in three sections, which allow the framing and interpretation of dynamics and challenges regarding Section 1, "Territorial Specificities and Tourism Development"; Section 2, "Sustainability and Challenges for the Inland Tourism Products"; Section 3, "Culture and Heritage in the Construction of Differentiated Tourism Products."

In Section 1, "Territorial Specificities and Tourism Development," which includes five chapters, approaches are established on the dynamics of tourism in territories with territorial and/or geographical specificity, investigating the contributions operated by the tourism sector and its relations that promote the construction of differentiated destinations and the use of existing resources. Elena De Uña-Álvarez and Montserrat Villarino-Pérez establish, in Chapter 1, an approach to the diversity of ecocultural resources held by rural areas and their strategic importance for their development. They acknowledge the concerns of the EU in this matter and develop the problem of the interior territories in Spain. They analyse the importance of tourism in the development strategies for the interior territories and the role of public and private actors, their logic of action, and projects. Studies are framed in the inland territory of Galicia (NW Spain) focusing on strategies put forward in the Sacred Riverside rural area (in the Galician language, "Ribeira Sacra", hereinafter RS) and the Courel Mountains UNESCO Global Geopark (hereinafter CMGG). The Estrela Geopark research team (Hugo Gomes, Emanuel Castro, Patricia Azevedo, and Fabio Loureiro) address the importance of the Unesco World Geoparks in the construction of collective strategies for sustainability, highlighting the contributions of science, education, and tourism in the construction of collective territorial strategies and the promotion of the communities' well-being. Develop the strategy of Estrela Geopark in the promotion of different actions focused on natural and cultural heritage that aims to stimulate socio-economic activity through the creation of projects related to the tourism sector, to disseminate and promote visits and tourist circuits in the territory. Emphasise the importance of education as the basis of geotourism and a pillar for the sustainability of heritage and local communities. Gonçalo Fernandes and Ines Cravino develop, in Chapter 3, an approach to border territories and their current significance in the face of growing mobility and the valorisation of natural and cultural resources affecting the promotion of tourism. They take as their geographical space to study the border in center region between Portugal and Spain and reveal the trajectory of the oldest European border, its development problems, and the current dynamics of cooperation and tourism promotion. In this context, they describe the dynamics of tourism as a propeller of new territorial relations and economic dynamics between the communities of the two countries, which present more elements of union than of separation, and demand strategies of cooperation for their valorisation and sustainability. In Chapter 4, Francisco Silva establishes approximations and conceptual reflections on the dynamics of the interior territories and peripheral island regions. This chapter focuses on the case study of tourism in the Azores archipelago in Portugal. It seeks to contribute to an understanding of this reality so that similarities, bridges, and possible strategies may be followed by other destinations located inland, the other geographical extreme of islands. Reflects on how the promotion of authenticity, uniqueness, and differentiation has contributed to the development of tourism in low-density peripheral territories, namely inland and island regions.

Meltem Altınay Özdemir and Emre Ergün, in Chapter 6, discuss how creativity in tourism increases originality and innovation, which can affect the competitive advantage of companies and destinations. Reflect the importance of creativity and its relationship with culture in the creation of tourist experiences bringing diversity to tourists, preserves the cultural heritage of destinations, and re-evaluates a place's identity. The study examines the current offers and future potential of creative tourism in Fethiye, a coastal destination in Mugla, Turkey, based on the aforementioned theoretical overlaps in creative tourism.

In Section 2, "Culture and Heritage in the Construction of Differentiated Tourism Products," four chapters are included to address sustainability issues in inland territories, demonstrating how academic results are considered in the development of sustainable tourism policies, and evidencing dynamics in the development of tourism close to the communities, the valorisation of endogenous resources, and the promotion of experiences. Forms of tourism consumption with specific characteristics are made explicit, recognising local traditions and cultures and seeking to preserve heritage and ways of life. Processes that reproduce traditional practices are analysed, valuing customs and proposing (re)discoveries that commune with traditions, local ways of life, and ancestral practices of living in the countryside and nature. The valorisation of recreational activities, leisure, and sports with an intense connection with nature, as well as the valorisation of culture in the sustainability of rural areas and its promotion, allow us to understand tourism in inland areas and the emergence of proposals that contribute to sustainable tourism, with consumers who value the authentic and seek recallable experiences. Fatima Carvalho and Silvia Fernandes develop a chapter (7) on how academic results, published in theses, scientific articles, and reports, can contribute to the promotion of sustainable tourism development strategies and policies. The main objectives that structure the chapter are related to which themes are highlighted in the sustainability of tourism, the difference between the academic production in Portugal and Brazil, and how the themes of sustainability, environment, governance, and planning are related to tourism and its contributions to development. The developed work aims to establish what proportion of academic works, listed in the scientific repository of open access in Portugal (RCAAP), have proposed innovative governance strategies for sustainable tourism. Mafalda Marques and João Reis address the Slow Tourism in Low Density Territories: NUTS III - Beiras and Serra da Estrela. This chapter (8) analyses the adequacy of the tourist offer of low-density territories to the growing demand for slow tourism. Authors have the purpose to capture the value of the rural areas of the territory (Nut III – Beiras Se Serra da Estrela in Portugal) under analysis and provide them with a renewed image, taking advantage of a seemingly new concept – slow tourism.

This same theme is the object of analysis in Chapter 9, by Zeynep Çokal and Nilgün Demirel, which concerns the sustainability of local culture in Turkey. It addresses the importance of local culture and its intrinsic characteristics in the development of tourism and its differentiation in the face of globalisation and standardisation of behaviours. The study promotes the compilation of literature on slow tourism in the sustainability of local culture and the preservation of socio-cultural values and identities. To conclude this second section, Chapter 9, Fluvial resources and recreational and sports activities. Contributions of sport fishing to the recreational and tourist attraction of Serra da Estrela, by Gonçalo Fernandes and Elsa Ramos, aims to reflect on the importance of inland surface waters for the development of local communities, and their potential to foster tourism and leisure activities, promoting differentiated tourism products that promote this inland tourist destination. It seeks to inquire into new business opportunities, while simultaneously reflecting on ways of controlling uses and issues of preservation and environmental sustainability. The study is developed in the Serra da Estrela territory, framing the destination, and establishing an approach to the evolution of recreational fishing and its contributions to the strengthening of tourism and leisure, given the exceptional conditions of mass water supply and its environmental impact.

The last section (3), "Culture and Heritage in the Construction of Differentiated Tourism Products," is composed of five chapters that bring the patrimonial and cultural value of the inland territories to differentiated tourist approaches. Literature, music, archeology, and gastronomy are arguments in the promotion of these territories and in the promotion of new strategies for their valorisation as tourist resources. Differentiated approaches are emerging, recognising the importance of culture in the revitalisation of areas of low deprivation, with penalising consequencies at the demographic level and identifying in tourism dynamics for their recomposition and promotion of social and economic welfare, ensuring, simultaneously, the perpetuation and conservation of their heritage.

María Lechuga Sancho and Alicia Martín Navarro presents Chapter 10, "Culture and Heritage in the Construction of Differentiated Tourism Products," which corresponds to the development of bibliometric methodology, which seeks to assess the evolution of research on social responsibility in the tourism sector. Analyses are established on the evolution of productivity in this area of knowledge, and its quantitative and qualitative representation, making it possible to promote contributions to future research. Anabela Sardo and Rita Arala Chaves present in Chapter 11 the case of the Portuguese project Fundação Eça de Queiroz /'Eça de Queiroz Foundation', showing how this organisation, starting from the personal patrimony and work of a renowned Portuguese writer from the last half of the 19th century, managed to preserve part of his patrimonial assets, as well as his recognised literary work, for the pursuit of lasting and socially relevant ends. Those constitute, in the present case, projects of a cultural, commercial, and tourism nature that contributes to the development of the Portuguese statistical sub-region (NUTS III), Tâmega e Sousa, included in the statistical North, a territory close to the internationally recognised Alto Douro Wine Region, whose landscape has been a World Heritage Site by UNESCO since 2001.

Chapter 12, by Martín Gómez-Ullate, Javier Barra Sanz, and Manuel Rodríguez Palacios reflects on rural development, its heritage characteristics, and how music can influence it, promoting new cultural valorisations and socio-cultural dynamics. It reflects on indicators of quality of life and well-being, and on the sustainability of local communities, beyond the references of growth and richness. The study is supported by fieldwork in different European rural contexts and deepens the state of the art on the importance of music in rural communities and the promotion of tourism. Projects of relevant impact in their territories are presented, constituting cases of good practice. Milica Maksić Mulalić develops Chapter 13, entitled "Sustainability of Cultural Heritage in Inland Territories: the Case of Archaeological Heritage in Serbia." This chapter presents a case study of Serbia's Gamzigrad-Romuliana and Caričin grad archaeological sites, investigating the possibilities regarding the sustainability of World Heritage in inland territories and the existing policies for their preservation. It has as a central object of investigation and reflection the issue of how conflicts between cultural heritage protection and tourism development can be harmonised in inland territories.

The conflicts between cultural heritage protection and tourism development are analysed, and the information systematised. By investigating planning solutions and the coordination mechanisms of the actors and institutions included in the process, general recommendations are given for improving the sustainability of cultural heritage in low-density territories of relevant heritage interest.

Mario Ossorio in the last chapter (14) seeks to analyse the relevance of events in enogastronomy during the pandemic. The evolution of the concept of wine and food tourism and the economic consequences for rural territories and the wine industry are the basic objectives of this research. Furthermore, the different categories of these tourists are examined, emphasising the relevance of the experience within food services. It describes the Strada del Vino dei Campi Flegrei, isola d'Ischia e Vesuvio, a territory near the City of Naples, where well-known Italian wines are produced and several wine-centred events are organised, reinforcing the attractiveness of these rural areas. The importance of the combination of local food and wine generates its attractiveness and demand that becomes relevant to analyse for the tourism sector. The role of the customers' experiences connected to enogastronomy and the possibilities offered by a rural context are described. The chapter tries to investigate the socio-economic advantages of enogastronomic tourism for the territory and the effects on its community.

The diversity of themes addressed throughout the book and the contributions of authors from different countries, with different socio-cultural backgrounds, enrich the publication as a whole, resulting in a deepening of knowledge about inland and/or low-density territories, case studies, best practices, innovative products, and research forms/methodologies.

The various contributions gathered allow us to get to know the territorial dynamics that contribute to the valorisation of the ecocultural resources of these territories, to the preservation of ways of life, and to the conservation of heritage manifestations (material and immaterial) that constitute an argument for the promotion of tourism and its continuity and preservation.

The challenges that we register at the beginning of the second decade of the 21st century require a deeper understanding of the knowledge and dynamics that run through the inland territories, their rurality, and the constraints on their communities. The pandemic effects that affect our society and that have deeply conditioned tourism, whether in mobility, access to services, or in recreational and leisure practices, require a greater appreciation of inland destinations, the preservation of their culture and natural resources, the stimulation of the creativity, and the promotion of new projects and initiatives linked to local identities.

Thus, the chapters developed allow us to contribute to a better understanding of tourism and of its challenges and dynamics in inland territories, in the dissemination of products and strategies, and in the promotion of new projects, which we can systematise:

- Disseminate products and tourist practices
- Expand knowledge of tourist activities in peripheral areas
- Systematise knowledge about inland tourism
- Spread good practices and strategies for promoting tourism in low-density territories
- Disseminate models of territorial governance and innovation
- Provide knowledge on differentiating tourist activities
- Stimulate tourism entrepreneurship activities
- Value the ecocultural heritage as a tourist asset
- Build scientific resources for new appropriations of inland territories

We are in the presence of a publication of interest to Higher Education Institutions (Teachers and students), tourism management organisations (national and local), municipal tourism departments, local development associations, tourism entrepreneurship, and all the development agents of the inland territories.

Finally, I would like to thank all the authors for their contributions and commitment, allowing us to enrich this international publication by disseminating their research and projects and promoting a deeper and more up-to-date understanding of the dynamics and challenges of tourism in inland territories.

Gonçalo Poeta Fernandes
CITUR, Polytechnic Institute of Guarda, Portugal

Acknowledgment

I would like to acknowledge the assistance of all academics and researchers who contributed to the successful completion of the book, in particular, the authors for their valuable contributions and commitment in the development of this publication and all the colleagues who took part in the review process, and made it possible. Without their support and interest in promoting these territories and disseminating its investigations, this publication would not have been possible to bring to light. We also would like to express our sincere gratitude to Norberto Santos for her insightful foreword.

The theoretical developments, the territorial analyses, case studies and good practices in the valorization of inland territories, through new dynamics or ways of valorizing resources and heritage, are undoubtedly valuable contributions for these territories, a better understanding of them and the paths to follow.

Gonçalo Poeta Fernandes
CITUR, Polytechnic Institute of Guarda, Portugal

Section 1

Territorial Specificities and Tourism Development

Chapter 1
Fostering Ecocultural Resources, Identity, and Tourism in Inland Territories (Galicia, NW Spain)

Elena De Uña-Álvarez
University of Vigo, Ourense, Spain

Montserrat Villarino-Pérez
University of Santiago de Compostela, Spain

ABSTRACT

Inland territories hold a great diversity of ecocultural resources, increasingly constituted in tourist products for local development. Their role in improving the socioeconomic conditions and wellness of local communities, as well as in promoting tourism and sustainability, depends on the involvement of public and private actors. The relationships and the collaboration of local actors are essential in that regard. The study of aforementioned processes takes place in the inland territory of Galicia (NW Spain). The methodology of research relied on in-depth interviews. Due to the key role of the local actors, the interviews focused on their professional and life experiences. The analysis of the answers establishes the definition and the appraisal of the main resources, attached to territorial identity, and highlights the engagement and involvement of the actors in the territorial dynamics that foster the promotion of the ecocultural resources for tourism.

INTRODUCTION

Inland territories hold a great diversity of ecocultural resources, increasingly constituted in tourist products. They are often rural and mountainous areas with sparse and ageing population. Regarding their surface area (the 80% of the EU territory), the important meaning of their identities, and the challenges that they face, the European agenda (European Parliament, 2018; European Committee of the Regions,

DOI: 10.4018/978-1-7998-7339-6.ch001

2020) highlights the need to foster local development through more holistic, proactive, and place-based strategies, defining the crucial role that tourism is expected to play. As it is the case with other European areas, inland Spanish territories are today less developed areas, sparsely populated, and in demographic decline. They face big challenges in order to reverse or at least minimize the hindrances that this situation brings about (Serra et al., 2014; Nieto Masot et al., 2020). Tourism provides an opportunity to set into motion their ecocultural resources and improve their socioeconomic conditions as well as the wellness of local communities. In this context, the boost to carry out new arguments and initiatives -promoting tourism and sustainability- depends on the involvement of both public and private actors. Such processes are framed in the inland territory of Galicia (NW Spain) focusing on strategies put forward in the Sacred Riverside rural area (in the Galician language, "Ribeira Sacra", hereinafter RS) and the Courel Mountains UNESCO Global Geopark (hereinafter CMGG).

The chapter concerns in particular the mobilization of ecocultural resources, considering new economic and sociocultural initiatives that emphasize their enhancement, the reaffirmation of territorial identity, the consolidation of territorial cohesion, and the generation of conditions for sustainable development. Inland territories have always supplied goods and services in addition to those of the agricultural type, stemming from their environmental and cultural features (geodiversity, biodiversity, and lifestyle). They identify the main components in resources-oriented strategies (Féret et al., 2020) through territorial trajectories of diversification, restructuring, innovation, and branding (e.g. Ilbery, 2014; Camarero & Oliva, 2016; Gómez-Ullate et al., 2020). The narratives provided by local actors reflect the relationship between nature and culture rooted in specific places and times (Buttimer & Seamon, 1980). They also highlight their appraisal of the material and the immaterial resources that belong in a territorial project (Colletis-Wahl & Pecqueur, 2001), bring into light -based on personal and collective experiences- the role that local identity and social cohesion play in attracting exogenous development opportunities (Pike et al. 2006). Furthermore, they account for the configuration of alliances (Byrd, 2007; Sharpley, 2014) especially aimed at creating tourist products (Williams & Lew, 2015; Richards, 2020). In order to gain insight into these processes, the methodology of research was qualitative, relying on in-depth interviews conducted in the years 2011, 2012, and 2019.

Driven by the acknowledgment of the importance of their function, 17 local actors were selected because of their engagement with territorial actions that affect the discovery, the preservation, and the activation (or the creation) of ecocultural resources. The analysis of in-depth interviews outlines different ideas, experiences, and practices interwoven through adaptation, adjustment, and innovation processes. Strongly attached to territorial identity, such processes make up the ecocultural values for tourism. They distinguish those Galician inland territories that enjoy increasing capacities and collective self-esteem, related to the expectations, the participation, and the organization of local actors, all of them coalescing into a synergetic current flowing among the latter. Nevertheless, it must be noted that the narratives of the local actors involved in the territorial dynamics pinpoint the difficulties to achieve socioeconomic advantages and visibility without performing actions that encompass both bottom-up and top-down processes.

The main objective of this chapter is to describe and analyze the role played by local resources, territorial identity, and tourism in the evolution of adaptive or new dynamics that arise from times of crisis (in the first decade of the 21st century). These dynamics strengthens the importance of ecocultural resources, incorporates a wide range of intangible values, broadens the spectrum of the tourist offer to encompass multifaceted perspectives, and spreads the image of the territory as a tourist destination. Three specific objectives are of particular concern: to explain the role of local actors in territorial dynamics; to describe and assess, through the actors' narratives, the strategies that underlie territorial projects based on ecocultural resources; and to furnish information regarding future actions to audiences so diverse as institutions, managers, associations, and local residents.

INLAND TERRITORIES, RESOURCES STEWARDSHIP, AND TOURISM

Since the 1980s, research on inland territories has been focused on the processes of change concerning both synergies and conflicts in the adaptive or creative responses to various global and local pressures (e.g. Lowe et al., 1999; van der Ploeg & Renting, 2000; Woods, 2007; Bosworth et al., 2016; Cejudo & Navarro, 2020). Based on the attractiveness of ecocultural resources, tourism is increasingly an alternative option to build a future. Inland tourism can drive territorial transformations, especially in times of crisis, by promoting local initiatives (Hall et al., 2013; Fernandes, 2020) attached to specific ecocultural resources shared by local communities, and contemplated as living context. In Spanish inland territories (more than the 80% of the territory, where the 12% of total population lives) fostering ecocultural resources -attached to local identity and activated through different tourist modalities- seems like a winning plan of action (Ruiz Pulpón & Cañizares Ruiz, 2020). This argument appears in the proposals of associations such as the Forum of Spanish Regions with Demographic Challenges ("Foro de Regiones Españolas con Desafíos Demográficos" [FREDD]) or the Spanish Association of Mountain Municipalities ("EsMontañas"), both of them established in 2013. The vernacular knowledge and resources draw a particular attention, with an increasing social and institutional awareness of the local identities. At the same time, the success of sustainable projects to revitalize inner areas requires the support of supra-local levels (Sánchez-Mesa Martínez, 2019), whether it is at the European (LEADER or LIFE programs), the national (Rural Development Programs), the regional (planning from Spanish Autonomous Communities), and provincial governance entities (known in Spanish as "Diputaciones").

Tourism provides an opportunity to build innovation strategies, managed by the Rural Development Groups (Spanish, Grupos de Desarrollo Rural [GDR]) and other collectives of local actors. According to the last reports (2021) of the Spanish Observatory of Rural Tourism and Ecotourism, the COVID-19 pandemic has brought about a significant rise of tourism flow to inland destinations -rural and nature-after the lockdown as a getaway (Spanish, "escapada"), avoiding gatherings of many people. These tourists dedicate the 67% of their expenses in destination to purchasing local products, and they show a favourable disposition to activities such as trekking, guided tours, enjoying local gastronomy, and resting. The Autonomous Community of Galicia, located in the northwest corner of the Iberian Peninsula, is one of the FREDD members. It spreads through a surface of 29.576 km^2 and has a total population of 2.701.819 inhabitants (Instituto Galego de Estatística [IGE], 2020), the 32% of which are residents of the seven main urban areas. Since the socioeconomic crisis of the 1970s (Rodríguez-Pose, 2000), emigration (Martínez-Filgueira et al., 2017), depopulation and ageing (Hernández Borge, 2011), together with land abandonment and the drop of income levels (Corbelle-Rico et al., 2012; Sineiro-García et al., 2014; Peón Pose et al., 2020) have been especially significant in the inland territories (Ourense and Lugo provinces). Nevertheless, several studies show positive experiences in these territories through initiatives devised, for instance, by building up from the identification of opportunities for gastronomy products (Domínguez García et al., 2010), animating heritage as a sign of identity (Paül, 2013), or fostering inland tourism (Cànoves et al., 2017). In all of them, it has proved fundamental the involvement of the local actors in order to overcome the structural weaknesses.

Ecocultural resources can become the main tourist attraction of a territory, fostering the empowerment of the local communities to build their own future (Wallace & Russel, 2004). The insertion of this type of resources in a tourist product is born out of awareness, cooperation, and innovation. The synergies thus created contribute to the emergence and the strengthening of the territorial identity in the framework of a project where memory, history, heritage, and landscape interact, a project that needs also of the mobilization of the local actors as agents of change. The transformation arises from a reassessment of the relationship between nature and culture (Pilgrim et al., 2010) with different objectives. The recognition of ecocultural resources occurs in the framework of a shared history, of a distinguishable and owned territory, belonging to a territorial heritage that transcends the nature-culture duality (Fredheim & Khalaf, 2016). This acknowledgment implies the beginning of a path that opens up opportunities where a constellation of actions shapes new initiatives and management approaches with the capacity to improve the quality of life (Dissart & Seigneuret, 2020). A tourist product based on available ecocultural resources can promote the territory's sustainability in the environmental, economic, and sociocultural dimensions. By being committal to small-scale operation founded on emotion and experience (Martínez Quintana, 2020), tourism can foster the stewardship, protection, and sustainable management of these resources.

In this way, Global Geoparks are new contexts for the empowerment of local communities through the stewardship and sustainable uses of ecocultural resources (Farsani et al., 2011; Henriques & Brilha, 2017). To become a member of the UNESCO Global Geoparks Network [GGN], created in 2014, a territory needs to have sharp boundaries and contain a singular geologic and geomorphologic heritage. To this, it adds the existence of a development project and an active involvement of the local communities. Key factors in this regard are the promotion of tourism and education. According to the foundational statements of the International Geoscience and Geoparks Programme (UNESCO, 2015), Geoparks are "laboratories for sustainable development." This approach sustained the foundation of the European Geoparks Network [EGN] in the year 2000, as an outcome of a LEADER project (taking part Germany, Spain, France, and Greece). Thus, the significance of Spanish Geoparks (Orús & Carcavilla Urquí, 2020) transcends geoconservation and includes aspects linked to scientific, cultural, educational, and socioeconomic values. Tourism is one of the most important options for the preservation of ecocultural resources and the promotion of sustainable development.

FOSTERING ECOCULTURAL VALUES BY ENGAGING LOCAL ACTORS

One Territory, Two Models

The morphology of inland Galicia alternates between mountain ranges and enclosed valleys (Yepes Temiño, 2002). In the eastern sector, the Miño River and its main tributary, the Sil River, flow along canyons (where is bounded RS); to the NE are located the Courel Mountains (Figure 1). This inner area, far from the Galician coast, is thinly populated and has a low degree of urbanization. The RS and the CMGG belong to a territory characterized by its imbrication in nature and the interdependency of common local actors (owners, associations, GDR, and other entities), even though they present two different models regarding protection and management of their resources. Thus, the territory's promotional slogan reads *the taste of a landscape* (RS) and *500 million years walking* (CMGG).

The RS (20 municipalities) extends throughout 2.523 km^2 encompassing the canyons of the Miño-Sil system (Lugo and Ourense provinces). The steep riversides, which slopes reach up to 400 m, connected

Figure 1. Location of inland territory
(Base map from SITGA and CMGG. Own elaboration)

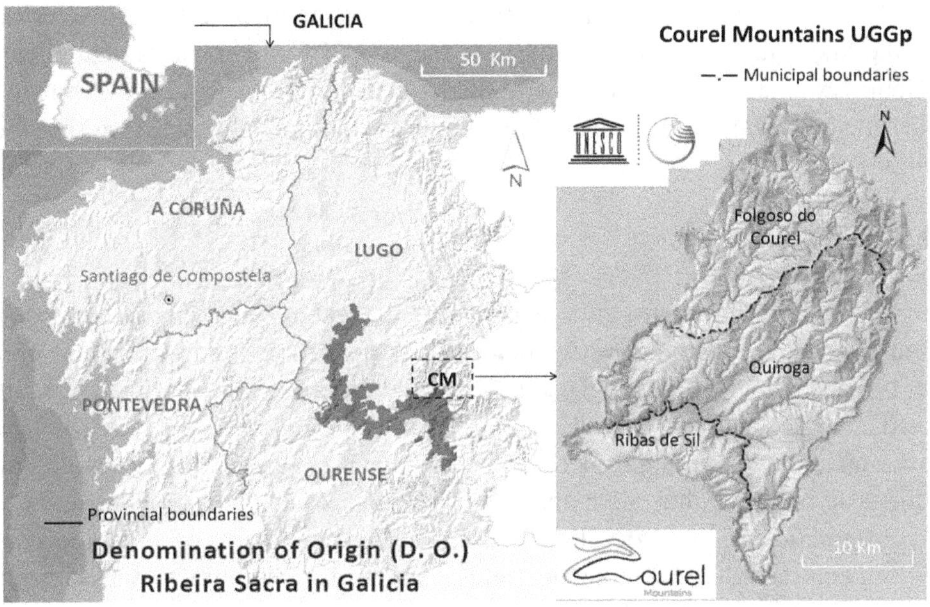

with the upper plains (altitude between 600-700 m) shape a singular waterscape. The population density (23 inhabitants/km^2) is far below the Galician average (91 inhabitants/km^2). Most of inhabitants lives in the towns of Monforte de Lemos (Lugo, 18.347 inhabitants) and Castro Caldelas (Ourense, 1.204 inhabitants) (IGE, 2020). Coincident with the viticulture space of the Designation of Origin RS (1995), one of the main tourist references of the territory is the "heroic viticulture" performed in terraces, attached to the taste of a millenary cultural landscape. The high density of monasteries, hermits, towers, and ethnographic assets, together with the value of the forests, have altogether made the Galician Administration (Xunta de Galicia, 2017) to acknowledge the RS an Asset of Cultural Interest (Spanish, Bien de Interés Cultural [BIC]) in the category of Cultural Landscape. With the understanding that such a landscape reflects an interaction more than a thousand years old between nature and mankind, the RS candidature to be included on the World Heritage List is due for consideration at the 45th session of the UNESCO Committee, scheduled for June 2021. Nature tourism, rural tourism, and ecotourism are the modalities that animate the promotion and knowledge of ecocultural resources.

The CMGG extends throughout 578 km^2 encompassing three municipalities (Lugo province). It is a mountainous area (maximum height 1641 m) with deep valleys, ravines and waterfalls shaped by the fluvial network of the Sil River and includes a sector of the Sil canyon. The population density is extremely low (9 inhabitants/km^2) above all in the highest areas (5 inhabitants/km^2), with the 42% of the total population living in the municipality of Quiroga (IGE, 2020). The CMGG (designated on the 11th of April, 2019) has been recognized as a singular and unique territory together with other 16 Geoparks that comprise the Spanish Forum of Geoparks. Its candidature to the list of GGN involved the Association for the Territorial Promotion of the Courel Mountains, created in the year 2016, among whose members are the councils of the three municipalities and the entity managing the European LEADER projects (Rural Development Group RS-Courel). The CMGG (53% of which belongs to the Natura

2000 network) presents undoubtable natural values (e.g. linked to geoforms, glacial areas, waterfalls, or caves) counting with the 40% of the terrestrial flora from Galicia and with the presence of the brown bear. It is to be highlighted also its cultural assets (e.g. Roman mining, mills, ovens, or traditional villages). This is the first Galician territory included in the list of GGN. One of its municipalities, namely, Folgoso do Courel, was designated “Tourist Municipality” by the Galician government in February 2020. Geotourism and scientific tourism are the most important modalities to fostering the sustainable use of ecocultural resources.

Narratives of Local Actors

The analysis of the in-depth interviews aims at gathering relevant information highlighted in the answers of six qualified informants (I) from diverse sociocultural spheres (e.g. education and journalism) and eleven owners (O) and managers (M) engaged in territorial processes. Eight of them are autochthonous. Once conducted, recorded, and transcribed, the interviews were analyzed with the AtlasTi software. Applying this procedure, there is an initial encoding by automatic assignment of text excerpts to key codes. Afterwards, the tasks of debugging and manual encoding by revising and structuring the text excerpts. Appropriate configuration in Excel of all the excerpts corresponding to each key code facilitated the analysis of the contents regarding the following fields: territorial assets (attractiveness/resource, heritage/cultural landscape, identity, territorial quality), territorial processes (tourist product, dynamics, crisis, initiatives, innovation, development policies and plans), and lastly, effects and impacts (networks/associations, diffusion/promotion, obstacles, perspective of future). Every comments provided here none include personal data, giving priority to the summary of narratives. The synopsis has a reference according to the actor’s category, their sample number, and the year of interview conducted.

Recognition of Resources, Identity, Life Quality

All the local actors are highly motivated to put forward and approach projects promoting the territorial enhancement. They feel satisfied, and assert that doing something in the territory where they live, being able to change things for the better, is one of the most important things in life. After a decade working on territorial projects, managers affirm that they are very satisfied starting new ideas and initiatives, which raise great interest for the local community. The main ecocultural resources identified by local actors represent the native culture, traditions, and feelings. The actors highlight specific features of those territorial resources and define their role as tourist attractions that have become identity signs of the territory. The recognition of ecocultural resources encompassing aesthetic, scientific, historic, socioeconomic, and quality of life values.

They mention as the main resources in RS: the fluvial canyons, the monasteries and hermits of Romanesque style, the heroic viticulture, which is a completely handcrafted process, and the Balneary of Aguas Santas (I, 05, 2012; M, 17, 2019); while in CMGG the recumbent folds, other natural assets and the mountain culture are the most outstanding (M, 14, 2019). The answers also indicate the key role of the Tourism Paradors (singular Spanish hotel accommodation in Monasteries, Castles, and other historic buildings) both in the RS and in the area that would later give birth to the CMGG due to their area of influence. The Parador of San Vicente do Pino (Monforte de Lemos, Lugo, opened in 2003), housed in a Neoclassical Monastery and the palace of the Counts of Lemos, provides economic benefits and contributes to foster ecocultural resources through diverse activities explaining the paths, the monuments, the wineries, and the catamaran routes (M, 10, 2011). In the same way, the Parador of Santo Estevo (Nogueira de Ramuin, Ourense, opened in 2004), placed in a Baroque-Romanesque Monastery within autochthonous woods, work hand in hand with local groups and offer to customer experiential tourist products (M, 03, 2012).

Developing a sense of appreciation for the resources is a process attached to territorial identity. Setting into motion tourist resources is perceived as a factor in the construction of an idea of what is of one's own, contextualized by awareness of and appreciation for the territory, including recognition from the inside and from the outside. This process affects material and immaterial resources, taking part in the dissemination of local tourist products. The answers underline here the meaning of the territorial identity because only when we see ecocultural resources as one's own that we see it as something valuable (M, 06, 2011) and it helps in protecting the assets and in generating employment (I, 09, 2012). This argument remains over time with the need of the internal and external recognition, considering that it is fine that others appreciate the resources, but being as good that people go abroad to see how other areas to discovery the values of the resources (M, 14, 2019). Such a recognition deploys emotions and affective connections with the ecocultural resources, an aspect commented upon by the foreigner's local actors, asserting the exceptional quality of life in the territory. Being perfect because not only they are connected with the land by a physical, psychical, and spiritual sensation (O, 11, 2011) but also exist a life quality away from noise, tranquility, and natural beauty (M, 16, 2019).

Ecocultural Resources, Territorial Processes, Tourism

Despite the difficulties that may exist, the trend of the territorial processes since the 1980s has been to a coalescing of local wills and interests to develop territorial projects. The first initiatives to foster tourism consist of day trips, most of all focused on the Romanesque cultural resources, animated by some qualified informants (I, 02, 2011). Began small actions that expand slowly the ecocultural values of the territory. The local actors also point to the need of public funding in order to facilitate the activation of the resources that can lead to local development (I, 08, 2012). The project of the Geopark bring into value the geoheritage, having a high social impact the creation of the Campodola lookout (M, 14, 2019). The evaluation of the experiences over time is positive. Twelve years ago, there was a strong rivalry between Ourense and Lugo provinces; "now, nobody argues anymore" (M, 17, 2019). Lastly, the local actors try to develop joint actions together the RS and the CMGG (e.g. meat and chestnut products, enotourism and mountain tourism).

The years 2008, 2009, and 2010 have been hard. The impact of the crisis implied a reorientation of the territory, with the image of an uncrowded and less expensive destination. The drop in demand impinged on the accommodations network (most of which were small-sized). Native people returned and people who had never been in the territory begin to visit it (M, 13, 2019). The initiatives of rural tourism, nature tourism, and enotourism increased with a proximity offer. The discovery of the value of a common heritage contributed to strengthening social cohesion and to the visibility of the territory. Despite the existence of particular attachments in the riverside, the plains, or the mountain areas, the significance of the resources transcended the political division in different provinces. Sustaining the territory throughout the hard times was an achievement that can be explained on the premise that a crisis scenario was not something unheard of in the area. In fact, the agricultural labour in the riversides of the fluvial canyons "went through multiple moments and waves of splendour and decay" (O & M, 15, 2019); after the 1930s the Spanish civil war stopped the "heroic viticulture", and, subsequently, most of terraces were abandoned; then the Denomination of Origin for wine (1995) led to a revival. However, during the "hard years" the rural tourism houses did not close: they adapted to the crisis situation (M, 17, 2019), making a diversification in the offer by multiple tourist activities.

The local entities agglutinating these processes were the Regulatory Board of the Designation of Origin of wine and the Tourism Consortium (with a Tourism Dynamization Plan in 2006-2009 for 22 municipalities of the territory), together with the GDR management of the European funds. Besides them, an assortment of associations contributed positive synergies in a way that foster the activation of ecocultural resources (e.g. trekking, balloon travels, catamaran travels, and restaurant business). The Tourism Dynamization Plan needed the participation of a great diversity of actors, trying to stimulate tourism in order to revitalize an area afflicted with severe demographic problems. In the beginning of the 21st century second decade, people that remain in villages were the elders and the "nostalgic ones" but also "urban people" who buy and rehabilitate houses (I, 02, 2011). At the same time, infrastructures for accommodation, communications, services, and activities were created. Thus, the territory understood as a tourist destination. Some initiatives have enjoyed noteworthy success, such as the creation of the Wine Centre (Monforte de Lemos, 2009), which functions also as a tourist information centre, defined as "a great idea of the local government" (M, 13, 2019). This and other tourism-related initiatives were beneficial for the entire environment of the territory, diversifying the offer through small tourism businesses, some of them highly innovative for the standards of the area (e.g., catamaran visits including food, rest, and bathing, golf sessions). As for family tourism in the Thermal Spa & Golf Resort of Aguas Santas (Pantón, Lugo, opened in 2006); this tourist establishment, built next to an old thermal spa (closed in the mid-1950s) is, besides a balneary, "a family-oriented hotel" (M, 16, 2019).

The funding of the projects by the European Union and the autonomous government had a controversial effect. It implied a large investment and an employment creation (I, 05, 2012). But several GDR contended for the assignment of those funds. One of the weaknesses is that every holder of a public office has a localist view of the development policies, plans, and assistance. The narrative from the GDRs managers asserts that there were difficulties with the local government, inasmuch as the idea of to use LEADER funds doing something different through new collaborative projects for territorial development involves a joint vision above the municipalities (M, 14, 2019); bureaucracy is another obstacle, especially mentioned in the process of the funds assignment, related to rural development plans which splits into 17 Spanish autonomous communities (together with different norms and limitations). These weaknesses are overbalanced by territorial brand image and diffusion, making of it a winning territory. In this context, the actions of the GDRs and, likewise, the Tourism Consortium are appraised favourably. Working with council majors from different political parties and with the Galician government (Xunta de Galicia), the local actors acknowledge that it is very difficult and the Tourism Consortium is "doing a very good work" (M, 13, 2019). The Rural Development Groups have also done a good work, promoting new projects established by young people (M, 17, 2019). The funds dedicated to development, both the European and the regional ones, played a key role in supporting nature tourism, rural tourism, and enotourism. They have also provided funds for tourism businesses located along the Winter Way (Spanish, "Camino de Invierno," a variant of the Way of Saint James). These processes started at a time when the territory had no external projection yet.

After 17 years analysing and promoting the value of their resources with European LEADER and LIFE funds, the CMGG was born promoted by the GDR RS-Courel. The existence of the Geopark, itself a brand of territorial quality, reflects processes of discovery and social recognition in which several actors are involved, both public and private. Possible oppositions to the project were considered in the elaboration of a strategy towards the future. The candidature to become a Geopark achieved the internal recognition of ecocultural values, proceeded in the background, working step by step, and "surprised the Xunta because projects of this type are usually promoted from above" (M, 14, 2019). The year 2003 saw the inauguration of the first geoviewpoint for the Campodola-Leixazos recumbent fold (Global Geosite) and the signposting plan was started. Since 2006, the promotion of appreciation for the ecocultural resources was pushed forward through expository speeches, didactic booklets, conferences, sessions, and school activities. In the year 2010, the Scientific Station of the Courel, located at the village of Seoane and run by the University of Santiago de Compostela, started its activity with the objective to support research on and teaching about the Galician mountains, including projects contributing to the local development. The Courel Mountains, through the GDR, were involved in national cooperation projects such as Geojob: the geodiversity as a job field (2009-20119, "Geoempleo: la geodiversidad como yacimiento de empleo"); or Museums, naturally (2011, "Museos, naturalmente") which provided the context where the Quiroga Geological Museum was created, the first one of its type in Galicia. Such dynamics is characterized as a process of continual learning in sessions attended by people from other Spanish Geoparks. In the words of local actors, the CMGG project is a challenge, and a specialized working team must be hired to defend and make its development plan. It is provided with the infrastructures and basic services (accommodations, hostels, and camping). They are optimistic toward the future because the start-up of a project including everything related to the mountain "is brilliant" (M, 17, 2019) and will be able to provide further territorial sustainability.

TOWARDS LOCAL EMPORWERMENT

The interactions The narratives of the local actors assert that the diversity of the ecocultural resources in the RS and the CMGG has contributed to the transformation and the reinvention of the territory, accompanied by the restructuration of traditional tourist products and the emergence of new ones. Both cases count with their own development strategies, founded upon a collection of singular ecocultural resources and configured in a common trajectory building up from the interactions of the local community with external agents. Some of the interviewed actors confirm the necessity of public programs in order to effectively promote sustainable strategies, and they also point to the existence of endogenous obstacles.

Material and immaterial assets are clearly intertwined within a strategy for the attraction of tourism, and the involvement of the local community in productive and non-productive projects is encouraged. The actors point to the importance of exchanging ideas and experiences in order to develop further strategies towards the future. In the RS, the territorial assets are defined not only on the basis of the world and the landscape of wine. The protected spaces of the Natura 2000 network (the Sil and the Cabe canyons), the BIC (such as the Monastery of Santo Estevo de Ribas de Sil, transformed into a Tourism Parador), the historic ensembles (e.g. the Castro Caldelas town), the Winter Way, and other identifying elements such as wineries, the Wine Centre, and ethnographic assets, are all of them equally relevant tourist attractors. The material and the immaterial values of the resources are embedded into a unique cultural landscape representative of the territorial identity. The interviewees highlight also the value of singular services related to the tourist use of the ecocultural resources (e.g. catamaran, balneary-spa, or tourist train). In the CMGG there is a manifold of natural resources among which can be found unique instances of tectonic deformations (recumbent folds), exokarst (Mouras valley), caves (Buraca das Choias), glacier systems (A Seara), autochthonous forests (A Regueira), and orchid micro-reserves in limestone. The appraisal of geodiversity and biodiversity is related to the presence of other ecocultural resources such as traditional villages, viewpoints, museums (geological and ethnographic), and interpretation centres (for caves). Currently, the CMGG collaborates with the Galician Association for Territorial Stewardship. To its promotion as a tourist destination integrating nature and culture, it is adjoined the accreditation as a destination for scientific tourism obtained in March 2019 (Observer Science Tourism, 2021). Such an acknowledgment substantiates the value of the Geopark for scientific tourism based on knowledge, participative tourism related to citizen science, and research tourism.

The trajectory of the territorial processes as described by the local actors has a bottom-up orientation in regard to innovation and networking. In the RS, the response to a time of crisis was the diversification of the tourist products. This situation, shared in the context of collaborative work by the managers of local entities, contributed to the idea of a future CMGG. The initiatives to protect and activate the ecocultural resources (attending to local priorities) are focused on a strategy based on the notion of economically sustainable development (considering the sociocultural dimension). Such initiatives define strategies addressing local development, directing both the territory's functional diversification and the creation of territorial quality branding. As for the evolution of the processes, the importance of tourism as a driver of change can hardly be overemphasized. In this sense, the actors point to the role played by the local entities, namely, the Regulatory Board of the Designation of Origin RS (since 1996), the Tourism Consortium (since 2006), and the Rural Development Groups, which agglutinate different expectations and interests, in order to use new opportunities. The extent to which the strategies deployed for the empowerment of the territory become successful is based on the strengths created by the synergies among the local actors. But the answers to the interviews also show that were it not for the support of external agents, it would not have been possible to sustain the territory in a time of crisis and to reorganize the options for the future afterwards. These conditions are the reflection of a territorial dynamics tuned to the objective of sustainability and empowerment through neo-endogenous development (Ray, 2006; Marango et al., 2021). The contributions made by European funds, especially through the LEADER program, as well as by national and regional funds, have been decisive. Always according to the opinion of the local actors, the synergies between bottom-up and top-down processes have generated positive territorial outcomes.

The interactions between the local actors drive the selection and the definition of a collection of ecocultural resources as a common heritage attached to territorial identity that can be set into motion in order to facilitate the territory's sustainability. In this process, as it was advanced by Lowenthal (1998), the recognition of value relies on the perception of tangible and intangible elements within the scope of local knowledge, usage, and practices (Di Meo, 2008; Pillet Capdepón, 2012; Gray, 2019), generating a discourse plenty with symbolic roots, stemming from the material concept of the objects and reaching to the social concept of territory, grounded on territorial identity. The activation of the ecocultural resources encompasses the collective know-how and is dependent on the ability to bring into view what is invisible about a form of existence (Davallon, 2016; Fernandes et al., 2021) with different degrees of appreciation. In fact, the results of the interviews show that the ways in which the resources are included in tourist products have both common and distinguishing features in the RS and the CMGG. Among the first it is to be found the catamaran travels, the network of viewpoints for contemplative activities, all sorts of routes, museum visits, and the promotion of local gastronomic products (e.g. the Geolodays, in collaboration with the Spanish Geological Society). Among the second it must be highlighted in the RS the winery visits, the journeys in the tourist train, the historical-artistic routes, and the wine fairs (RS Tourism, 2021), whereas in the CMGG, which includes a subarea of the Designation of Origin of the RS wine, the expectations (Courel Mountains Geopark, 2021) are high about trekking, cave interpretation, nature routes, and scientific or educational activities.

CONCLUSION

The Organization for Economic Cooperation and Development (2020) states that a new model for investigation about tourism destinations is rising. It implies a reassessment of the tourism sector toward the future, contemplating the new processes of diversification, innovation, and promotion, focusing on the study of new products, new management modes, and new programs that appear at the light of the necessary drop in costs. The processes of reorientation and network dynamics that protect vulnerable territories and make them resilient are understood as key interactions that must be investigated, preferably through a place-based approach.

The local actors in the RS and the CMGG define distinguishable destinations for new types of tourists, promoting the mobilization, use, and collaborative management of their ecocultural resources. The actors' awareness and appraisal of these resources emerge in a sociocultural context. They are agents for the protection of the ecocultural resources and for the revitalization of local entrepreneurship in an inland territory. The activation of the resources, supported by the identity of one's own territory, is framed by the system of local interactions animated by the opportunities detected in the supra-local levels. The local actors refer to a complex of ideas, experiences, and practices mirroring their capability to keep, activate, and bring into awareness the assets of the territory. It includes internal and external wellness, beauty, emotion, and tranquility. The collaboration among the actors, together with innovation aimed at capturing opportunity niches, encourages the use of the ecocultural resources for an inland tourist modality that, besides being healthy and sustainable, is attached to territorial identity.

In the context of inland territories, the RS and the CMGG models display a high potential to protect and promote the ecocultural resources uses with strategies aimed at creating or reinforcing local empowerment. Both in the RS and in the CMGG, the narratives of the local actors pinpoint the power of the ecocultural resources to distinguish the areas as singular territories. The mobilization of resources in several tourist modalities takes place through products that share and spread territorial identity. Tourism becomes embodied in the culture of the territory. Nevertheless, just as stated by the local actors, at the beginning and throughout the evolution of these territorial strategies remains the need for external funding.

The challenges faced by inland territories have become urgent and complex. Their influence and the responses that they generate, especially in inland Galicia (NW Spain), will open up future research venues.

ACKNOWLEDGMENT

This research was supported by the National Research Project "Inland tourism in Spain: challenges, valuation and strategies before changeable situations and of crisis to stimulate products and tourist destinations; dynamics of cases", funding by the Ministry of Economy and Competitiveness of the Government of Spain [CSO2016-74861-R].

REFERENCES

Bosworth, G., Rizzo, F., Marquardt, D., Strijker, D., Haartsen, T., & Thuesen, A. (2016). Identifying social innovations in European local rural development initiatives. *Innovation, 29*(4), 442–461. doi:10.1080/13511610.2016.1176555

Buttimer, A., & Seamon, D. (Eds.). (1980). *The Human Experience of Space and Place*. Routledge.

Byrd, E. T. (2007). Stakeholders in Sustainable Tourism Development and their Roles. *Tourism Review, 62*(2), 6–13. doi:10.1108/16605370780000309

Camarero, L., & Oliva, J. (2016). Understanding Rural Change: Mobilities, Diversities and Hybridizations. *Social Studies, 13*(2), 93–112. doi:10.5817/SOC2016-2-93

Cànoves, G., Blanco-Romero, A., Prat, J. M., & Villarino, M. (Eds.). (2017). Turismo de interior en España. Productos y dinámicas territoriales [Inland tourism in Spain. Products and territorial dynamics.]. València, Spain: Universitat de València, Servei de Publicacions, Colección Desarrollo Territorial.

Cejudo, E., & Navarro, F. (Eds.). (2020). *Neoendogenous Development in European Rural Areas: Results and Lessons*. Springer International Publishing. doi:10.1007/978-3-030-33463-5

Colletis-Wahl, K., & Pecqueur, B. (2001). Territories, development and specific resources: What analytical framework? *Regional Studies, 35*(5), 449–459. doi:10.1080/713693825

Corbelle-Rico, E., Crecente-Maseda, R., & Santé-Riveira, I. (2012). Multi-scale assessment and spatial modelling of agricultural land abandonment in a European peripheral region: Galicia (Spain), 1956–2004. *Land Use Policy, 29*(3), 493–501. doi:10.1016/j.landusepol.2011.08.008

Courel Mountains Global Geopark. (2021). *500 millones de años caminando* [500 million years walking]. Retrieved January 19, 2021, from https://www.courelmountains.es/

Davallon, J. (2016). Penser le patrimoine selon une perspective communicationnelle [Thinking about heritage as a communicative fact]. *Science and Society, 99*(99), 15–29. doi:10.4000ds.5257

Di Meo, G. (2008). *Processus de patrimonialisation et construction des territoires* [Processes of heritagization and construction of the territoires]. Retrieved December 15, 2020, from https://halshs.archives-ouvertes.fr/halshs-00281934

Dissart, J. C., & Seigneuret, N. (Eds.). (2020). *Local Resources, Territorial Development and Well-being*. Edward Elgar Publishing. doi:10.4337/9781789908619

Domínguez García, L., Swagemakers, P., & Fernández, X. S. (2010). Mountain Food Production: Promising niches for rural development in Galicia, Spain. In *Proceedings of the 9th European IFSA Symposium: Building sustainable rural futures* (pp. 1607-1616). Vienna: Universität für Bodenkultur.

European Committee of the Regions. (2020). *EU strategy for Rural Revival* (NAT-VII/006). The EU's Assembly of Regional and Local Representatives.

European Parliament. (2018). EU Agenda for Rural, Mountainous and Remote Areas (2018/2720 RSP). *Official Journal of the European Union, C*, 11.

Farsani, N.T., Coelho, C., & Costa, C. (2011). Geotourism and Geoparks as Novel Strategies for Socio-economic Development in Rural Areas. *International Journal of Tourism Research, 13*, 68–81. doi:10.1002/jtr.800

Féret, S., Berchoux, T., Requier, M., & Abdelhakim, T. (2020). Framework providing definitions, operational typology and review of EU strategies for rural areas. *Working Document from SHERPA: Sustainable Hub to Engage into Rural Policies with Actors*. Retrieved February 16, 2021, from www.rural-interfaces.eu

Fernandes, G. (2020). Ecocultural Heritage, Tourism, Resilience and Landscapes Dynamics in Border Mountains of the Iberia Central Range. In A. Rocha, A. Abreu, J. V. de Carvalho, D. Liberato, E. Alén González, & P. Liberato (Eds.), *Advances in Tourism, Technology and Smart Systems, 171* (pp. 489–504). Springer. doi:10.1007/978-981-15-2024-2_43

Fernandes, G., Castro, E., & Tracana, R. B. (2021). Tourism as a Strategy for Geo-Education and Cultural Valorization of Territories: Promoting Ecocultural Values and Geoconservation in Estrela Geopark, Portugal. In F. Brandão, Z. Breda, R. Costa, & C. Costa (Eds.), *The Role of Tourism in Achieving Sustainable Development Goals* (pp. 117–137). IGI Global. doi:10.4018/978-1-7998-5691-7.ch007

Fredheim, H., & Khalaf, M. (2016). The significance of values: Heritage value typologies re-examined. *International Journal of Heritage Studies, 22*(6), 466–481. doi:10.1080/13527258.2016.1171247

Gómez-Ullate, M., Rieutort, L., Kamara, A., Santos, A. S., Pirra, A., & Gotay, M. (2020). Demographic Challenges in Rural Europe and Cases of Resilience based on Cultural Heritage Management. *European Countryside, 12*(3), 408–431. doi:10.2478/euco-2020-0022

Gray, M. (2019). Geodiversity, geoheritage and geoconservation for society. *International Journal of Geoheritage and Parks, 7*(4), 226–236. doi:10.1016/j.ijgeop.2019.11.001

Hall, C. M., Harrison, D., Weaver, D., & Wall, G. (2013). Vanishing Peripheries: Does Tourism Consume Places? *Tourism Recreation Research, 38*(1), 71–92. doi:10.1080/02508281.2013.11081730

Henriques, M. H., & Brilha, J. (2017). UNESCO Global Geoparks: A strategy towards global understanding and sustainability. *Episodes, 40*(4), 349–355. doi:10.18814/epiiugs/2017/v40i4/017036

Hernández Borge, J. (2011). Despoblación y envejecimiento: Galicia hacia el año 2020 [Depopulation and ageing: Galicia towards the year 2020]. *Revista Galega de Economía, 20*, 121–140.

Ilbery, B. (2014). *The Geography of Rural Change*. Routledge. doi:10.4324/9781315842608

Instituto Galego de Estatística. (2020). *Statistical Open Data of Galicia*. Retrieved January 15, 2021, from http://www.ige.eu/web/index.jsp?paxina=001&idioma=es

Lowe, P., Ray, C., Ward, N., Wood, D., & Woodward, R. (1999). The Forces of Change Affecting Rural Areas. In *Participation in Rural Development* (pp. 1–13). European Foundation for the Improvement of Living and Working Conditions.

Lowenthal, D. (1998). Fabricating Heritage. *History & Memory, 10*(1), 5–24.

Marango, S., Bosworth, G., & Curry, N. (2021). Applying Neo-Endogenous Development Theory to Delivering Sustainable Local Nature Conservation. *Sociologia Ruralis*, *61*(1), 116–139. doi:10.1111oru.12315

Martínez-Filgueira, X., David Peón, D., & López-Iglesias, E. (2017). Intra-rural divides and regional planning: An analysis of a traditional emigration region (Galicia, Spain). *European Planning Studies*, *25*(7), 1237–1255. doi:10.1080/09654313.2017.1319465

Martínez Quintana, V. (2020). *Eco-Cultural Tourism: Sustainable Development and Promotion of Natural and Cultural Heritage*. doi:10.5772/intechopen.93897

Nieto Masot, A., Cárdenas Alonso, G., & Engelmo Moriche, A. (2020). Spatial Analysis of the Rural-Urban Structure of the Spanish Municipalities. *International Journal of Geo-Information*, *9*(4), 213. doi:10.3390/ijgi9040213

Observer Science Tourism. (2021). *Montañas de Courel* [Courel Mountains]. Retrieved January 18, 2021, from https://www.observersciencetourism.com/es/destino/monta-as-de-courel

Organisation for Economic Cooperation and Development. (2020). *The territorial impact of COVID-19: Managing the crisis across levels of government*. Retrieved December 14, 2020, from https://www.oecd.org/coronavirus/policy-responses/the-territorial-impact-of-covid-19-managing-the-crisis-across-levels-of-government-d3e314e1/

Orús, A. H., & Carcavilla Urquí, L. (2020). Twenty Years of Spanish Geoparks: Analysis and Future Prospects. *Geoheritage*, *12*(4), 87. doi:10.100712371-020-00510-2

Paül, V. (2013). Hopes for the countryside's future. An analysis of two endogenous development experiences in South-eastern Galicia. *Journal of Urban and Regional Analysis*, *5*(2), 169–192. doi:10.37043/JURA.2013.5.2.5

Peón Pose, D., Martínez-Filgueira, X. M., & López-Iglesias, E. (2020). Productive vs. residential economy: Factors behind the recovery of rural areas in socioeconomic decline. *Revista Galega de Economía*, *29*(2), 67–44. doi:10.15304/rge.29.2.6744

Pike, A., Rodríguez-Pose, A., & Tomaney, J. (2006). *Local and Regional Development*. Routledge. doi:10.4324/9780203003060

Pilgrim, S., Samson, C., & Pretty, J. (2010). Ecocultural revitalization: Replenishing Community Connections to the Land. In S. Pilgrim & J. Pretty (Eds.), *Nature and Culture: Rebuilding Lost Connections* (pp. 235–256). Earthscan Publications. doi:10.4324/9781849776455

Pillet Capdepón, F. (2012). El turismo de interior en la España peninsular: El patrimonio territorial como destino turístico [Inland Tourism in Peninsular Spain: Territorial Heritage as a Tourist Destination]. *Boletín de la Asociación de Geógrafos Españoles*, *59*, 345–365. doi:10.21138/bage.1461

Ray, C. (2006). Neo-endogenous rural development in the EU. In P. Cloke, T. Marsden, & P. Mooney (Eds.), *The handbook of rural studies* (pp. 278–292). Sage Publications. doi:10.4135/9781848608016.n19

Ribeira Sacra Tourism. (2021). *Descubre Ribeira Sacra* [Discovering Sacred Riverside]. Retrieved January 12, 2021, from https://turismo.ribeirasacra.org/

Richards, G. (2020). Designing creative places: The role of creative tourism. *Annals of Tourism Research*, *85*, 102922. doi:10.1016/j.annals.2020.102922

Rodríguez-Pose, A. (2000). Economic convergence and regional development strategies in Spain: The case of Galicia and Navarre. *European Investment Bank Papers*, *5*(1), 88–115.

Ruiz Pulpón, A. R., & Cañizares Ruiz, M. C. (2020). Enhancing the Territorial Heritage of Declining Rural Areas in Spain: Towards Integrating Top-Down and Bottom-Up Approaches. *Land (Basel)*, *9*(7), 216. doi:10.3390/land9070216

Sánchez-Mesa Martínez, L. J. (2019). Planning Territorial Policies against Inner Areas Depopulation in Spain: Keys for sustainable management of cultural and environmental resources. *Il Capitale Culturale*, *19*, 53–81. doi:10.13138/2039-2362/1980

Serra, P., Vera, A., Tulla, A. F., & Salvati, L. (2014). Beyond urban-rural dichotomy: Exploring socioeconomic and land-use processes of change in Spain (1991-2011). *Applied Geography (Sevenoaks, England)*, *55*, 71–81. doi:10.1016/j.apgeog.2014.09.005

Sharpley, R. (2014). Host perceptions of tourism: A review of the research. *Tourism Management*, *42*(1), 37–49. doi:10.1016/j.tourman.2013.10.007

Sineiro-García, F., Vázquez-González, I., & García-Arias, A. I. (2014). Mountain Family Farms in Galicia, Spain: Challenges and Strategies. *Mountain Research and Development*, *34*(4), 375–385. doi:10.1659/MRD-JOURNAL-D-14-00015.1

Spanish Observatory of Rural Tourism and Ecotourism. (2021). *Tendencias turísticas: Ecoturismo y turismo rural* [Tourist trends: Ecotourism and rural tourism]. Retrieved January 22, 2021, from https://www.escapadarural.com/blog/ecoturismo-y-turismo-rural/

UNESCO. (2015). *International Geoscience and Geoparks Programme*. Retrieved December 16, 2020, from http://www.unesco.org/new/en/natural-sciences/environment/earth-sciences/international-geoscience-and-geoparks-programme/

Van Der Ploeg, J. D., & Renting, H. (2000). Impact and Potential: A Comparative Review of European Rural Development Practices. *Sociologia Ruralis*, *40*(4), 529–543. doi:10.1111/1467-9523.00165

Wallace, G., & Russel, A. (2004). Eco-cultural tourism as a means for the sustainable development of culturally marginal and environmentally sensitive regions. *Tourist Studies*, *4*(3), 235–254. doi:10.1177/1468797604057326

Williams, S., & Lew, A. (2015). *Tourism Geography. Critical understandings of Place, Space and Experience*. Routledge.

Woods, M. (2007). Engaging the global countryside: Globalization, hybridity and the reconstitution of rural place. *Progress in Human Geography*, *31*(4), 485–507. doi:10.1177/0309132507079503

Xunta de Galicia. (2017). *Resolución de la Dirección General del Patrimonio Cultural* [Resolution of the General Directorate of Cultural Heritage]. Retrieved December 16, 2020, from https://www.xunta.gal/dog/Publicados/2017/20171229/AnuncioG0164-261217-0001_es.html

Yepes Temiño, J. (2002). *Geomorfología de un sector comprendido entre las provincias de Lugo y Ourense* [Geomorphology of a sector comprised between the Lugo and Ourense provinces]. A Coruña, Spain: Publications of the Laboratorio Xeolóxico de Laxe, Serie Nova Terra.

ADDITIONAL READING

Bernard, H. R., & Ryan, G. W. (2010). *Analyzing qualitative data: Systematic approaches*. Sage Publications.

Cloke, P., Marsden, T., & Mooney, P. (Eds.). (2006). *The Handbook of Rural Studies*. Sage Publications.

de San Eugenio Vela, J., Nogué, J., & Govers, R. (2017). Visual landscape as a key element of place branding. *Journal of Place Management and Development, 10*(1), 23–44. doi:10.1108/JPMD-09-2016-0060

Freeman, R. E. (1984). *Strategic management: A stakeholder approach*. Pitman.

Harril, R. (2004). Residents' attitudes toward tourism development: A literature review with implications for tourism planning. *Journal of Planning Literature, 18*(1), 251–266. doi:10.1177/0885412203260306

Hoggart, K., & Paniagua, A. (2001). The restructuring rural Spain? *Journal of Rural Studies, 17*(1), 63–80. doi:10.1016/S0743-0167(00)00037-1

Stock, F. (2009). Identity, image and brand: A conceptual framework. *Place Branding and Public Diplomacy, 5*(2), 118–125. doi:10.1057/pb.2009.2

Waligo, V., Clarke, J., & Hawkins, R. (2013). Implementing sustainable tourism: A multi-stakeholder involvement management framework. *Tourism Management, 36*, 342–353. doi:10.1016/j.tourman.2012.10.008

KEY TERMS AND DEFINITIONS

Adaptation: A territory's adjustment to existing conditions in both the global and the local contexts.

Courel Mountains: Eastern mountain range in inland Galicia, NW Spain, province of Lugo, recently designated Global Geopark by UNESCO.

Ecocultural: A quality of the elements in which nature and culture are embedded together.

Inland: Territories far from the coast and from the main urban areas.

Revitalization: Territorial process of dynamization and development involving environmental, sociocultural, and economic dimensions.

Sacred Riverside: A rural area in inland Galicia, province of Ourense, NW Spain, "Ribeira Sacra" in the Galician language, candidate to be included in the World Heritage List.

Sinergy: Complex interaction among several factors or people that contribute to generate new conditions for the territory.

Chapter 2
The Contribution of UNESCO Global Geoparks in the Promotion of Sustainable Tourism:
The Example of Estrela UGGp

Hugo Gomes
Associação Estrela Geopark (AGE), Portugal

Emanuel de Castro
Associação Geopark Estrela (AGE), Portugal

Patrícia Azevedo
Associação Geopark Estrela (AGE), Portugal

Fábio Loureiro
Associação Geopark Estrela (AGE), Portugal

ABSTRACT

The Estrela UNESCO Global Geopark, which encompasses the highest mountain in mainland Portugal, is a geographically defined territory with resources of notorious geological interest, with precious ecosystems, identity, and ways of life. Its geodiversity, associated with traces of glacial origin, the richness of its natural and cultural heritage, make this a unique territory, whose structured approach will strengthen its knowledge, the importance of its functions, and its tourist, heritage, and cultural potential. The Estrela UGGp has the mission of contributing to the protection, valorisation, and dynamics of the natural and cultural heritage, with special emphasis on geological heritage, with a view of disseminating scientific knowledge, promoting tourism and the sustainable development of serra da Estrela. Overall, the Estrela UGGp fosters the development of an integrated development strategy, combined with geoconservation, education, and tourism, supporting the construction of development strategies for the well-being of the community.

DOI: 10.4018/978-1-7998-7339-6.ch002

UNESCO AND GEOSCIENCES PROGRAMS

UNESCO, in addition to its literacy campaigns and its World Heritage Sites, is also known for working in such a way that scientific and technological knowledge is a right of humanity, putting the human being at the center of priorities, that is, ensuring that science and knowledge are put towards the well-being of all the inhabitants of planet Earth, promoting their sustainable development.

Since its foundation, UNESCO has sought to achieve development and peace through scientific knowledge and the science produced, putting it at the service of all, on the assumption of contributing to the resolution of the real problems of societies, making them more informed and aware and, in this way, better prepared to achieve their development.

UNESCO's International Geosciences and Geoparks Program, created on the 17th of November 2015, is today a new paradigm of valorisation, promotion and development of territories, anchored in the importance of their geological heritage. In the centre of its concerns, UNESCO Global Geoparks advocate a holistic view of the territories, the implementation of concerted actions between the different development agents, a conservation and valorisation strategy for their sites of geological interest and a territorial-based development policy that is effectively integrated and participatory.

Having said that, UNESCO Global Geoparks are well-delimited territories, holders of relevant geological heritage, linked with a sustainable development strategy that has in the promotion of Geosciences, education for sustainable development and geotourism, their main pillars (Hose, 2012) . Also, these must also value, promote and preserve biodiversity, cultural heritage, crafts and gastronomy, among others, seeking to improve the living conditions of populations, promoting, in an integrated way, endogenous values (Stueve et al., 2002).

Each UNESCO Global Geopark is unique and tells part of the long history of Earth Planet (4600 Million years), using its geology as a tool for the development of its communities, through the implementation of strategies based on Science, Geoconservation, Education for Sustainable Development and Tourism.

UNESCO Global Geoparks (UGGp), as holistic territorial strategies for the promotion, enhancement and protection of natural heritage, in particular geoheritage, play an important role in the sustainable development of territories. At the same time, UNESCO classified territories can generate new job opportunities, new economic activities and additional sources of income, especially in rural regions. These strategies encourage the production of new products and enhance regional handicrafts, involving geotourism activities and "geoproducts", focusing on improving the economy of local communities.

In these pandemic times, tourism in rural areas gains expression and followers, due to the less densified modalities of accommodation and the search for outdoors activities, which leads to the emergence of products leveraged on the territory's endogenous resources, such as: health and well-being tourism, nature tourism and heritage interpretation for the valorisation of nature and landscapes. As such, tourist practices are increasingly complex, in the way they are sought, the means involved, the needed experiences, and the relationship with the territory. Tourism, in its territorial relationship and transformations generated, promotes local resources, develops products to which services are associated and generates new jobs.

The frequent divergence between the global needs of nature protection and the local needs of socio-economic development has been, in many situations, an obstacle to the management of natural places, in particular those of low density, as it can disqualify their perception and the activities established by their population. They are places of opportunity, reinforcing new approaches, generating territorial cohesion, promoting self-esteem and the identity of local communities, which are fundamental for differentiating tourist experiences. These positions require specific and concerted attention from the various interlocutors, especially the management entities, capable of creating conditions for the sustainability and qualification of these territories and their valorisation.

The promotion of different actions focused on natural and cultural heritage aims to stimulate socio-economic activity through the creation of projects related to the tourism sector, to disseminate and promote visits and tourist circuits in the territory. Education is the basis of geotourism and a pillar for the sustainability of heritage and local communities. The visitor develops attitudes according to the available information and didactic and scientific interpretation. In this way, geotourism plays a fundamental role in sustainable local development, while leisure activities enhance environmental education (Hose, 2000, 2005; Farsani et al., 2011).

THE NEW TOURIST PARADIGMS OF THE 21ST CENTURY

Tourism is a relevant economic and social activity, which appropriates territories and is structured through their attractiveness. Today, the importance of this activity for the economy it is undeniable, due to its ability to generate revenue and wealth, to the increase in the number of flows and also for its propensity to generate new jobs and thus, to generate better living conditions. In fact, tourism is no longer just an economic activity, but a social phenomenon.

Throughout the history of Tourism, we have heard of great changes. For years, this activity was enjoyed only by a public with financial capacities far above average and in this way, the sector was also considered luxury, both for its characteristics and for the public it managed to reach. However, after the *Grand Tour*, tourism started to be perceived differently, by promoting the discovery of the pleasure of travelling as a noble and enriching personal experience. Travelling was now no longer geared towards war expeditions, missionaries or pilgrims, nor towards scientists, but rather towards travellers who loved culture, the landscape and above all tourists who travelled for pleasure and to make the most of their free time. However, these were only the first changes that were perceived in tourism. This is a sector that is constantly changing and needs rapid updates and adaptations in terms of tourist tastes and profiles.

In the 21st century, we are facing a different tourist profile, one that perhaps we would not have imagined a few years ago. Today the tourist is increasingly more autonomous, more curious, more adventurous and above all more open to new experiences, as the memories of these moments contribute to their return to the same place in the future. Thus, Tourism becomes part of the everyday life of territories, since they become its main experience.

The territories have in their essence, a set of characteristics that differentiate them and make them unique, of which we highlight their geography, geology, culture and, mainly, their identity. It is from these characteristics that differentiating tourism products are developed, as we can take advantage of the best that each territory has, so that we can promote it effectively. Therefore, and taking into account the constant changes in recent years, in relation to existing tourist offers and also due to the tourist profile, we face a differentiating tourism today. This means that we find several tourist products properly structured and we focus more on diverse target-publics. There is an increasing work of the responsible tourism entities, to differentiate the country's offer, as a territory of all and for all, a territory that is much more than the Sun & Sea, known for Nature Tourism, Health and Wellness Tourism, Scientific and Educational Tourism and more recently, a country with strategies at the level of Sustainable and Accessible Tourism (Nunes, 2010). In this sense, it can be said that the work of tourism is constant and will never be finished, because what is certain today, tomorrow may be a banality, which arouses little interest. Therefore, it is up to those who work in tourism, to think it and structure it always taking into account the identity of each of the territories, for which the work and emergence of Global Geoparks, a classification established by UNESCO, which has tourism one of its most important foundations, has contributed to.

UNESCO GLOBAL GEOPARKS AND ITS RELATION WITH TOURISM

UNESCO Global Geoparks are eminently territorial and holistic classifications, which are based on territorial approaches such as the enhancement, preservation and promotion of biodiversity, cultural and geological heritage, gastronomy and scientific research. The concept of "Geopark" emerged at the end of the 20th century in Europe.

A UNESCO Global Geopark is a well-delimited territory with a remarkable geological history which, due to its relevance and significance, constitutes a common legacy that should be safeguarded and valued for future generations, with the aim of enhancing, promoting and developing territories, anchored in the importance of geological heritage and its significance. (UNESCO, 2015).

This UNESCO classification focus on the implementation of a sustainable development strategy, based on the conservation of geological heritage, in association with the remaining elements of natural and cultural heritage, with a aim of improving the life of the populations living within, promoting endogenous values in an integrated manner. The basic philosophy behind the creation of geoparks is centred on the development of networks which allow an exchange of experiences and a joint promotion of the concept, and of each of the network members. The three pillars of this UNESCO classification are the Conservation of Geological Heritage, Education for Sustainability and Tourism Development. The creation of UNESCO Geoparks gave visibility and notoriety to Geosciences, encouraging interest in a more specific knowledge, by those who visit the territory. The classification of a territory as a UNESCO Global Geopark has direct impacts on tourism, contributing to the dissemination of the territory, giving it scale and greater notoriety.

Aware of the value and relevance of its Geological Heritage, UNESCO Geoparks, whose mission is to contribute to the protection, valorisation and promotion of natural and cultural heritage, in a perspective of deepening and disseminating scientific knowledge and promoting sustainable development of the territories from their endogenous resources, can contribute to develop more sustainable territories, through new tourism strategies, based on the concept of Geotourism, not only as a product, but especially with a new paradigm of tourism action. In a broader sense, this is a "bottom-up" approach to territorial development, which assumes the natural heritage as an effective resource for the promotion and valorisation of the landscape and a resource with effective tourism potential.

Nowadays, Geotourism is seen as a strategy for territorial enhancement whose main objective is to experience the endogenous resources, and not only the geological ones, promoting their understanding, environmental and cultural valorisation, with the local community being the main beneficiary. In this sense, Geotourism is an important tool for the dissemination and interpretation of the Geological and Geomorphological Heritage because it allows the knowledge about geosciences to reach a greater number of people, triggering a greater openness to the enhancement of the Heritage and the whole territory, because we can only value what we really know. In fact, the integrated interpretation of heritage and landscape translates into a new tourist experience, making Geotourism a territorial strategy and not only a product based on geology.

Tourism is, in fact, a relevant economic and social activity, which appropriates territories and is structured through their attractiveness, contributing to new functions and to the emergence of other activities, so the increased visibility provided by the UNESCO brand leads to various tourism phenomena, such as increased demand and appreciation by tourists, a growth in revenues by the various companies linked to the tourism sector, a greater number of flows and last but not least, the creation of a greater number of jobs. The relationship between Geoparks and Tourism is unequivocal, either by the wealth they generate, or by the improvement of living conditions they provide to local people, allowing them to live and work in the place where they grew up, especially because the basis of this classification is a strategy for promoting sustainable tourism, enhancement of endogenous potential, an eminently differentiating and structuring factor of destination and tourism product.

SUSTAINABLE TOURISM: CHALLENGES AND OPPORTUNITIES

According to the World Tourism Organization, sustainable tourism should safeguard the environment and natural resources, ensuring the economic growth of the activity, i.e., capable of meeting the needs of the present and future generations. Therefore, tourism development should be guided by "*managing natural resources, and should be favoured and encouraged by national, regional and local public authorities*" (WTO, 2007).

Tourism activities must be designed and planned in such a way that the natural heritage, made up of ecosystems and biodiversity, is protected, and endangered species of wild fauna and flora are preserved. Those involved in tourism development, especially professionals, must allow limitations to be imposed on their activities when they are carried out, particularly in sensitive areas such as mountains.

Sustainable tourism emerges as an alternative to mass tourism, as it is concerned with the number of people who will visit the receiving regions and locations. In this sense, tourism management should consider environmental, cultural and social issues, seeking to minimise the impact of activities and ensuring that local communities are economically and socially included in these strategies. But is this approach a reality or a utopia?

Nature tourism, rural tourism and ecotourism are considered modalities of tourism that are more sustainable, because they develop activities in harmony with the site. However, sustainable tourism must be developed by all, because failure to meet this new reality will condemn many destinations economically, since they will lose their attractiveness due to the negative impacts caused by, for example, the destruction of natural values.

But tourism, namely the so-called "mass tourism", also has its negative consequences, both environmentally, economically and socially. Among these consequences, the following stand out for their potential to generate problems: intensification of building activity, overcrowding in many holiday resorts, direct competition with other land uses, the loss or disruption of local traditions, changes in social structure and local lifestyles, impacts on the landscape, changes in ecosystems, and the contamination of natural resources. All these aspects can be aggravated by seasonal tourism (WTO, 2006; Cunha, 2003).

Tourism is no stranger to constant changes in society, politics, technology, culture and the economy, and new trends are emerging accordingly to both Demand and Supply. Among the most significant trends is a greater awareness of the environment and the new paradigm of sustainability, and in Tourism, the growing relevance of "Nature Tourism" and the use of Protected Areas for its development, requires a new way of being and acting on the part of Tourism Management.

Today, Tourism is one of the most relevant economic and social activities in the World, in the European and, particularly, National context. Its contribution to the creation of wealth and improvement of citizens' well-being is felt in multiple aspects, penetrating and integrating diverse areas of society. "*By grouping by affinities the reasons that lead people to travel, it is possible to identify a variety of types of tourism*" (Cunha, 2003, 2011). That is, the need to seek the different, to live unique and singular experiences or simply to respond to cultural, physical or mental appeals, contribute to the emergence of new tourism products and new leisure activities that make up the tourism offer. At the base of these "needs" is a multitude of factors in constant change, which allows us to speak of trends, not only on the Demand side, but also, consequently and imperatively, on the Supply side. In this way, the changes in mentalities, behaviour and lifestyles that underpin the change are being felt at the level of consumer preferences and demands, with more rigorous evaluation criteria. Mass Tourism" is rapidly being replaced by "Alternative Tourism" which values personal experience. Among the most significant trends is the growing environmental awareness and the importance of defining an appropriate use of the available resources in order to preserve them, rather than degrade them. This duality in Tourism is even more sensitive since the new destination areas, distinguished by their unique ecological resources and/or landscape value, should be subject to protection and valorisation.

Sustainability in tourism is an inescapable path, which is being taken with growing awareness by the sector's agents, on a global scale. It means taking into account the needs of visitors, the sector and the communities and its environmental, economic and social impacts now and in the future. Sustainable Tourism must make appropriate use of environmental resources, respect the socio-cultural authenticity of communities and ensure that economic activities are viable in the long term. It also requires the informed participation of stakeholders and constant monitoring of its impacts, while maintaining a high level of satisfaction among tourists (WTO, 2007).

In 2015, the United Nations approved the 2030 Agenda, which consists of 17 Sustainable Development Goals (SDGs). Tourism has directly and indirectly contributed to all SDGs, having been included mainly in Goals 8 (sustainable economic growth), 12 (sustainable consumption and production), but also contributing to the achievement of targets of other SDGs.

At the national level, the vision for the Tourism Strategy 2027 is based on the affirmation of "*Tourism as a hub for economic, social and environmental development throughout the territory, positioning Portugal as one of the most competitive and sustainable tourist destinations in the world*". Portugal, as a quality tourist destination, bases its competitive advantage on the principles of sustainability, on the diversified offer and on the valorisation of its distinctive and innovative characteristics, placing people at the centre of its strategy. The current epidemiological pandemic of the new Coronavirus (COVID-19) has caused a negative impact on the economy and society. However, and according to the WTO - "One Planet Vision" - the responsible recovery of the sector, after the COVID-19 pandemic, will allow it to resume activity in an even stronger and more sustainable way (Agência de Turismo de Portugal, 2017).

At the European level, the *European Ecological Pact* was also presented, which aims to transform the European Union into a modern, more efficient and competitive economy regarding the use of its resources. The action plan of this "Pact" stresses the importance of the commitment and involvement of all sectors of activity, including Tourism (Council of Europe, 2021). In the same strategy, Tourism of Portugal is preparing to present the *More Sustainable Tourism Plan*, which provides a set of actions that introduce an effective sustainable component throughout the tourism value chain.

In this context, UNESCO Global Geoparks are strong strategies that reflect the commitment to strengthen the role of Tourism in building a better world for all. In fact, these territories distinguished by UNESCO present as main challenge the necessary balance between the promotion of tourism activity and the preservation of their identity and heritage values. In parallel, these are a real opportunity to build a new paradigm for tourism, based on scientific knowledge, enhancement of endogenous resources and community involvement, moving away from a vision of mere territorial consumption to become a community phenomenon and integrated into the territoriality of destinations. The UNESCO Geoparks are, therefore, places for innovation, promotion and differentiation of tourism, revealing themselves as new paradigms for the XXI century.

GEOTOURISM AS A TERRITORIAL DEVELOPMENT STRATEGY: THE CASE OF THE ESTRELA UGGP

The Associação Geopark Estrela (Estrela Geopark Association) is the manager of the Estrela UNESCO Global Geopark brand and has as its mission "*to contribute to the protection, enhancement and promotion of natural and cultural heritage, with special emphasis on geological heritage, in a perspective of deepening and disseminating scientific knowledge, promoting tourism and the sustainable development of the Estrela UNESCO Global Geopark territory*" (Castro et al., 2017; Gomes et al, 2019).

Tourism is one of the pillars of any UNESCO Global Geopark and, as such, actions were selected that aim to, among others: boost tourism in the Estrela UGGp; strengthen partnerships; contribute to the continuity of the identity of Estrela; contribute to the creation of a strong tourism brand, based on heritage and culture; encourage an increase in the number of visitors and average expenditure/day per visitor; as well as contribute to overcoming tourist seasonality.

Within the scope of the tourism strategy, the Estrela Geopark invests in the consolidation of the interpretative routes network, which aims to make known the existing natural and cultural heritage, with particular focus on geological heritage, promoting the enjoyment, knowledge and valorisation of the endogenous resources of the Serra da Estrela territory. The Interpretative Routes of the Estrela Geopark include the 9 municipalities that integrate this classification (Figure 1). This is a journey through its heritage and geosites classified by UNESCO, constituting an opportunity to promote a tourism of and for ALL! As such, the extension of this network is intended, consolidating the promotion and knowledge of the territory.

The promotion of the concept and the Estrela UNESCO Global Geopark brand is only possible through synergies with local agents, who are on the ground every day in contact with the populations of this territory. For such, this association has defined a network of partners in order to amplify this dissemination, as well as to strengthen connections and promote the territory. This network has been divided into 4 segments: Institutional, Educational, Business Partners and Local Producers. The aim is to extend this network even further so that the whole territory can be covered by the Estrela UGGp brand.

The demands of tourism are becoming more and more accentuated in these days, so the existence of qualified technicians who can suppress the needs inherent to this area is necessary. In this perspective, the Associação Geopark Estrela bets in the consolidation of training in the area of Tourism and UNESCO Global Geoparks, through trainings based in what the mission and the values of this Geopark are, as

Figure 1. Territory of the Estrela UNESCO global geopark
(Source: Associação Geopark Estrela).

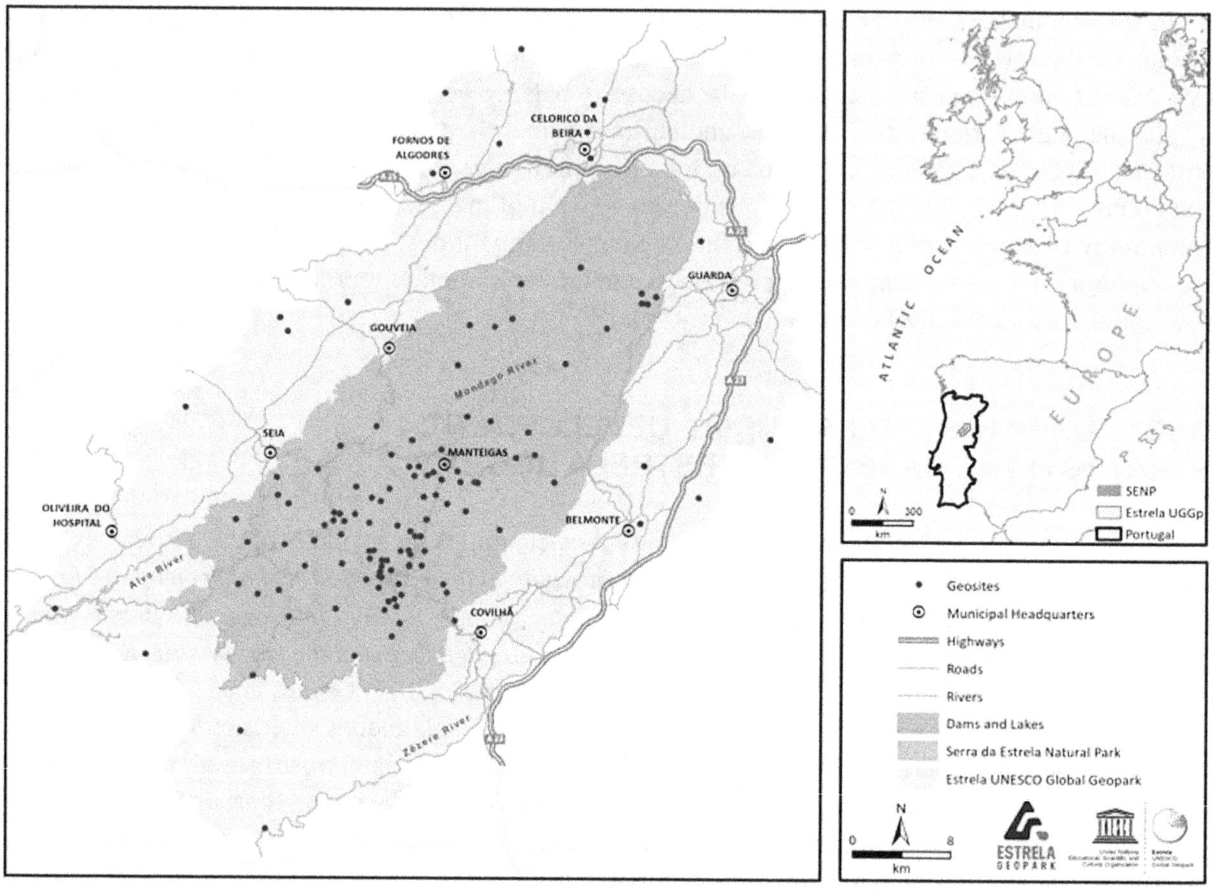

well as on all the patrimonial elements related to Estrela. It is necessary to counteract seasonality and offer a variety of resources, based on what are tourist activities *per se*, but also on all the elements that can viewed as heritage and used for tourism development.

It is imperative that in a developed and conscious society, sustainable tourism is practiced assiduously, stimulating the increase of the number of visitors and the average expenditure per day and per visitor, but allowing the preservation of the resources. In this way it will be possible to increase tourist affluence in a sustained way, so that all visitors can enjoy these resources in the same way. The focus on the dissemination and interpretation of all the heritage, as well as the incentive to tourist training are also crucial, thus allowing a more enriching experience to be provided to visitors, which can translate into an increase in the length of their stay.

However, to talk about Tourism in a territory classified as a UNESCO Global Geopark, is unequivocally addressing the issue of Geotourism. Since its initial definition, (Hose, 1995) where it refers to Geotourism as "*the provision of interpretative services and facilities that enable tourists to acquire the necessary knowledge to understand the Geology and Geomorphology of a place beyond mere aesthetic appreciation*" (Rodrigues, 2009), the concept has clearly evolved, moving away from the vision exclusively focused on geological values and incorporating the other heritage elements and endogenous resources of territories. In fact, the practice of tourism sustains and enhances the identity of a territory, taking into account its geology, environment, culture, aesthetic values, heritage and the well-being of its residents (Arouca Declaration, 2011) is the concept that best defines Geotourism today, representing a key lever for the promotion of territorial development in UNESCO Global Geoparks.

As an important pillar for the sustainable development strategy, Tourism seeks to promote the valorisation of heritage, the development of new products and services and encourage the practice of local arts and crafts, thus promoting economic growth and the creation of new job opportunities and added value for the territories. In this perspective, Tourism, and in particular Geotourism, emerges as a strategy around which various products are developed, such as nature, wellness, scientific and educational tourism (Castro et al., 2018). Still on this theme, Geotourism can be stimulated in these UNESCO territories through a series of strategic lines, of which we highlight:

- the strengthening of partnerships and networking, both within each geopark and its local agents, and between the different geoparks (169 spread throughout the world in 2021);
- the enhancement of territorial identity, thus creating unique experiences for visitors and fostering a sense of belonging for locals; the creation of a strong tourism brand, based on heritage and culture;
- encouraging visitation and increasing tourist flow, always taking into account the carrying capacity of the territories and the preservation of heritage values;
- increase in tourism supply throughout the year, avoiding problems associated with tourist seasonality;
- a structured approach to sustainable tourism supply that is built around the intrinsic characteristics of the Serra da Estrela territory.

In fact, since the first references to Geotourism, this concept is increasingly emerging as a holistic and territorial-based strategy, from which sustainable tourism initiatives are promoted, based on endogenous resources identity. In this context, the prefix “GEO” means territory and not geology, as is often referred to. Therefore, we are not facing a tourism product based on geology, but an integrated approach that fits into the very dimensions of a UNESCO Global Geopark.

Some of these ideas justify that, in the context of Portugal, the concept of Geotourism gains greater expression when associated with the UNESCO Global Geoparks, since through its networking and the initiatives developed internally, these give a strong contribution to the territorial development of sectors of the country with greater economic and demographic problems.

It is necessary to reinforce the idea that Geotourism is not a tourism for a market niche, Geotourism constitutes itself, a strategy for knowledge dissemination about the territory and can be used in several areas, not only in tourism but also in education (Hose, 2000). In this sense, this has been an approach implemented by the Estrela Geopark, in order to diversify the existing tourism products, the reduction of seasonality and the development of a sustainable approach based on the construction of new tourist experiences.

The UNESCO Geoparks in Portugal are a great example of this, because they take advantage of the territory in the best way, giving to who visit it, a completely different perspective from the usual. These territories classified by UNESCO work Geotourism in different perspectives, whether scientific, touristic, educational or simply with a vision of awareness and valorisation of its heritage elements (Figure 2).

Figure 2. Estrela UGGP interpretative panels
(Source: Associação Geopark Estrela)

The Portuguese Geoparks promote Educational Programmes, Interpretative Routes, small lectures to the Communities, Conferences that address issues that are often of concern to these territories, but above all work on enriching strategies for the territories, taking advantage of what the Earth itself gives us, its geology, which is complemented by its geography, its culture, its identity, i.e., a Geopark enriches the territory by exploring its heritage values.

In Portugal, in 2021, there are five UNESCO Global Geoparks, each with its own characteristics, which differentiate them, but which unequivocally work in common areas. As with others, this is a classification that requires much networking, cooperation and mutual help, because this is a relatively recent classification that needs to gain strength and notoriety that is only possible with networking between all Geoparks (Zouros, 2004). In this sense, the five UNESCO Geoparks in Portugal have been working with the Tourism of Portugal, whose goal is to make these territories properly structured and sustainable authentic tourist destinations.

In fact, there has been an effort to strengthen the UNESCO Geopark brand as differentiating tourist destinations, through a network work that aims essentially at the empowerment of local tourism agents, the structuring of products, based on science and the valorisation of its heritage of international relevance, and structured communication that enhances the positioning of these territories in the national and international context.

In short, the classification of Estrela as a UNESCO Global Geopark allowed the emergence of new tourism approaches in the territory, through a territorial-based strategy that has materialized in Geotourism itself. This approach has allowed a greater involvement of communities and tourism partners, the valorisation of endogenous products and the activation of new tourism products. At the same time, we are currently witnessing a greater awareness of the value of heritage and its valorisation, through public policies that reinforce the specificity and identity of its landscape, pushing back old processes of deterritorialisation. All in all, we are facing a new sustainable vision of tourism activity in Estrela, articulated, networked and with effective added value for their communities, the primary objective of a UNESCO Geopark, which tends to be transposed to the tourism approaches of today.

CONCLUSION

The XXI century poses prominent challenges in the relationship between territories, their resources and the use that is made of them by various activities, among which we highlight, in the context of this chapter, Tourism. In fact, the relationship between the latter and its territoriality has been the subject of debate, especially in the search for answers about the sustainability of this relationship.

It is worth pointing out that tourism sustainability is much more than a mere environmental or ecological approach. This presupposes a territorial vision, where natural resources and communities contribute to a development based on the intrinsic characteristics of each place. In this context, the UNESCO Global Geoparks have emerged to meet these challenges, emerging as opportunities to build a new paradigm of tourism sustainability. From the geological and geomorphological heritage and its landscape, various areas of sustainable development are worked, among which we highlight Tourism, for its inductive capacity, but also by the transversal approach that presents in the context of the destinations themselves. Thus, since the beginning of this century, Geoparks have grown in number, but especially in the relevance they present as territories and strategies for sustainable development.

From this approach, Geotourism has seen an obvious link with UNESCO Global Geoparks, although it is not reduced to the latter. Despite a clear relationship the approach can be often confusing. As we have clarified throughout the article, Geotourism should be perceived essentially, as a tourism development strategy and not in its reductive vision of a tourism product strictly related to geological or scientific visitation. This is a much more holistic, integrated and territorial approach that includes geological heritage, its relationship with non-geological heritage, endogenous resources and the identity of the territories, which together contribute to the differentiation and uniqueness of each of the destinations, where landscapes and inhabitants are not mere tourist scenarios, but active players in the construction of different tourism products.

Based on this integrated vision, the Estrela Geopark has sought to consolidate a new tourism vision for the Estrela territory, enhancing the value of its resources as the main asset for this territory, promoting networking, the emergence of new products, strengthening synergies between different territories and a sustainable approach, which is not exhausted in the meaning of the word itself. Thus, Tourism emerges as a strategy to leverage new approaches of territorial development, which guarantee a higher added value for populations and territory, which benefits from the favourable positioning that UNESCO itself gives to these places. On the other hand, the transversal approach, imposed by the Geoparks to Tourism, relating it with Science, Education, Community Development and Communication itself, emphasize a differentiator development model, emerging as an opportunity for this century.

In short, to speak of UNESCO Global Geoparks is to transpose the geographical limits of its territory to the concept and the approach that results from them. Here, the close relationship between this classification and its territoriality allows to look at Tourism not only as a consumer of space, but essentially as a factor of promotion and enhancement. In fact, if we think that these territories are in themselves the main development resource, we understand the true meaning of sustainability and the reason why the 161 UNESCO Geoparks around the world and the five in Portugal have contributed to the new look on Tourism and in particular on its relationship with the territorial identity itself, *sine qua non* condition for a differentiating tourism offer, with low territorial impact, in line with the carrying capacity of the territories and with the tourist experience, the great motivator for tourism travel and demand.

REFERENCES

Agência de Turismo de Portugal. (2017). Estratégia 2027 - Liderar o Turismo do Futuro. Author.

Castro, E., Loureiro, F., Gomes, H., & Vieira, G. (2018). O Património Geológico do Geopark Estrela e a sua Valorização. In Geopatrimónio - Geoconhecimento, Geoconservação e Geoturismo. CEGOT.

Castro, E., Vieira, G., Loureiro, F., Patrocínio, F., Firmino, G., Gomes, H., & Fernandes, M. (2017). Managing the Estrela Aspiring Geopark: A framework for promoting geoheritage conservation. *Conference Ciência 2017.*

Conselho da Europa. (2021). https://www.consilium.europa.eu/pt/policies/green-deal/

Cunha, L. (2003). *Introdução ao Turismo* (2nd ed.). Editorial Verbo.

Cunha, L. (2011). Autenticidade e Inovação: factores de renovação dos destinos turísticos maduros. Cogitur Journal of Tourism Studies, 4(4), 9-28.

Declaração de Arouca, (2011). Congresso Internacional de Geoturismo. In *Geotourism in Action*. Arouca.

Farsani, N. T., Celeste Coelho, C., Carlos Costa, C., & Carlos Neto, C. (2011). *Geoparks and Geotourism. New Approaches to Sustainability for the 21st Century*. Universal Publishers.

Gomes, H., Loureiro, F., Cezar, L., Castro, E., & Vieira, G. (2019). Interpreting Late Pleistocene paleoenvironments through the geosites of the Estrela Geopark (Central Portugal). In *International Meeting on Paleoclimate: change and adaptation – book of abstracts*. Instituto Terra e Memória - Centro de Geociências da Universidade de Coimbra.

Hose, T. A. (1995). Selling the Story of Britain's Stone. *Environmental Interpretation*, *2*, 16–17.

Hose, T. A. (2000). European "Geotourism" – geological interpretation and geoconservation promotion for tourists. In Geological Heritage: its conservation and management. Academic Press.

Hose, T. A. (2005). Geoturismo europeo. Interpretacíon geológica y promoción de la conservación geológica para turistas. In D. Barretino, W. A. P. Winbledon, & E. Gallego (Eds.), *Patrimonio geológico: conservación y gestión*. Istituto Tecnológico Geomineiro de España.

Hose, T. A. (2012). 3G's for modern geotourism. *Geoheritage*, *4*(1-2), 7–24. doi:10.100712371-011-0052-y

Nunes, P. A. (2010). *Desporto, Turismo e Natureza - O Turismo de Natureza como pólo de atração turística complementar ao produto Sol e Mar na sub-região do Litoral Alentejano*. Universidade Técnica de Lisboa - Faculdade de Motricidade Humana.

Rodrigues, J. (2009). Geoturismo – Uma Abordagem Emergente. In C. Neto de Carvalho & J. C. Rodrigues (Eds.), *Geoturismo & Desenvolvimento Local* (pp. 38–60p). Idanha-a-Nova.

Stueve, A. M., Cooks, S., & Drew, D. (2002). *The Geotourism Study: Phase I. Executive Summary*. Travel Industry Association of America.

UNESCO. (2015). *Statutes of the International Geoscience and Geoparks Programme and Operational Guidelines for UNESCO Global Geoparks*. UNESCO.

WTO – World Tourism Organization. (2006). Parte I – Introdução e Conceitos Básicos. OMT, Introdução à Metodologia da Pesquisa em Turismo.

WTO – World Tourism Organization. (2007). *A Practical Guide to Tourism Destination Management*. WTO.

Zouros, N. (2004). The European Geoparks Network. Geological heritage protection and local development. *Episodes*, *7*(3), 165–171. doi:10.18814/epiiugs/2004/v27i3/002

Chapter 3
Borders as a Space for Mobility, Cooperation, and Tourism:
Dynamics, Resources, and Policies at the Central Portugal and Spain Border

Gonçalo Poeta Fernandes
https://orcid.org/0000-0003-1278-0369
CITUR, Polytechnic Institute of Guarda, Portugal

Inês Gomes Cravino
CFAEBI, Portugal

ABSTRACT

The border is framed as a political/administrative element, as a space of contact of cultures and geographies, as a line of separation and permeability, generator of mobilities of different scales. During the last decades, the Iberian border has experienced depopulation dynamics and socio-economic reorganisations that have transformed the ways of working, the models of social organisation, occupation, and land use. In this context of growing mobility, tourism has become an important activity for the border due to its ability to generate employment and foster economic and social development. The mobilization of natural and cultural resources assumes significance in the valorization of these spaces, in line with the current policies of cross-border cooperation promoted by the EU and the efforts of the two Iberian countries. The border between central Portugal and Spain is taken as a study object, highlighting the existing dynamics and forms of cooperation, given the heritage values, capable of generating new attractions and functions in the oldest European border.

INTRODUCTION

Borders can generally be understood as regions where economic and social life are directly and significantly affected by the proximity of an administrative division, whose political, financial and cultural differences imply territorial specificities and pose their own challenges to management. Borders, as political boundar-

DOI: 10.4018/978-1-7998-7339-6.ch003

ies and contact zones, represent two basic implications: one refers to the material barrier effect through the creation of means of controlling border flows that can have immediate social and economic effects; a second is a result of the perception of the border based on historical features of identity and linguistic and cultural diversities, which in themselves generate differences and separation. Although a physical border may change its meaning and functions, the effects of immaterial barriers can persist in the form of imaginary barriers, embodying feelings and estrangement that penalise these spaces, requiring the construction of cooperative relationships (Lois & Caballo, 2015; Trigal, 2013). The control of the movement of people, goods and capital between states is the most relevant function of border management, accompanied by its assertion of sovereignty in terms of its administrative boundaries. The persistence of national borders has been based on the fact that they are crucial for the reproduction of territoriality and national identity narratives (Medeiros, 2010; Paasi & Prokkola, 2008). In this context, and considering a line that separates, divides and controls the existence of areas/territories on both sides of the border line, it implies that the effects disappear with distance, besides the emergence of new meanings and dynamics depending on their accessibility, resources held, forms of settlement and economic activities.

Tourism development is seen as an instrument for the diversification of economies in peripheral and often underdeveloped regions. From both nature and tourism perspectives, travel destinations should not be limited by political borders and these, because of their heritage and cultural hybridities, can represent destinations for visitation (Campesino & Jurado, 2014; Fernandes, 2014; Timothy, 2006; Trigal, 2005). There is a growing significance of tourism and leisure activities, due to the increase in mobility, cooperation and the dissemination of border resources and heritage, as well as the incorporation of technologies in the qualification and promotion of these territories. The creation of the Schengen space enhances and expands mobility, without border controls, generating positive effects on tourism and leisure activities among the citizens of colliding countries. These spaces present, due to their history and natural heritage, resources of great value, of which the international rivers, the classified natural areas, the castles and fortifications, the historical villages, the traditions and festivities, the gastronomy and the presence of different cultures stand out at the central Iberian border. Given this panorama, this chapter seeks in an articulated manner to reflect on the meaning of the border, considering the dynamics of formal and informal relationships around this line of territorial specificity. The border is framed as a political/administrative element, as a space of contact of cultures and geographies, as a line of separation and permeability, generator of mobilities of different scales, potentiated by the dimension of the border in the tourist process.

The Luso-Spanish border is taken as an object, in particular, the border between central Portugal and Spain, highlighting the existing dynamics and forms of cooperation for the construction of a space of resilience, given the existing constraints and heritage values, capable of generating new attractions and functions in the oldest European border.

THE BORDER: A COMPLEX, SYMBOLIC AND CHALLENGING CONSTRUCTION

The term "border" establishes the legal line of division between states, indicating their differentiation, a space in which cultures and policies generate tensions, in a dynamic relationship that requires its own measures and logics of governance. They represent fields of forces that oscillate between cooperation and conflict and draw a geographical space with distinct social and cultural values, according to their administrative and functional attributes, with a dynamic that recreates itself, generating, depending on the moment, different permeabilities, flows and socioeconomic relations by their communities. The perceptions of the past, associated with the constraints and barriers that these spaces signified, nowadays reveal a growing relationship, whether by means of the approximations generated, or by the expansion and improvement of physical and technological communications, or even by existing cooperation policies and programmes, as can be seen on the Iberian border.

Each border carries its own identities and imaginaries as a result of its geography, its historical and political evolution and the cultural elements of its communities. They are spaces of exchanges, cultural hybridisms and specific heritages as a result of their functions, flows and social interactions generated over time (Dominguez *et al.*, 2013; Sofield, 2006).

Structural exchanges consist of flows of goods, services or information arising from specialisation and complementarity in production and in the relationship models between communities and territories. Cyclical exchanges have their origin in the "border effect" and income differential, linked to the proximity of different legal, socio-cultural and economic systems (Leloup & Stoeffel, 2001, p. 74).

The strategy of considering the border area as regional is also employed by its communities, and assumes two dimensions which, manifestly, interact in a way contrary to the perception and potential of these spaces: one connoting the border region as a composition of an international territory and the other perceiving the border as a regional space, not including the elements which compose the border of the adjacent country. The people who inhabit each side of the border are considered to be part of a distinct nation, and that nation is not only a political and economic entity, but something that reproduces meanings and identities in a system of cultural representation. We can understand that there is, in these approaches, the possibility of tension/rivalry that tends to disappear, particularly on the central and southern borders of Europe, through a process of cooperation and territorial construction (Medeiros, 2010; Sahlins, 1996).

It should be stressed that the border does not disappear as an identity with the physical separation line, but can acquire new meanings for those who live it every day. According to Sahlins (1996) and Timothy (2015), border populations increasingly participate in the construction and redefinition of the meaning of the border and, consequently, in its organisation. Indeed, the peripheral border approach forces it to understand how individuals and communities appropriate their borders and identities in order to use them as political and development arguments (Calderon, 2017; Medeiros, 2018; Sahlins, 1996). Border regions present a double peripherality, both in terms of their geographical positioning and in relation to the decision-making centres, thus presenting social, economic and structural disadvantages (Fernandes, 2014; Lange, 2011). Moreover, these regions were for decades marginalised in the face of the decision-making centres and limited by a real and legal barrier (the *raia*), aggravating their socio-economic condition which, in the 1960s the *Iberian Raia* as the name of "border of underdevelopment" (Lange, 2011; 2012; Cabero Dieguez *et al.*, 1995; Cabero Diéguez & Caramelo, 2001).

Given the reduced dynamics, reflected by demographic and socioeconomic indicators below the national average, these regions have been one of the main sources of concerns about development and cohesion of the European Union, leading to he development of support and aid programmes capable of reducing the identified development deficits and the problem of their marginalisation in the face of differentiated and competing contexts of proximity (Fernandes & Natário, 2020; Medeiros, 2009).

In the last three decades we have witnessed strategies of de-fronteirisation between Portugal and Spain, which have promoted and facilitated mobility between the two sides of the border, and an increase in institutional relations (formal and informal), taking advantage from European Union support and cooperation programmes and the openness of communities and political representatives to the creation of channels for dialogue and rapprochement. Of these processes, the following advances stand out for their importance:

- Increased access and ease of movement;
- Increased institutional relations;
- Creation of specific support programmes for territorial cohesion;
- Increased socio-cultural relations and activities;
- Community awareness of the need to increase formal and informal partnerships and relationships;
- Increased community support infrastructures;
- Political effort to establish collaborative bases;
- Decrease in the perception of the border as a barrier.

Since the implementation of the Schengen Agreement, the free movement of people has been permitted between the signatory countries, thus fostering the development of the European integration project (Decoville & Durand, 2018). The Schengen area guarantees unrestricted travel within a territory of 26 countries, home to more than 400 million citizens. The Schengen Agreement that set out the gradual abolition of checks at common borders. A Europe without internal borders brings huge benefits to the economy as well, which shows how tangible, popular and successful the Schengen achievement is and the importance it has for the societies and cross border territories (European Commission, 2015. Fosters the mobility between countries and boosts many social and economic activities, generating financial and cultural flows that enrich the regions.

Through commercial exchanges, relationships have been fundamentally altered by the incorporation of diverse social connections, cooperative and sometimes in solidarity, which allow their communities to maintain themselves and generate functions for economic and social well-being. The enhancement of heritage and respective identities allows promoting new attractions, supported beyond commercial relations in tourism and leisure practices that tend to be widened and diversified. Cooperation tends to be a vitalising factor, generating instruments that reduce the weaknesses that these lines tend to represent and assume. The development of the conditions for enhancing the territory for a diverse set of natural and cultural heritage resources should be promoted in a framework that ensures the sustainability of the values, natural resources and identity of the territories (Figure 1). In these processes, employment opportunities and socio-professional integration of communities are promoted, as shown by the use of tourism resources, the generation of services associated with natural and cultural resources and the cooperative strategies of local governance and the impulses of the respective communities (population and companies) (CPC, 2014; Fernandes & Natario, 2020, Fernandes *et al.*, 2012). *The income generated by tourism movements between borders can be a catalyst for permanent peace solutions, as both parties realise the benefits and the risks involved. Such an understanding can change the attitude and behaviour of local people and businesses over time when the feeling of sharing from the same pool becomes stronger* (Kozak & Buhalis. 2019, p.3)

In this context, tourism as a development factor in border spaces has been increasingly recognised over time (Fernandes, 2014; Kovácks *et al.*, 2020; Prokkola, 2010), due to the fact that many peripheral border areas hold great potential for the development of tourism and leisure activities. Their heritage is

Figure 1. Conceptualisation of the relations of mobility, cooperation and promotion of tourism and leisure activities on the border
(Source: Authors elaboration)

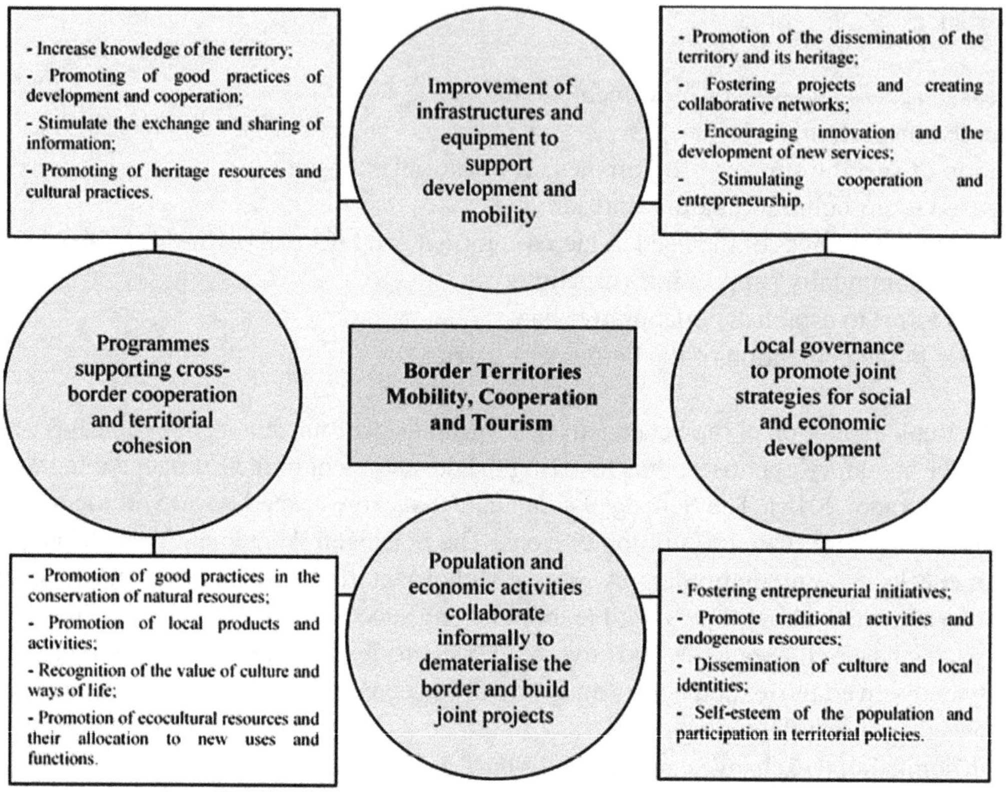

preserved and with authenticities that provide differentiating experiences (a consequence of their relative isolation) and allow the diversification of economies in these peripheral and often underdeveloped regions (Dominguez *et al.*, 2013; Timothy & Saarinen, 2013;), so the border constitutes a privileged area for action and support by European Union policies and programmes.

Relations between cross-border regions are categorised according to the situation between the colliding countries: alienation, coexistence, cooperation (mutual assistance), collaboration (cooperation) and integration. Proactive, resilient and integrative cooperation is relevant in the construction of a spatial context marked by differences, which guide forms of appropriation and require specific forms of governance. The temporary movement of people outside their usual place of residence to cities that are adjacent to the dividing line between two countries has various motivations, such as business, leisure, visits to family and/or friends, social events and shopping, among others (Moral *et al.*, 2016). The border territory can act or be converted, in an analogous way, to other well-known tourist and cultural routes or itineraries: historic routes, pilgrimage routes, routes of historic towns, among others. This "appropriation" includes the movements of travellers and tourists and tourism products, suported by the history and heritage at the border. This border tourism would have, in a way, a transversal and integrative character with other tourism products (Jurado-Almonte & Pazos-Garcia, 2018).

THE BORDER AS A SPACE FOR RESILIENCE AND COOPERATION

Borders are places where political actors clash, cultures mix and economies converge (Timothy, 2006). Since the entry into force of the Lisbon Treaty in 2009, the European Union has had the opportunity to support, coordinate or complement the actions of member states in their cross-border development. European integration and territorial cooperation policies have deeply influenced the way people in border regions live, both in terms of perceptions and actual spatial practices (Decoville & Durand, 2018)

Resilience is generally understood as the ability of a system to absorb disturbance and reorganise itself while undergoing change, in order to maintain essentially the same function, structure, identity and feedbacks (Walker *et al.*, 2004). Alongside the general understanding of the term, the literature distinguishes different types of resilience (Grove, 2018). It includes multiple system states that experience disturbance and can be absorbed before the system changes its structure (Biggs, Schluter & Schoon, 2015). Resilience, according to Grove (2018), focuses on identifying and improving conditions that allow for systemic adaptations following perturbations, whether the adaptation is a matter of moving back to a previous state or moving forward to a new and potentially improved configuration.

We may consider that border spaces, lands and people live their daily life of invading geographical, linguistic, fiscal and political (or obedience) limits. They live in a kind of territory where, from here to there and vice versa, complicities, rather than oppositions and rivalries, predominate. In fact, the conflict associated with the separation is complemented (or interspersed) by the conjugation of interests between residents from one side and the other, with the exploitation of context advantages, as in the case of smuggling, favoured by price differences, tax burden and prohibitions (Ribeiro & Valente, 2013).

In many cases, geographical positioning has resulted in the marginalisation of border peoples, in time and space, as it is also often a reflection of the unequal distribution of power in the economy and society (Sofield, 2006; Hidalgo, Dominguez & Benzinho, 2014). In an interdependent or even integrated border area, forms of socio-economic behaviour and practice are undoubtedly different from those in which a border relationship is in tension or hostility. Thus, borders assume different roles and meanings as a result of their history, political orientations or geographical characteristics, promoting themselves as spaces of porosity/permeability or containment/barrier. Cross-border cooperation has been an important instrument to achieve the objectives of the European Spatial Development Perspective (ESDP), as well as to implement the objectives of the Lisbon and Rotterdam efforts to achieve sustainable growth and territorial cohesion in Europe. Particularly in achieving the goal of cross-border cooperation (ESPON, 2007; Medeiros, 2018), more fluid cross-border relations have been promoted between regional and local governments and administrations along the Spanish-Portuguese border.

Cross-border cooperation processes at the Iberian level, despite their undeniable dynamism, are not progressing and developing as quickly as in other regions of Europe. The undoubted improvements created by the programmes in terms of infrastructure and equipment, formal and informal approaches between entities and agencies and the funds available have led to significant improvements for cooperation and cohesion. Nevertheless, development has effects on a national scale, causing border regions to continue to experience problems of depopulation and economic inertia, which keep border effects and associated context costs (Calderon, 2017; Fernandes, 2019). Therefore, the European Union considers them as special places for the application of public policies oriented towards development, where the challenge of mitigating internal differences and strengthening cohesion and cooperation between communities and governance models (Espon, 2017; Medeiros, 2016, 2018).

Figure 2. Mobility and cooperation in the development of border territoriesand tourism
(Source: Authors elaboration).

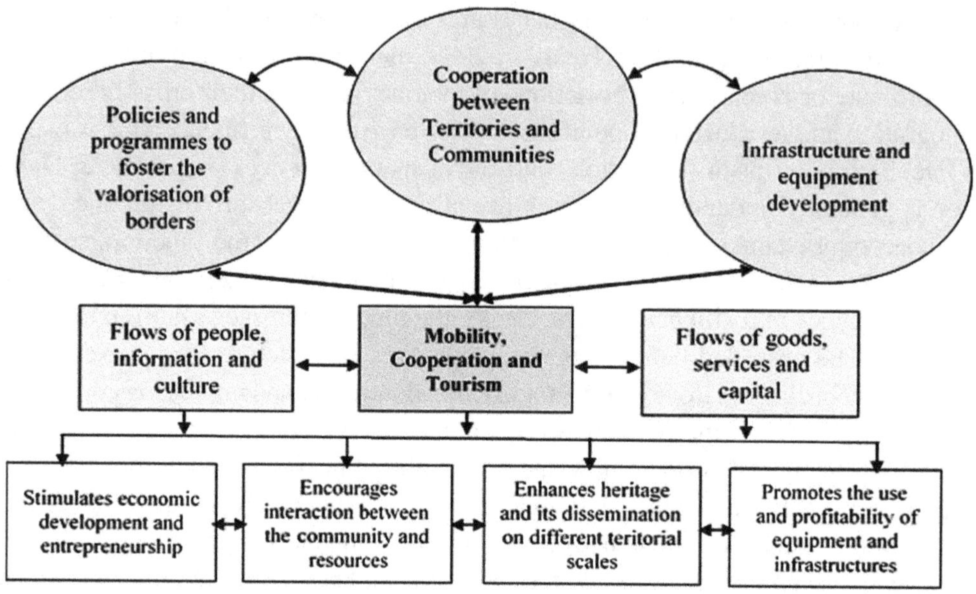

Cross-border mobility and consumption, which includes all border crossings, regardless of destination, trip purpose or distance travelled, corresponding to a broader approach than traditional tourism concepts (Mayer et al., 2019). The usual activities, including shopping and utilitarian trade, are related types of cross-border consumption. For many tourists the opportunity for cross-border shopping might be considered to be a primary tourist attraction. Although the people who undertake them are not usually defined as 'tourists', their activities can contribute to the broader notion of tourism by extending the consumption of goods and services associated with the territories of movement (Timothy & Teye, 2005).

Thus, cross-border cooperation assumes, nowadays, one of the central roles of European Union policies. Taking into account the design "Europe without borders", where the emphasis on complementarity and cooperation prevails over competition (Lima, 2012). Tourism has proved to be a strategic factor in territorial cooperation, especially in inland areas, highlighting in particular its ability to bring together measures that have direct or indirect impact on the four existing axes of action (Promoting competitiveness and employment; Environment, heritage and natural environment; Accessibility and territorial planning; Fostering cooperation and economic and social integration). In this way, the different editions of the INTERREG initiative, first, and the tourism programmes and other programmes at the regional and municipal levels, have co-financed projects and socio-economic initiatives related to tourism, in order to promote cooperation and establish new markets for these regions (Jurado-Almonte & Pazos-García, 2018).

In this context, with economic, social and technological changes, and the growing interdependence of economies as a result of the integration process, the traditional role of the border has changed, and through cooperation support has been generated to mitigate the effects caused by these changes and to manage cross-border areas (Medeiros, 2016). However, development weaknesses and the lack of an effective integration strategy in national governance continue to be denoted, which can be seen in the continued loss of population, decline in economic activity and loss of services, among others situations.

THE BORDER BETWEEN CENTRAL PORTUGAL AND SPAIN: A SPACE FOR MOBILITY, DISCOVERY, TOURISM AND ECOCULTURAL VALORISATION

The concern over the development of the Luso-Spanish border became more significant with the accession of the two Iberian countries to the European Community, through reconfigurations related to intrinsic weaknesses, remoteness from major centres of economic activity, political decision-making and demographic devitalisation. *Portugal and Spain are among the oldest nation states in Europe, and each has a strong sense of national unity and mission. After the April 1974 revolution in Portugal and the death of Franco in 1975, a democratic transition took place in both Iberian countries. There occurred then their return to international incorporation, from which they had been relatively isolated during the dictatorship era* (Martins, 2007, p.12). Over the years, social and economic deregulation has been deepened because the development strategies that these areas have followed have not achieved the desired effectiveness. In particular, European programmes have not achieved the objectives of development and territorial cohesion (Arencibia & González, 2013; Fernandes, 2019; Fath & Hunya, 2001). The specificity of social, economic and territorial problems in these regions has been recognised, along with the need to create more socially and territorially equitable interventions for their cohesion (Hidalgo et al., 2014). In recent years, there has been a dynamic favourable to the "de-fronteirisation" by new structures and the recovery of resources and communities of interest (Calderón, 2017). In addition, many initiatives have emerged and sought to encourage cooperation between regions in order to counter their depopulation and weak productive structure (Fernandes, Natario & Braga, 2014; Lange, 2012;). The unquestionable improvements created by the programmes in terms of infrastructure and equipment, formal and informal approaches between entities, agencies and the funds available, have led to significant improvements in cooperation and cohesion. These developments have effects on the national scale, with border regions experiencing problems of depopulation and economic inertia (Calderón, 2015; Hortelano, 2015), thus keeping border effects and associated context costs in place.

The relations in the cross-border space have acquired a growing importance and increased their influence on the European political, economic and cultural evolution. In the case of the border of Portugal and Spain, it is possible to speak of a porosity that is historical, with approximation of the territory(ies), either through public policies (community, national, regional and local) or through private investments. There is a concern of the regional governmental entities in the promotion of cooperation, in the development of projects and common initiatives and, particularly, in the promotion of entrepreneurship. It is worth highlighting the POCTEP BIN-SAL Empreende projects (2016-20), which streamlined investments and entrepreneurship in order to enhance endogenous resources and services associated with tourism and territorial cohesion, involving municipalities, development associations and HEIs.

In general, it can be considered that the border between Portugal and Spain presents (in both countries) very similar characteristics, as it evolves progressively, where regions and municipalities establish partnerships and the border increases its permeability, strengthening the closeness of people and forms of institutional and business cooperation (Figure 3). This border is multi-peripheral in terms of its geographical position, its relationship with the decision-making centres, its capacity to mobilise and valorise resources, and its effective territorial integration and investments, thus presenting structural constraints. Taking this perspective allows us to identify conditions that contribute to the capacity of systems to adapt and evolve and to build resilience. Governance and management issues and the discussions that accompany building resilient tourism destinations continue to be the subject of research in the tourism literature. At the same time they attract increasing attention in fields such as planning, ecosystem management and

Figure 3. The Iberian Peninsula and the border area between central Portugal and Spain (Source: Authors elaboration)

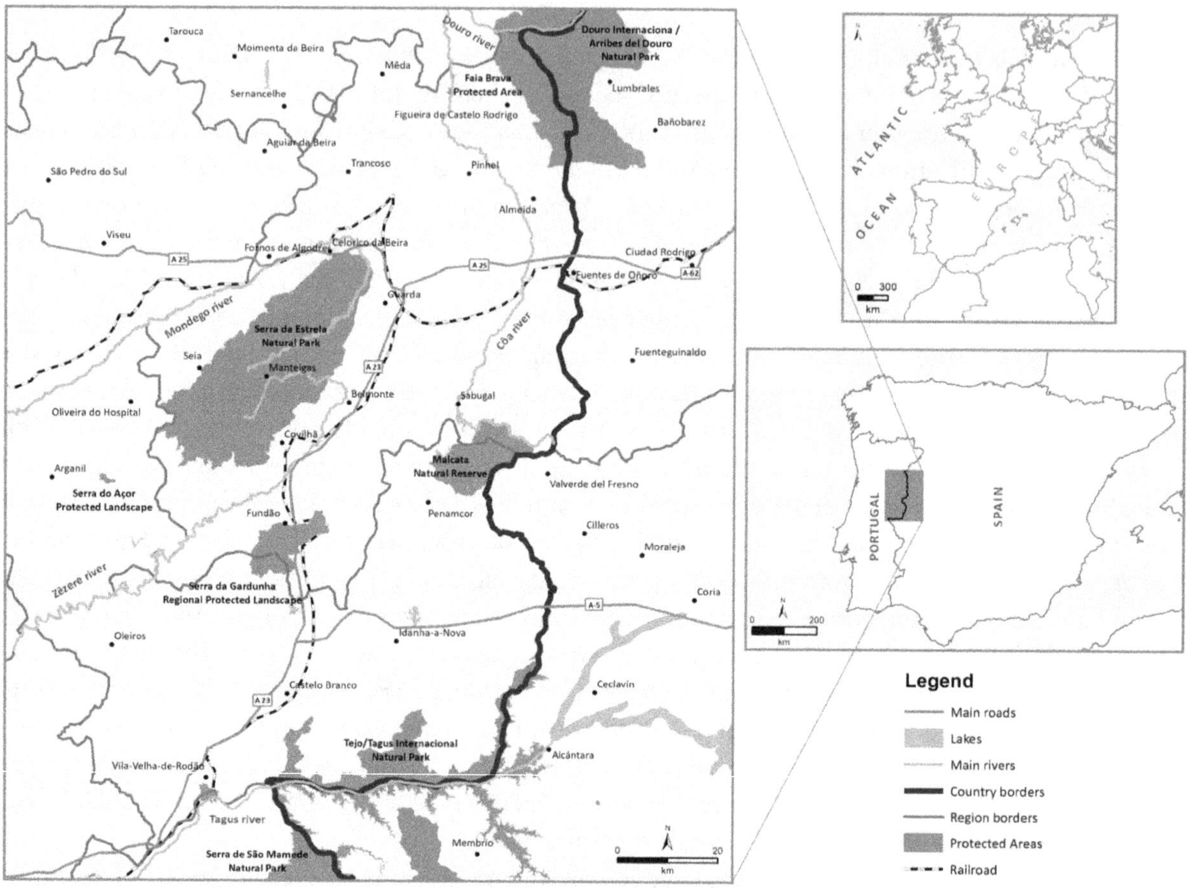

the protection of areas with special status (e.g. National Park, Geopark, UNESCO World Heritage Site). The increased permeability of the border influences the development of tourism infrastructure, associated tourism services and the transformation of the territories on both sides. Political factors can also create problems for cross-border cooperation in tourism by establishing different social and ideological systems that affect not only tourists but also local tourism stakeholders. In this context, it is important to be aware that tourism development has a transversal impact on society-economy-environment, which consequently drives the need to create synergies, which, in border spaces, are faced with greater governance demands. These imply that the actors involved develop concepts of change, grounded in complex and potentially unfavourable territorial relations in the face of the different costs of existing contexts. It is important to distinguish, as Hernández-Ramírez (2017) points out, cross-border tourism and tourism at the border. The first takes as destination a territory larger than the administrative border itself. The territory to be visited would cover a more or less wide space, of transition and boundaries, somewhat lax on both sides of the border and where the fact of the border was felt. Globally, the tourism challenge for destinations is to become robust enough to withstand disturbances and flexible enough to recover or reinvent themselves, contributing to their resilience. This implies a more evolutionary and adaptive perspective, considering and managing tourism destinations as complex adaptive systems (Hartman, 2018).

In this context, safety and security issues, compounded by the threat of international terrorism or public health crises (such as that experienced with Covid-19), result in increased surveillance at border crossings around the world. It is important to remember that for border control purposes, airports and seaports are also classified as borders and that following this kinds of global alert, most countries have increased a series of procedures and controls to restrict or limit the movement of people. These measures have incalculable repercussions on the world economy, with tourism being the sector most affected, in view of the restrictions on mobility and, in particular, the closing of borders, which once again take on the function of a barrier. These positions and measures have given rise to tensions between governments and the tourism industry, which seeks to facilitate cross-border tourism flows. An example of this is the growing practice of governments issuing their citizens with "travel advice" or creating lists of countries not to visit, warning them against travelling to certain countries or destinations due to perceived threats or risks.

Under favourable conditions, cross-border territories contribute to tourism development in the framework of neighbourhood relations and interests. As an activity to be driven, tourism is supposed to be a factor of regional development of the peripheral (border) parts of each country (Kabakchieva & Vasileva, 2020). The interest in alternative forms of tourism (e.g. ecotourism, rural tourism, river tourism, gastronomic tourism and sports tourism, among others) is recognised and should be considered as potential fields for cross-border tourism development. These products and some niche markets are becoming increasingly popular, and imply the discovery of a preserved natural environment, along with new or unexplored tourist destinations (Kovács & Nagy, 2013). These forms of tourism can play an important role in the development of border regions which, in the past, were far removed from tourism flows as destinations. Depending on their characteristics, they may generate regional brands. In the literature there is a threefold typology in the spatial relations between borders and tourism: the boundary line is far from the tourist area, the tourist destination is adjacent to the border on one side, or the tourist areas extend or lie on the border (Gelbman & Timothy, 2010). In this context, the Iberian region is in the third category, due to the diversity of associated resources around the borderline. Both sides of a border are visited and the assets of at least two neighbouring countries are exploited. This includes tourism derived from specific qualities of a border (e.g. differentiated prices of goods and services or distinct cultural identities and practices) and the goal of visiting to experience the area "on the other side", including tourists and tourist infrastructure, attractions and spaces that extend or are adjacent to a border.

The connotations that this space receives are diverse and change according to the events, actors involved and policies defined. Due to its meaning and function, the border was historically characterised by a strong militarisation but weak development policies, presents a historical and natural heritage of great integrity and presents an increase in equipment and accessibility, which enhance its exploitation as a visiting space.

It is emphasised that places value their ecocultural assets in order to increase their competitive profile. Heritage stands out as a differentiating resource for the territories and a strategic element in the design of new products aimed at capturing specific segments of tourism demand, fosters the collective identity of the receiving communities, strengthens feelings of mutual aid and cooperation and encourages conservation and sustainable use actions. Thus, heritage valuation becomes a reference factor for local development, capable of generating territorial balance as a result of the attraction it generates and the activities it brings about in its economic and social appropriation, among which tourism and leisure activities take on relevance. At the same time, heritage symbols (castles and bastions, for example), which have constituted life and history on the border and which also represent tourism resources, are strengthened and preserved. This is why the border itself can become a destination and the main attraction for tourists (Mayer et al., 2019; Timothy, 2005, 2006).

Thus, in addition to the known tourism products (rural, ecotourism, agrotourism, gastronomy etc.) that can be found in border areas, it is possible to develop a new product, which can be known as "border tourism", "cross-border tourism" or "tourism on the border", and which satisfies a very diverse demand, eager to consume the history and heritage of border areas for tourism (Jurado-Almonte & Pazos-García, 2016). Border tourism is characterised by the fact that the destination is located at the same border and that tourist activity focuses on visiting spaces/attractions that belong to different states. The tourist visit is an experience of rediscovering, contemplating and participating in the cultural singularities, as well as the historical heritage, which is sometimes marked by cultural and political confrontations of the past and local hybridisms. Campesino Fernández (2016) distinguishes and emphasises what are potentialities and tourism products for the case of la Raia, or the Iberian border. He states that, for the most part, the natural and cultural heritage of the border remains a "potential resource with attractive and expected profitability" that has not been transformed into tourism products. However, its history has generated a remarkable historical and cultural legacy. The restrictions arising from the presence of national borders have always had a negative impact on the development of tourism in these regions, generating inertia and mistrust to effectively use the potential of these territories. The main problems that can be blamed for the weak dynamism of tourism on this border stem from the following situations:

- Lack of studies on the tourism sector and of the resources that support the activity capable of developing guidelines for the development of strategies and the construction of a differentiated tourism product;
- Perception of conflicts between tourism and the cultural identities of communities, showing a lack of integration with local culture and ways of life;
- Lack of an integrative cross-border network (between public institutions, tourism support agencies and other organisations (including private entities) to support common efforts to manage benefits and invest in new tourism products and markets;
- Insufficient publicity and promotion of the attractions of border regions and of the means of access and stay, namely transport, hotel networks and tourist facilities, routes, festivities and traditions
- Joint marketing strategy for diversified products and services;
- Dependence on external tour operators, transport and travel companies that originate outside the region and discard working niche markets or specific products, which should tend to have a global market scale.

The structural backwardness in the Luso-Spanish border regions led to the creation of the Operational Programme for the Development of Border Regions (INTERREG) Portugal-Spain, with the Community objective of moving towards greater economic and social integration and convergence within the European Community. It was only with the creation of the INTERREG initiative that the Community took on very clearly the need to develop border areas and stimulate cross-border cooperation between neighbouring populations separated for centuries (Lange, 2012; Policarpo & Mogollón, 2015; Oliveira, 2005). However, cross-border cooperation has not achieved its main goals. There are some difficulties derived from the limitations caused by the strong role played by the States (countries), the great differences and asymmetries that exist, and also their own historical inertia, which continues to feed resistance and fears on both sides of the border (Fernandes, Natario, & Braga, 2014, 2016; Salgado, 2010). In this context, it is necessary to systematise a set of factors and opportunities to be considered for border development:

- Important financing instruments (INTERREG V-A Programme 2014-2020), which will have a special impact on the strengthening of local development structures and the modernisation of economic activities already in place;
- Heritage sites that requires greater awareness and dissemination, namely religious and military monuments and historical infrastructures (bridges, roads and viaducts, among others);
- Enhancement of the cultural and natural heritage which, in addition to its legal protection, will be subject to qualification processes, safeguarding its historical and cultural value, affecting the new functions and uses for which tourism is privileged. To highlight the concerted strategy around Wine, Gastronomy and the UNESCO World Heritage Site and the BIOFRONTEIRA project, framed within the Protecting the Environment and Promoting Resource Efficiency of Cross Border Cooperation;
- Presence of two international rivers, with great ecological value and classified as international natural parks (Douro and Tagus), generates an attraction potential to develop aquatic activities, navigation, active tourism, sport fishing and bird watching, among others;
- Articulation of the cities in constructive strategies to promote the discovery and valorisation of the border heritage – Cencyl Cities Network (The heritage of the Cencyl Cities – includes 440 elements classified as Goods of Cultural Interest and National Heritage);
- The historical border element, with cultural hybridisms, differentiated and complementary commerce and services and the possibility of visiting the old Portuguese and Spanish border posts, taking advantage of the offer and services on both sides of the border;
- The existence of some brands for the promotion of regional products and the promotion of tourism in this territory. To highlight the actions promoted with the name Lat. 60, the gastronomy associated with the wines and meat products of some provinces and municipalities;
- Financing and support for the creation and development of business initiatives related to rural and inland tourism (active leisure and itineraries), which articulate tourist resources and products with the projection of the border.

Challenges and positions arise, both in the consolidation of cooperation and in the strategic development capacity of the border as a destination and in the development policies, which generate social cohesion and equity. In this sense, joint policies and strategies enable to:

- Promote a network of classified natural spaces, including UNESCO World Geoparks
- Enable compatibility in the uses for leisure and tourism of the important water reservoirs in this territory, namely in the reservoirs of the Tagus and Douro river basins;
- Create articulated spatial planning instruments on both sides of the border, establishing joint rules for the protection and enhancement of natural and cultural heritage;
- Strengthen associations and cross-border cooperation between public administrations, non-profit institutions and entrepreneurs;
- Change mentality towards a more tourism-oriented local culture in which quality and different tourism products are promoted.

The relationship between borders and tourism development is complex and manifests itself in many ways. A physical border can form a barrier to tourism flows, or can be crossed almost unnoticed. The permeability of the border, the barrier effects caused by the border, such as regulations for the movement of people and goods, will directly influence tourism and the development of flows of goods and people (Kabakchieva & Vasileva, 2020; Prokkola, 2008; Timothy, 2006;). The tourism activity between the border itself and the territories beyond it is a necessary line of research, on the basis of the new emerging cross-border identities and functions, showing the integration between bordering territories (Mansvelt Beck & Hortelano Mínguez (2016).The potential of tourism– resources of the cross-border area – is significant and diversified, and in most cases, well preserved and capable of generating different attractions (markets) throughout the year. The capacity to foster new activities and services is recognised, generating important socio-economic dynamics for the revitalisation of the border and its communities.

CONCLUSION

In peripheral cross-border regions, we have witnessed a loss of economic dynamism and a negative population growth that has weakened the territories and their economies. In these scenarios, heritage, natural and cultural resources may contribute to the strengthening of tourism, which in turn makes it possible to diversify economic structures and strengthen territorial competitiveness and social equity.

Cooperation programmes and institutions have played a key role in directing financial resources to these territories. However, data on population and employment trends show that the infrastructure, facilities and institutions created have failed to stabilise the population and attract investment to these territories. Cooperation depends externally on the Structural Funds. The old border is still there, present and alive in the minds of the vast majority of its inhabitants through its cultural, linguistic, emotional and identity dimensions and in the sense of belonging to a national community, which is distinct and different from the others.

In Europe, the importance of tourism as an economic factor has increased substantially. Especially in less developed peripheral and cross-border regions, tourism can help to diversify economic structures and strengthen competitiveness. The EU supports the development of tourism in these regions through its regional policies and funding programmes. In this context, cross-border cooperation between adjacent countries and regions plays an important role, particularly in removing border obstacles, promoting material and non-material exchange and strengthening territorial cohesion. The EU's cross-border policies help considerably in encouraging cross-border cooperation and developing integrated border regions. In general terms, borders and administrative boundaries are understudied components of tourism mobility, being traditionally perceived as mobility territories for other destinations and not as tourist destinations themselves. The historical recognition of borders, their political relations, cultural hybridisms, historical heritage and natural resources bring new interests to these spaces and their functioning.

It becomes fundamental to understand the globalised nature of contemporary tourist flows and the different degrees of permeability or impermeability of borders, given that this determines tourist mobility and, to a large extent, the generated/economic, social and environmental effects of tourist activities in different destinations. The reality we see at the beginning of this second decade of the millennium (due to the effects of Covid-19) is striking in the significance that borders assume for the world economy and how they have a function as a control valve for the flows of people and goods. Cross-border regions share a common historical heritage, related traditions and similar natural assets. In all cases, cooperation favours the creation of a diversified and differentiated tourism product, contributing to the development of economies and the construction of an image of territorial value.

The heritage, material and immaterial, of central Portugal with Spain, associated with its history, cultural diversity and natural resources generate relevance and international recognition. They constitute arguments for the construction of a growing tourist function in these territories and foster attractiveness and entrepreneurship. The expansion of tourism stimulates the economy, preserving values and identities and generating new uses and functions. Furthermore, it promotes the consolidation of relations between communities and their mobility along the border, allowing exchanges and the desired sustainability.

REFERENCES

Arencibia, A., & González, C. (2013). *La Cooperación Transfronteiriza*. Amarú Ediciones.

Biggs, R., Schlüter, M., & Schoon, M. (2015). *Principles for Building Resilience: Sustaining Ecosystem Services in Social-Ecological Systems*. Cambridge University Press. doi:10.1017/CBO9781316014240

Cabero Diéguez, V., Campesino Fernández, A. J., & López Trigal, L. (1995). El conocimiento de las franjas fronterizas. Aportación de los geógrafos españoles. *Boletín de la Asociación de Geógrafos Españoles*, (21-22), 93–106.

Cabero Diéguez, V., & Caramelo, S. (2001). The Evolution of the Spanish-Portuguese Border and the Process of European Integration. Anais… Congress of the European Regional Science Association, 41.

Calderón, F. J. (2015). Reviewing the Spanish-Portuguese border: Conflict,interaction and cross-border cooperation. *Estudios Fronterizos, nueva época*, *16*(31).

Calderón, F. J. (2017). Fronteras intraeuropeas, desactivación fronteriza, cooperación transfronteriza e instituciones: El caso de La Raya ibérica [Borders within Europe, border deactivation, cross-border cooperation and institutions: The Iberian Raya case]. *Estudios Fronterizos (México)*, *18*(36), 78–101. doi:10.21670/ref.2017.36.a04

Campesino Fernández, A.-J. (2016). Paisajes del agua y turismo fluvial en la Raya ibérica. In Paisaje, cultura territorial y vivencia de la geografía. Libro homenaje al profesor Alfredo Morales Gil. Universidad de Alicante. doi:10.14198/LibroHomenajeAlfredoMorales2016-05

Campesino Fernández, A-J. & Jurado, J. M. (Eds.). (2014). *Turismo de Frontera (III)*. Productos turísticos de la Raya ibérica. Edición de la III Jornada de Turismo de Frontera, Huelva, 3-4 de abril, 2014, Servicio de publicaciones, Universidad de Huelva.

Consultores de Polítcas Públicas (CPC). (2014). *Análise da situação económica, social e ambiental da zona transfronteiriça de espanha e portugal*. Doc. de Trabalho para a programação 2014-2020, Programa de Cooperação Transfronteiriça Espanha Portugal.

Decoville, A., & Durand, F. (2018). Exploring cross-border integration in Europe: How do populations cross borders and perceive their neighbours? *European Urban and Regional Studies*, 1–24. doi:10.1177/0969776418756934

Dominguez, L., Aliste, J., Ibáñez Martinez, M. L., Natário, M., & Fernandes, G. (2013). *Estudio Socio-economico de la Frontera entre Portugal y España*. Edita Riet.

ESPON. (2007). Cross-Border Cooperation. Cross-Thematic Study of INTERREG and ESPON activities. Luxembourg: INTERACT and ESPON.

ESPON. (2017). Shrinking rural regions in Europe Towards smart and innovative approaches to regional development challenges in depopulating rural regions. Luxemburg: Monitoring Committee.

European Commission (2015). *Europe without borders. The Schengen area*. doi:10.2837/78709

Fath, J. & Hunya, G. (2001). Cross-Border economic co-operation on present and future EU borders, a summary of findings, Enlargement. *Research Bulletin*, 2.

Fernandes, G. (2014). *Regiões de fronteira e desenvolvimento turístico: desafios, estratégias e utopias para a BIN-SAL 1492. In A JANGADA DE PEDRA – Geografias Ibero-Afro-Americanas*. Universidade do Minho.

Fernandes, G. (2019). Rural depopulation, social resilience and context costs in the border municipalities of central Portugal. Dichotomies of social reorganization vs absence of public policies. *Economía Agraria y Recursos Naturales - Agricultural and Resource Economics*, *19*(1), 121-149.

Fernandes, G., & Natário, M. (2020). Entrepreneurship in Border Territories: Context Costs, Resilience, and Logics of Action of the Companies and Entrepreneurs – Study of the Border Municipalities of Beira Interior in Portugal and Salamanca in Spain. In J. Dantas & L. Carvalho (Eds.), *Handbook of Research on Approaches to Alternative Entrepreneurship Opportunities* (pp. 129–152). IGI Global. doi:10.4018/978-1-7998-1981-3.ch007

Fernandes, G., Natario, M., & Braga, A. (2014). *Entrepreneurship and structural constraints to development of the bordering municipalities of beira interior norte. In Proceedings Renaissance of 11w Regions of Southern Europe*. APDR.

Fernandes, G., Natario, M., & Braga, A. (2016). Empreendedorismo e dinamismos socioeconómicos em territórios de fronteira. Uma abordagem à região transfronteiriça da bin-sal (beira interior norte-salamanca). In *Las fronteras en la investigación peninsular: temáticas y enfoques contemporáneos*. Publicaciones Universidade de Santiago de Compostela.

Fernandes, G., Natário, M., Daniel, A., & Braga, A. (2012). Evolução demográfica em municípios de baixa densidade no interior centro de Portugal. In *Respuestas de la Geografía Ibérica a la crisis actual. XIII Coloquio Ibérico de Geografía*. AGE e APG.

Gelbman, A., & Timothy, D. J. (2010). From hostile boundaries to tourist attractions. *Current Issues in Tourism*, *13*(3), 239–259. doi:10.1080/13683500903033278

Grove, K. (2018). *Resilience*. Routledge. doi:10.4324/9781315661407

Hartman, S. (2018). Resilient tourism destinations? Governance implications of bringing theories of resilience and adaptive capacity to tourism practice. In Destination Resilience – Challenges and Opportunities for Destination Management and Governance. Routledge. doi:10.4324/9780203701904-5

Hernández-Ramírez, J. (2017). Turismo en la frontera: Patrimonialización y cooperación transfronteriza en una periferia de la Unión Europea. *Etnográfica (Lisboa), 21*(2), 385–409. Advance online publication. doi:10.4000/etnografica.4940

Hidalgo, A., Dominguez, J., & Benzinho, C. (2014). *Costes de Contexto Tranfronterizos en el ámbito Empreserial- Territorio BIN-SAL, Diputacion de Salmanca.* OAEDER.

Hortelano, L. A. (2015). El Patrimonio territorial como activo turístico en la «raya» de Castilla y León con Portugal. *Cuadernos de Turismo*, (36), 247–268. doi:10.6018/turismo.36.230981

Jurado, J. M., & Pazos-García, F. J. (2016). Población y turismo rural en territorios de baja densidad demográfica en España. Boletín de la Asociación de Geógrafos Españoles, 71, 247-272. doi:10.21138/bage.2282

Jurado-Almonte, J. M., & Pazos-García, F. J. (2018). La Frontera como Recurso Turístico. Posibilidades del Bajo/Baixo Guadiana. *Revista de Estudios Andaluces*, (35), 1–34. doi:10.12795/rea.2018.i35.01

Kabakchieva, D., & Vasileva, V. (2020). Cross-Border Territories and Tourism. Published in Collection of Scientifical articles from Innovation and Entrepreneurship Conference, Ajax Publishing.

Kovács, I. & Nagy, G. (2013). Tourism in peripheric regions – The possibilities of thematic routes. *Seria Geografie*, *23*(2), 220-228.

Kovács, Z., Vida, G., Lados, G., & Tóth, B. (2020). The Potentials of Cross-Border Tourism Development in the Lower-Tisa Valley. *GeoJournal of Tourism and Geosites*, *28*(1), 360–375. doi:10.30892/gtg.28128-475

Kozak, M. & Buhalis, D. (2019). Cross–border tourism destination marketing: Prerequisites and critical success factors. *Journal of Destination Marketing & Management*, *14*, 1-9. doi:10.1016/j.jdmm.2019.100392

Lange, E. (2011). EGTC – a 'Breath of Fresh Air' for Practical Cross-Border Cooperation. Tracing the Legal Cross-Border Co-operation from the European Level to the Galicia-North of Portugal Cross-Border Region. *Regional Studies Association Annual International Conference.*

Lange, E. (2012). De 'Países Subdesenvolvidos' à 'Fronteira do Subdesenvolvimento' – Contributo para um Debate Conceitual do Desenvolvimento. *Anais... 18th APDR Congress - Innovation and Regional Dynamics*, 47-59.

Leloup, F., & Stoffel, S. (2001). Intégration régionale et frontière dans le Mercosur: Entre théorie, principes et réalités. *Mondes en Developpement*, *3*(115-116), 73–80. doi:10.3917/med.115.0073

Lima, F. (2012). As dinâmicas territoriais no espaço de fronteira na fachada atlântica peninsular: a eurocidade Valença/Tuy. *Cadernos curso de doutoramento em Geografia FLUP*. https://ler.letras.up.pt/uploads/ficheiros/9963.pdf

Lois, R. C. & Carballo, A. C. (2015). La frontera hispano-lusa en la actualidad: una visión geográfica. *Revista de historiografía*, *23*, 191-21.

Mansvelt Beck, J. & Hortelano Mínguez, L. A. (2016). La apertura de la frontera: ¿nuevas identidades transfronterizas? *Boletín de la Asociación de Geógrafos Españoles*, *72*, 271-284. doi:10.21138/bage.2340

Martins, R. C. (2007). Fronteira e função: o caso europeu. In Mare Oceanus. Atlântico: Espaço de Diálogos.

Mayer, M.; Zbaraszewski, W.; Pieńkowski, D.; Gach, G. & Gernert, J. (2019). Cross-Border Tourism in Protected Areas Potentials, Pitfalls and Perspectives. In *Geographies of Tourism and Global Change*. Springer Nature.

Medeiros, E. (2009). *The Cross-border cooperation in Raia Ibérica Geographical synthesis of INTER-REG-A territorial impacts*. CEG.

Medeiros, E. (2010). Old vs Recent Cross-Border Cooperation: Portugal-Spain and Sweden-Norway. *Area, 42*(4), 434–443. doi:10.1111/j.1475-4762.2010.00940.x

Medeiros, E. (2016). Territorial Cohesion: An EU Concept. *European Journal of Spatial Development*, (60), 1–30.

Medeiros, E. (2018). Should EU cross-border cooperation programmes focus mainly on reducing border obstacles. *Documents d'Analisi Geografica, 64*(3), 467–49. doi:10.5565/rev/dag.517

Moral, S., Cañero, P., Jimber, J., & Orgaz, F. (2016). Turismo fronterizo como motor de desarrollo de la frontera. Una revisión de la literatura. *International Journal of Scientific Managment Tourism, 2*(2), 249–265.

Natário, M., Braga, A., & Fernandes, G. (2018). Determinantes do desempenho empreendedor e inovador nas regiões transfronteiriças. *Innovar: Revista de Ciencias Administrativas y Sociales, 28*(70), 39–54. doi:10.15446/innovar.v28n70.74445

Oliveira, T. (2005). Tipologia das relações fronteiriças: elementos para o debate teórico-práticos. In *Território sem limites* (pp. 377–408). Editora da UFMS.

Paasi, A., & Prokkola, E.-K. (2008). Territorial Dynamics, Cross-border Work and Everyday Life in the Finnish–Swedish Border Area. Space and Polity, 12(1), 13-29. doi:10.1080/13562570801969366

Policarpo, F. & Mogollón, R. (2015). A actividade empreendedora na fronteira portugal-espanha alentejo- extremadura. *Anais... XVII seminário luso-espanhol de economia empresarial, sl3e'15.*

Prokkola, E.-K. (2008). Resources and barriers in tourism development: Cross-border cooperation, regionalization and destination building at the Finnish- Swedish border. *Fennia, Helsinki, 186*(1), 31–46.

Prokkola, E.-K. (2010). Borders in tourism: The transformation of the Swedish–Finnish border landscape. *Current Issues in Tourism, 13*(3), 223–238. doi:10.1080/13683500902990528

Ribeiro, M. T., & Valente, I. M. (2013). *Portugal Europe: Spaces Border and Crossborder Cooperation «Portugal Europe: Spaces Border and Crossborder Cooperation». In Comunicatin the EU policies beyond the bordes*. Oradea Uniersity Pressa.

Sahlins, P. (1996). *Frontières et identités nationales, la France et l'Espagne dans les Pyrénées depuis le XVIIe siècle*. Paris: Belin.

Salgado, A. R. (2010). La cooperación transfronteriza y sus consecuencias: Hacia la reestructuración territorial en Europa. *Investigaciones Regionales*, (18), 141–152.

Sofield, T. (2006). Border Tourism and Border Communities: An Overview. *Tourism Geographies*, *8*(2), 102–121. doi:10.1080/14616680600585489

Tarpai, J. (2010). Development of tourism destination management in the Ukrainian-Hungarian border region. *Jelenkori Társadalmi és Gazdasági Folyamatok*, *5*(1-2), 28-32.

Timothy, D. J. (2005). Tourism and Political Boundaries in an Age of Globalization. *Anais... Conference on Border Tourism and Community Development Conference*.

Timothy, D. J. (2006). Relationships between tourism and international boundaries. In Tourism and borders. Ashgate.

Timothy, D. J. (2015). Border and tourism. In Encyclopedia of Tourism online. Springer Publishing House.

Timothy, D. J., & Saarinen, J. (2013). Cross-border co-operation and tourism in Europe. In *Trends in European tourism planning and organisation* (pp. 64–74). Channel View. doi:10.21832/9781845414122-009

Timothy, D. J., & Teye, V. B. (2005). Informal sector business travelers in the developing world: A borderlands perspective. *Journal of Tourism Studies*, *16*(1), 82–92.

Trigal, L. (2005). La Raya Ibérica como espacio de atracción y cooperación. La Raya Ibérica Centro-Sur nº 21 – Papeles de Economia Española, economia de las comunidades autónomas, fundación de las cajas de ahorros, 11-13.

Trigal, L. (2013). Revisión de la terminología sobre la frontera y lo fronterizo. In *Condições de vida, coesão social e cooperação territorial* (pp. 149–167). Âncora Editora/Centro de Estudos Ibéricos.

Walker, B., Holling, S., Carpenter, R., & Kinzig, A. (2004). Resilience, adaptability and transformability in social–ecological systems. *Ecology and Society*, *9*(2), 5. doi:10.5751/ES-00650-090205

KEY TERMS AND DEFINITIONS

Border Tourism: Tourism activities developed around the border, by the citizens of the colliding countries or international travelers, promoting the border as a visiting destination and tourist experience in view of its natural and cultural heritage.

Context Costs: Operational costs as a result of different economic and administrative policies of the border territories, barriers resulting from past development strategies, cultural differences, and resistance to cooperation.

Cross-Border Cooperation: Cooperation promoted by border communities is a form of international cooperation, generating relationships of support, proximity, and common benefits, and building networks of economic and social stimulation, as well as resilience.

Cultural Hybridity: Reflect processes of exchange, sharing and representation of cultural traditions and practices that, originally from one country, are assimilated and integrated in other regions, promoting acculturations and hybridisms.

Schengen Mobility: Mobility provided by the creation of a space without borders, improving the circulation of citizens and their integration.

Chapter 4
Inland and Island Tourism, Contrasting Geographies Touching Each Other:
Azores, a Peripheral Archipelago

Francisco António dos Santos Silva
https://orcid.org/0000-0002-9742-7339
Centre for Geographical Studies, IGOT, Universidade de Lisboa, Portugal

ABSTRACT

Despite being opposite geographic regions, inland territories and small islands have very much in common in terms of economic and social constraints. Based on the case study of the Azores archipelago, the limitations and opportunities for tourism development in the small oceanic islands are analysed. This study is supported by empirical research involving the application of questionnaires to the main tourism stakeholders in the archipelago. The results lead to the conclusion that in the case of the group of small temperate islands, tourism has progressively asserted itself as strategic for the local economy, but the application of a sustainable development model must be considered as these territories are generally highly vulnerable, both environmentally and socially.

INTRODUCTION

The main phases of tourism growth have mainly been supported by mass tourism and, more recently, by short breaks, primarily in urban spaces or on the coast (Cooper, 2012). The massification of these territories and the diversification of demand have created opportunities for the development of many other destinations and segments of tourism (Richards, 2017). In a context characterized by demand diversification and high competitiveness, the promotion of authenticity, uniqueness and differentiation has contributed to the development of tourism in low-density peripheral territories, namely inland and island regions (Hall, 2010; Page, 2015).

DOI: 10.4018/978-1-7998-7339-6.ch004

This chapter focuses on the case study of tourism in the Azores archipelago. It seeks to contribute to an understanding of this reality so that similarities, bridges, and possible strategies may be followed by other destinations located inland, the other geographical extreme of islands.

In Portugal, the inland and islands territories, present several similarities, both regarding their social and economic constraints, as well as the importance of tourism as an engine and opportunity for the revitalization of these regions. The country is marked by important territorial asymmetries, in terms of spatial distribution of resources, opportunities and income, with inland regions and, in particular the Azores, as areas most affected by the risk of poverty (European Commission, 2019).

This study includes an empirical study supported by a survey, applied in 2012 to 8 tourism stakeholders in the region: i) Regional government (N=20); ii) Local government (N=39); iii) Local experts and researchers in tourism and regional planning (N=22); iv)Tourism and environmental non-governmental organizations (NGO) (N=36); v) Travel and tourism agencies (N=16); vi) Tourism recreation companies (N=59); vii) Tourist guides (N=43), and; viii) Tourist accommodation (N=67). In total, 302 validated answers were obtained, corresponding to a response rate of 54.5% of the operating universe, and a margin of error of 3.81% for a 95% confidence level. The survey was applied in both face to face and online formats.

The statistical data processing included measures of centrality and dispersion, and the variable *type of stakeholder* was considered independent, using the Kruskal-Wallis test and the "Mann-Whitney U" technique to analyse whether there were statistically significant differences.

The questionnaire was organised into three dimensions: i) potential and current tourism development; ii) tourism planning; and iii) sustainable tourism development. For many variables, the Likert scale, ranging from 1 (strongly disagree / not important) to 5 (strongly agree / very important) was used.

Tourists and resident stakeholder groups were not considered, as periodic questionnaires are applied to these groups in the region, however in order to gauge their opinion, data from the "Tourist Satisfaction Visiting the Azores 2019" (OTA, 2021) and the "Survey of Residents on Tourism in the Azores 2018" were used (SREA, 2019).

The empirical research on tourism in the Azores is complemented with a literature review on low-density peripheral territories and the economy on small islands territories. This allows meeting the main objective of the chapter, which is to present the bases for developing a model for these territories supported by the tourism sector, highlighting aspects that can be shared with inland regions.

GEOGRAPHY AND TOURISM OF PERIPHERAL TERRITORIES

Lagging Regions and Low-Density Territories

The reduction of regional inequalities and the promotion of territorial cohesion are essential strategies in the context of planning and sustainable development. Initially GDP per capita was practically the only criterion for assessing the level of regional development. Later, other variables began to be considered, such as the Human Development Index, constructed under the United Nations Development Program. At the regional level, other criteria have also been considered, such as accessibility, population density, demographic structure, geographical position, and size of the local economy.

Currently, despite continuing to be based on poor "low-income" areas, the classification of lagging regions has come to be associated with other concepts such as: peripheral, sparse, inland, and low-density (Pilati & Hunter, 2020). This association is justified by the fact that difficulties in terms of accessibility, distance from markets and their small scale are important constraints to development. In addition, many of these territories have an aging population and limited resources (Richards, 2017). Thus, many lagging regions are peripheral and low-density territories, among which inland and island regions are the most prevalent.

Their classification as lagging regions is of particular interest for the purpose of applying positive differentiation measures to these areas to promote territorial cohesion and more equitable and sustained development at both national and international levels. In the European Union, cohesion policy has come to be seen as a fundamental strategy for strengthening economic and social cohesion by reducing disparities in the level of development between regions. It is based on the assumption that a reconciliation between sustainable economic growth and social progress is essential (European Commission, 2017).

In Portugal, European policy and the application of structural funds have been coordinated with national strategies, with an emphasis on the National Program for Territorial Cohesion, whose main purpose is to promote solutions that counter "an unsustainable trajectory [due to the trend] of depopulation, aging and impoverishment of the inland regions, which represent around 2/3 of the national territory" (República Portuguesa, 2017, p. 3).

In order to define these regions, termed as low-density territories, a multicriteria approach was adopted, considering variables associated with population density, demography, accessibility, socioeconomic characteristics, settlement, and physical environment (Agência para o Desenvolvimento e Coesão, 2021). As a result, all the municipalities in the inland regions and several on the southern coast of the continent were included in this category.

The Portuguese archipelagos of Madeira and the Azores were not included in these programs since they are classified as outermost regions within the scope of the European Union's regional policy and thus benefit from specific programs. This classification has also been applied at the sectoral level, as is the case of tourism, whose strategic plan (Tourism Strategy 2027) defines "extending tourism activity to the entire territory and promoting tourism as a factor of social cohesion" as one of its main challenges (Turismo de Portugal, 2017).

Geography of Island Territories

The geography of the islands contains many realities. The classification of islands is not consensual and the difficulties in defining and establishing groups are important challenges reflected in studies on insularity and in the policies aiming to promote the development of these territories. Physical, political, socio-cultural, and geographical factors result in a significant and complex diversity of island territories.

Based solely on position and the geographic and demographic dimensions, three major groups can be defined for oceanic island territories: i) large islands with characteristics similar to those of the continents (e.g., Japan and Australia); ii) islands located close to the continents; and iii) scattered and small-sized oceanic islands away from the continental shelves, with a pronounced level of insularity, as is the case of Fiji with 332 islands, and the Azores with 9 islands (Silva, 2017).

The physical environment and climate are also important factors in the context of island geography. The political, economic and social factors are also considered highly relevant to the development of these territories. Small, isolated island countries with strong territorial dispersion have significant weaknesses that condition their development. Since the 1990s, the United Nations has placed particular emphasis on Small Island Developing States (SIDS), as they are a special case and share common challenges as far as their vulnerability is concerned (Grydehøj & Kelman, 2020; UN-OHRLLS, 2011).

In the literature, climate, especially the temperature of the sea, is another feature that distinguishes warm-water islands, which promote sun, beach and sea tourism from those of cold-water (Butler, 2006; Hall, 2010).

Development of Small Island Territories

The small oceanic islands have a set of common features that condition their development, of which the following are particularly noteworthy (Adrianto & Matsuda, 2004; Bernardie-Tahir, 2011; Briguglio, Persaud, & Stern, 2006):

- Small scale: economic structure constrained by the limitation of human, financial and natural resources and small-scale economies and know-how.
- Accessibility limitations: accessibility depending on air and sea connections, with increased costs, worse in the more isolated and dispersed territories that are further away from the main markets.
- Trade balance deficit: tendency to present significant structural imbalances in payments and to be highly dependent on foreign aid and remittances from emigrants.
- Limitation of resources: economies highly dependent on their natural resources and conditions for the development of key sectors such as agriculture, fisheries, finance, and tourism.
- Low resilience: particularly fragile territories from an environmental and socio-cultural point of view, with low carrying capacity and high exposure and vulnerability to natural risks and climate change.
- Territorial dispersion: many of the island regions consist of archipelagos with scattered islands, which hampers transport and the development of the local economy.

Considering these constraints, some island regions have adopted an inward-oriented development model to encourage local production. However, the small size of the domestic market, weak competitiveness and the scarcity of resources and know-how constrain the development of the economy.

The adoption of an outward-looking development model is an alternative for economies with the capacity to be specialized in competitive products and services endogenously based in foreign markets (fishing, other natural resources, tourism), or the creation of competitive advantages in the field of financial services or taxation.

From the second half of the twentieth century onwards, important transformations took place in many small island territories, associated with political factors, the expansion of commercial aviation and the globalization of the economy. As highlighted by McElroy (2003, p. 231), "the post-war history of small islands has been marked by two favourable development factors: the march of decolonization and the global spread of international tourism". Consequently, many of these territories began to invest in opening their economies more widely, promoting foreign investment and the development of strategic sectors, supported by natural resources, financial services, and tourism.

McElroy (2010) considers three dominant small island economy models, the MIRAB, PROFIT and SITE, and some hybrid combinations.

The MIRAB model refers to small islands where the economy is supported by the influx of two external sources of revenue: remittances from emigrants and international aid.

The so-called PROFIT formulation is "focused on the favourable intersection between flexible government policy and a dynamic private sector to carve out market niches: tax / insurance havens, offshore banking centres, duty-free manufacturing exporters and others" (McElroy, 2010, p. 316). Sustained by reduced jurisdiction and advantageous tax regimes "the financial services sector has become the blood stream of the economic machinery of many small island jurisdictions in today's global economy" (Saliba, 2007, p. 50). However, following the September 11, 2011 attack and the greater commitment to fighting corruption and reducing tax evasion, offshore financial centres began to be subjected to higher scrutiny and tighter regulation. Many of these offshore financial centres also set out to develop the tourism industry (Hampton, 2002).

The third model, referred to as SITE (Small Island Tourist Economies), applies to islands whose development is mainly supported by the tourism industry.

In recent decades, social and economic theory on insularity has evolved significantly, with a focus on finding opportunities within each of the realities, based on sustainability and resilience principles (Briguglio & Kisanga, 2004). In addition to the resources at their disposal, the ability of societies to organize and promote appropriate policies is essential to encourage the development of these regions. Hence, the qualification and settlement of populations are, in many cases, critical problems in need of a solution.

Small Island Tourism

The mental representation of insularity is generally associated with isolation and authenticity, whereby an image of fascinating, exotic and authentic spaces is created. The demand for island spaces has been strongly encouraged by large marketing campaigns which, since the 1980s, have promoted these new destinations by associating them with dream vacations in paradisiacal locations, conducive to romance, emotion and adventure. The image of a tropical idyll has been promoted, strictly associated with sun, beach, and sea tourism, organized mainly around tourist resorts and complemented by nautical tourism and, in some regions, by cruise tourism.

Due to the weaknesses of many of the island economies, tourism development has often been supported by investments from companies outside the region, to ensure infrastructure and a set of supporting resources such as hotels and restaurants. These tourist facilities are essential since, despite being attracted by the exotic and "virgin" nature of the islands, most tourists end up concentrating on heavily artificialized tourist sites (Terrasson, 2007).

Even this exogenous economic model (based on investment and external tourism demand), has had important effects on the local economy, with many island countries adopting tourism as the main export sector with a decisive weight in GDP, in some cases surpassing 30% (Maldives, Seychelles, Bahamas, etc. (UN-DESA, 2010). Some island destinations have become so popular that visitor numbers by far exceed their permanent population. However, in many cases, the growth of tourism has significantly increased the negative impacts. With limited resources and fragile environments, the impact of tourism is often unsustainable (Lombard & Labescat, 2010). "Issues related to the environmental impact of tourism, waste disposal and conservation of biodiversity can be far more sensitive for islands than for other tourism destinations" (Hall, 2010, p. 245). Islands can also be very vulnerable to the effects of climate change, especially to sea level rise (Veron, Mouchet, Govaerts, Haevermans, & Pellens, 2019).

The growth in tourist demand is higher in tropical destinations with paradisiacal beaches, namely the so-called warm-water islands (Baldacchino, 2006; Butler, 2006). However, most of the islands are in high latitudes, with cold climates and no potential for the development of bathing tourism. In this case, the climate and sea water temperature are not appealing for bathing tourism and are some of the characteristics that condition tourism development (Gössling & Wall, 2007). Nevertheless, in many of these territories, tourism has been gaining traction in the local economy by promoting the features that target niche markets and nature tourism, including adventure and tourism based on local history and culture (Baldacchino, 2006).

It is also important to consider a third group, the temperate islands (or Transition Islands), which consist of a relatively limited number of oceanic islands, located in the middle latitudes (Silva, 2013). These are destinations where nautical tourism and even bathing can be significant but highly seasonal and the image of these destinations tends to have nature as its main element. In the case of the Azores, the supply of sandy beaches is reduced.

Although these three groups of islands have distinct characteristics and potential for tourism, the fascination for island spaces is widespread, thus creating opportunities for tourism development as an important sector for these economies.

AZORES ISLANDS: TOURIST SUSTAINABILITY

Azores Archipelago: Geography and Tourism

The Autonomous Region of the Azores is composed of nine volcanic islands located in the North Atlantic Ocean, 1,500 km from the western coast of Europe and 3900 km from North America. This small Portuguese archipelago has a surface area of 23,221 km^2 and approximately 240 thousand inhabitants. The region presents a strong territorial dispersion, with the nine islands grouped, due to their geographical proximity, in three groups: i) Eastern, composed of the islands of Santa Maria and São Miguel; ii) Central, the islands of Terceira, Graciosa, São Jorge, Pico and Faial; and iii) Western, with the islands of Flores and Corvo.

Due to its small size, dispersion, isolation, limited resources and accessibility, the Azores present significant structural problems which are reflected in an uncompetitive economy and a GDP per capita below 75% of the European Union average, and the Portuguese region most affected by the risk of poverty (31.5%) (European Commission, 2019).

The region's economy is mainly based on services, with an emphasis on those provided by public administration, commerce, transport, and tourism. The agro-food sector (with a strong focus on milk production and agro-industry) and fisheries are also very important for the economy of the Azores (Azevedo, 2017).

Despite noticeable improvements, the average level of qualification of the adult population of the Azores remains low. More than 60% of the population aged 25-64 has low educational attainment levels, above the about 52% Portuguese average and well above the 22.5% from European Union, in 2017 (European Commission, 2019).

The data in Table 1 show the important regional disparity between the islands, with São Miguel, the largest island, standing out with the best indicators in terms of aging index, regional GDP, and population with higher education.

With a temperate oceanic climate and few sandy beaches, the Azores do not fit into the group of 'warm-water' islands, that are very attractive for Sun, Sea, and Sand Tourism (3S). Notwithstanding, tourism has become one of the main pillars of regional development. In 2015, this sector contributed to 14.1% of the GDP of the Azores and was responsible for 10% of employment in the region (SREA,

Table 1. Area, population, education, and GDP in the Azores

	Santa Maria	São Miguel	Terceira	Graciosa	São Jorge	Pico	Faial	Flores	Corvo	Total
Area (sq. km)	96.9	744.6	400.3	60.7	243.6	444.8	173.1	141	17.1	**2322.1**
Population(1)	5616	137307	55124	4208	8277	13651	14521	3628	464	242796
% Population(1)	2.31%	56.55%	22.70%	1.73%	3.41%	5.62%	5.98%	1.49%	0.19%	100%
Population density(1) (2)	58.0	184.4	137.7	69.3	34.0	30.7	83.9	25.7	27.1	104.6
Aging index(1)(4)	103.5	74.0	117.4	151.4	165.3	155.5	126.2	158.9	131.1	95.1
Regional GDP(3)	2.85%	58.11%	21.95%	1.44%	3.08%	4.85%	6.28%	1.22%	0.21%	100%
Higher education(5)	8.1%	10.4%	10.2%	7.2%	6.7%	7.9%	11.3%	8.5%	6.5%	10%

(SREA, 2021)

(1)2019;(2) people per sq. km; (3) 2018; (4) number of elders per 100 persons younger than 15 years old; (5) 2011

2021). These figures have significantly increased in recent years, particularly between 2014 and 2019, with the number of overnight stays growing to over 100%, more than double the national average (Table 2). This growth may be explained by five main factors: liberalization of airspace with low-cost companies operating from 2015; ii) new tourist demand dynamics seeking "alternative" territories; iii) Portugal's solid reputation as a tourist destination; iv) the quality of the destination; and v) Azores is a relatively young destination, situated between the growth and consolidation cycle, according to Butler's tourism life cycle model (Butler, 1980).

As a result of this consistent growth, in 2019, the Azores recorded a tourist saturation index of 400.3 - guests per population times 100, and a density of tourist demand of 418.5 tourists per km^2. These figures are significantly higher than the country's average (263.6 and 294.4, respectively), although much lower than those of the Madeira archipelago (583.0 and 1850.5 tourists per km^2).

However, the data in Table 1 point to some weaknesses, when compared with the national average and with the Madeira archipelago, namely the highest rate of seasonality, greater dependence on the national market and low RevPar. The main constraint to sustained tourism development in the region stems from the strong tourist concentration on a single island, which has progressively increased. In 2014, the island of São Miguel concentrated 65.3% of total overnight stays in the Archipelago and 57.1% of guests. These figures increased in 2019 to 68.9% and 64.2%, respectively. The expansion of the cruise product has also contributed to accentuating this concentration, and the port of Ponta Delgada is responsible for 87.5% of cruise passengers visiting the Azores (120 thousand passengers in 2019) (SREA, 2021).

Table 2. Tourism activity in the Azores and Portugal in 2019

	Overnight Stays	Guests	Foreign Guests (%)	Average Stay (days)	Seasonality Rate	RevPAR	Overnight Variation 2014-19
Portugal	70158964	27142416	60.5	2.58	36.3%	49.4	44.0%
Azores	2277800	771687	49.6	2.95	39.9%	38.1	103.3%
Madeira	7457196	1482239	78.2	5.03	30.8%	44.3	14.6%

(INE, 2020; SREA, 2021)

The image of the Azores is associated with insularity and the natural environment (landscape, heritage, sea, nature-based activities), which are the attributes most valued by visitors (OTA, 2021). The results of the questionnaires completed by tourists who visited the Azores in 2019 further demonstrate that they are 'very satisfied' with the Azorean cultural offer, and 'highlight' nature, safety, hospitality and the sea water quality as the most valued attributes of the destination (OTA, 2021).

The Regional Government vision in the tourism strategic plan is of an island destination in the middle of the Atlantic, "of exuberant nature, exclusive, of mystical beauty, where the visitor is welcomed as a special guest who, in safety, has at his disposal a variety of land and sea activities that will leave you dazzled and homesick" (IPDT, 2016).

It should also be noted that, according to the Survey of Residents on Tourism in the Azores 2018 (SREA, 2019), the opinion and attitude of residents towards tourism on the island is quite positive, with 97% agreeing that "tourism is good for the Azores", and 72.5% wishing tourism demand to grow.

Results of Empirical Research

It is possible to conclude from the questionnaires applied to the main tourism stakeholders in the Azores that, even prior to the considerable growth of tourism observed between 2014 and 2019 in the Azores, they all considered the activity to be one of the archipelago's most important (mean of 4.75 on a scale of 1 to 5, from *not very important* to *very important*). Additionally, tourism was deemed to have the highest potential for development in the region (4.52).

They highlighted the environment and natural landscape as the destination's main strengths. As for the factors that condition the development of tourism, accessibility was considered to be the most relevant. The highest level of dissatisfaction was attributed to the cost of air transport, both inter-island (mean of 1.54) and overseas (1.47), and transport coordination (1.98). The significant statistical differences among the various stakeholders were particularly noteworthy, mainly those from the Regional Government group, pointing to a more positive assessment.

The price-quality ratio of the services and the tourism planning and management policy were also important weaknesses highlighted by the stakeholders. Here, once again, significant statistical differences were found between the groups. As far as the quality of the services are concerned, the assessment mean was 3.12, and the most positive assessment (3.44) was given by the Travel Agencies group. The worst assessment was attributed to competitiveness of the destination, which yielded a mean assessment of 2.66. However, in relation to this variable, the Regional Government group provided a relatively positive assessment (3.40).

Conversely, a distinct and highly positive opinion was expressed in terms of the destination's potential for tourism development (4.26) and its capacity to harness meaningful visitor experiences (3.80).

Regarding the products with the highest potential for development, Nature-based Tourism was highlighted (mean of 4.5 and lowest standard deviation). This product was considered the region's main differentiation factor and the main feature of the tourist destination brand image (mean 4.35, median and mode 5). Nautical Tourism (cruises not included) (4.25) was also highly rated, followed by Touring Cultural and Landscape (3.71) and Health and Wellness Tourism (3.56) (Table 3).

The Sun, Sea, and Sand product is ranked in seventh position with a moderate potential (3.09), varying considerably from the importance it carries in the 'warm-water' islands, but significantly higher than in the 'cold-water' islands.

Table 3. Potential of tourism product development in the Azores

	Average	SD		Average	SD
Nature-based Tourism	4.50	0.64	**Sun, Sea, and Sand Tourism**	3.09	1.10
Nautical Tourism	4.25	0.78	**Cruises**	3.01	1.03
Touring Cultural and Landscape	3.71	0.95	**Meeting Industry**	2.96	1.02
Health and Wellness Tourism	3.56	0.96	**Golf**	2.91	0.98
Food and Wine Tourism	3.33	0.93	**City Breaks**	2.86	1.03
Cultural Tourism	3.23	0.96	**Religious Tourism**	2.80	0.98

(Silva, 2013)
Legend: SD - standard deviation || Scale: 1 – very low to 5 - very high

Cruise tourism is another product associated with coastal destinations, particularly islands, which also has a relatively low assessment score (3.01). This result is justified, on the one hand, by the fact that this product was in its early stages of growth in 2012, and on the other, as a result of the marginal position of the Azores in relation to the main cruise itineraries. In 2019, the number of passengers to disembark in the Azores stood at 20% in comparison with the archipelago of Madeira.

In the sustainability dimension, a high level of agreement was observed, with all the stakeholders considering that tourism development in the region should prioritise the guarantee of environmental, sociocultural and economic sustainability (mean of 4.64). They considered the promotion of greater population and company involvement in environmental management to be essential, with a mean score between 4.23 recorded for the Tourist guides group and 4.52 for Experts & researchers.

Stakeholders also agreed that tourism has contributed to local development, to improving the quality of life of the communities (mean of 4.00) and enhancing the supply of infrastructures and facilities for the local population (mean of 3.83). However, they disagreed that the citizens are part of the development policy-making processes and a considerable number disagreed that tourism development in the region is being channelled to all the islands (45%) and that it is contributing to territorial cohesion (62%).

Figure 1. The region should essentially promote tourism development in the islands with the greatest tourism potential
(Silva, 2013)

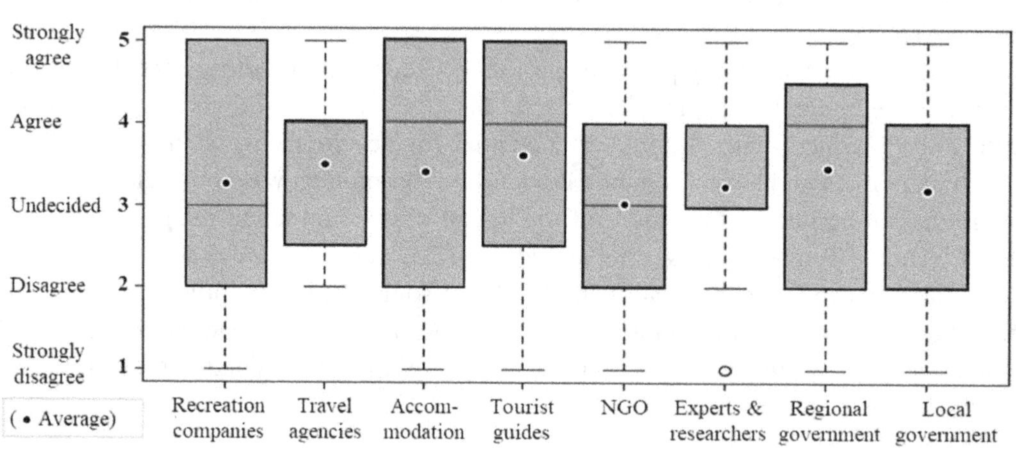

In the same context, stakeholders were asked whether they agreed that the development model should prioritise investment in the islands with the greatest tourism potential. The responses were highly dispersed (standard deviation 1.29), and the data extended across the full length of the assessment scale for almost all the groups of stakeholders. The NGOs, tourism and environmental non-governmental organizations, displayed the lowest levels of agreement (Figure 1)

Main Results

As a result of the tourism stakeholders responses and the data pointing to the trend of the sector, it may be concluded that this sector currently plays a central role in the economy of the Azores. In the opinion of both the stakeholders involved in the empirical research and of the residents (SREA, 2019) and visitors (OTA, 2021), nature is the main feature of the destination brand image. Further strengths that enhance the attractiveness of the destination are insularity, the uniqueness of the volcanic geomorphology, cultural landscape, hospitality and safety (IPDT, 2016).

However, isolation, territorial dispersion, accessibility and small-scale economy–related factors are substantial constraints to development. A breakdown of the main factors in the region, particularly in terms of their tourism potential, is presented below:

- **Economy**: uncompetitive, outermost, limited resources and small-scale market;
- **Accessibility:** very limited, dependence on air transport; competitiveness and cost-effectiveness of the transport sector highly conditioned by the small scale; few air connections and operating companies; strong territorial dispersion requiring high-cost investments in air and port infrastructures, thus increasing operating costs.
- **Physical Environment:** temperate oceanic climate, relatively rainy and unattractive winters for the tourism activity. As a result, the Azores have a relatively high rate of tourism seasonality (39.9% in 2019) (Turismo de Portugal, 2021). Extreme beauty and unique landscapes. Although there are few beaches, the omnipresence of the sea and richness of the marine fauna are conducive to the development of a variety of nautical activities, particularly diving and whale watching.
- **Culture and Hospitality**: the cultural pasture field landscape, some classified as UNESCO's World Heritage Sites. Tangible and intangible cultural heritage of these destinations is a major attraction, namely the festivities, religious and rural architecture and that connected to traditional activities, such as whaling and wine production. Hospitality and gastronomy are two of the Azorean society's most valued cultural characteristics by tourists during their visits to these regions (OTA, 2021). Coupled with safety, authenticity and tranquillity, they contribute greatly to the satisfaction and quality of the tourism experience.
- **Tourism**: sector with high growth between 2014 and 2019 and significant weight in the region's economy. However, several indicators such as the seasonality rate, RevPAR and over-concentration of the activity on only one island are weaknesses that need to be overcome;
- **Sustainability**: regional entities have carried out extensive work so that the destination is identified as a sustainable destination, highlighting the fact that it was the first region in Portugal and the first archipelago in the world to be certified as a sustainable tourist destination, awarded by the Global Sustainable Tourism Council.

SOLUTIONS AND RECOMMENDATIONS

Despite the low density and small-scale constraints common to other peripheral territories, and while not having the attractions of the tropical islands, the Azores have resources, a strong identity and singularities that enable tourism to be established as one of the main activities of the region.

However, the specific characteristics of the territory, the high concentration of tourist activity, a very rapid growth in demand between 2014 and 2019, and the crisis between 2020 and 2021 due to the COVID19 pandemic call for particular care when planning and defining the direction for development.

The territory of the Azores is both highly unique and vulnerable. Therefore, it is important that all the stakeholders resist the temptation of embarking on a path geared towards a rapid growth in demand to instead promote long-term prospective planning based on sustainability and resilience which, in turn, will contribute to the dynamization of other economic activities at a local level. Thus, the destination should prioritise the following:

- **Systemic Approach**: promoting a global identity of the destination, with a strong interconnection within the entire regional system, relying on tourism as a hinge for economic, social and environmental development across the territory;
- **Economy**: adopting a hybrid economic model (outward-looking and inward-oriented), considering an economy that is open to investment, in some products that are internationally competitive and open to tourism, while at the same time ensuring a small-scale external dynamic to promote the development of the local economy, based on local markets, authentic tourism in low-scale and highly personalized products and services, or strongly rooted in local culture, thus pursuing differentiating and creative standards of competitiveness for some of the local products.
- **Sustainability and Territorial Cohesion:** investing heavily in the enhancement of ecosystems, creating more reserves and implementing a cross-cutting management model to include the environment, tourism and culture; aiming for niche tourism; establishing and managing carrying capacities and promoting greater territorial cohesion;
- **Competitiveness:** aiming for market niches that enhance the main differentiating features of the region, by means of a more customized supply that will favour the local economy and higher quality services, to the detriment of massification and cost reduction; reducing seasonality by capturing new market niches, developing products, improving communication and investing in events;
- **Quality:** boosting investment in tourism infrastructures and facilities, investing in the qualification of human resources, promoting digital transition, investing in creative and smart tourism and monitoring the quality of the supply;
- **Brand Redefinition:** reinforcing the nature and sustainability brand, establishing an association with a triangle where the third vertex is the sea, and the centre is insularity. Simultaneously exploring a more trendy image, associated with SMART and Creative, where the traditional and the contemporary encounter a symbiosis and foster improvement of the quality and comfort of the infrastructures and facilities;
- **Accessibility**: reviewing investment priorities and support for improving accessibility, prioritizing internal mobility and encouraging visits to multiple islands;
- **Participatory Model:** the government should ensure the mobilization and involvement of the various stakeholders in the planning, monitoring and managing process of the destination;

Figure 2. Systemic approach: territory, society, economy and tourism

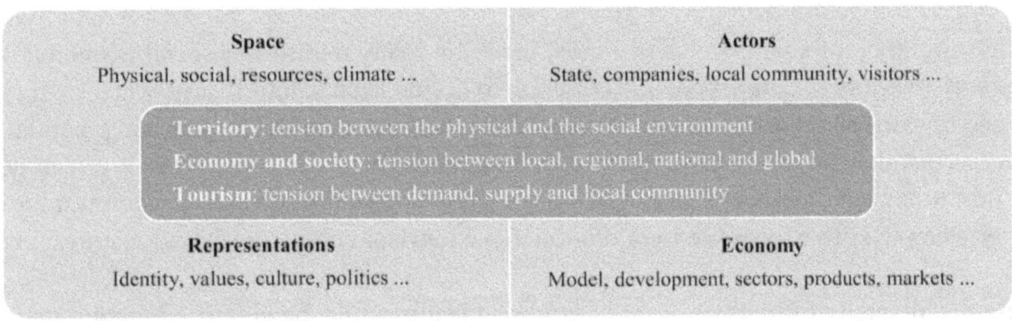

- **Markets**: aiming for greater segmentation in specialized markets, reducing dependence on the domestic market and continuing to explore the diaspora market;
- **Entrepreneurism and Human Resources**: investing in the enhancement of tourism professions and the qualification of entrepreneurs and managers, and stimulating innovation and entrepreneurship.

Several of these priorities are common to many of the inland low-density territories. We are faced with territories where traditional, slow and aged rural communities predominate, in stark contrast to the usually competitive, modern and frenzied worlds of multicultural visitors. These contrasts generate tension that can either give rise to conflicts or stimulate solutions that create synergies and win-win results.

In this context, territorial planning and governance are the gems of territorial management and essential for the game of forces to generate the much-needed dynamics to promote the sustainable development of these territories.

Tourism can play a pivotal role in their development and regional dynamics however, for sustainable development to be ensured, the adoption of a systemic approach to the territory is recommended, in which multiple dimensions are considered, namely: space, actors, representations and economy (Figure 2).

A systemic approach, departing from the region as a macro structuring system, will provide better conditions to safeguard the interests and values of the multiple local actors and, at the same time to find solutions that will capitalize and transform these territories into attractive spaces, both for visitors and the local community. It will also allow for a better response to European policies and programs promoted for territorial cohesion, when considering the territory in its entirety and its specific features, thus promoting greater sustainability and, consequently, the resilience of these regions.

FUTURE RESEARCH DIRECTIONS

Research around the geography of tourism often resorts to destination segmentation and studies on specific areas. However, there may be important gains for research with the development of studies that shift from the most common and traditional approaches. The case study presented herein does not fall within the geography of the territories covered in this book, but highlights the fact that islands are geographies that share several realities with inland territories.

CONCLUSION

Within the geography of small oceanic islands, there are many realities in social economic terms. As far as the tourism sector is concerned, it is common to divide islands into 'warm-water', which are very much geared towards sun, beach and sea tourism, and 'cold-water', with less tourism potential. It is also important to consider the group of temperate water islands, in which the Azores are included, where the most propitious conditions are furnished for tourism to have an expressive weight in the economy and society when the offer focuses essentially on niche tourism and nature-based, nautical and cultural products.

Despite this diversity of realities, a set of common features may be identified among many of these territories, such as an economic structure conditioned by the scarcity of resources and small production scale, significant accessibility limitations, territorial dispersion and high environmental and socio-cultural vulnerability. A knock-on effect of these factors may be observed in the development constraints, leading these economies to remain subject to a major dilemma: opening their economy to the world, but struggling to be competitive on a global scale.

The Azores are an example of such low-density peripheral territories with considerable economic constraints. Regarded as an outermost territory, the region has access to significant support from the European Union's structural funds and other measures to promote territorial cohesion at a national level.

Between 2014 and 2019, the number of tourists visiting the region more than doubled and it began to see a significantly higher density of tourism demand than the country's average, which reinforced the role of this sector as an engine of the regional economy. Its recent certification as a sustainable destination, the very positive attitude on the part of residents and the main attributes of the destination highlighted by visitors are very positive factors of the sector in the region.

However other data, and the very sharp growth in demand between 2014 and 2019, have presented a number of challenges and some negative impacts. It is therefore essential that the region ensures a tourism development model that respects the environmental and social carrying capacities, seeking to assert itself more in terms of quality and enhancement of the tourist experience, than in growth in demand.

This model should also seek to reduce the disparities between the islands in order to counter the trend of greater economic and social dynamism and the concentration of tourism activity on the island of São Miguel.

As already mentioned, the Azores are a peripheral region, but within this territory there are several centralities and peripheries, a reality that is shared with other regions, such as the inland territories (Pilati & Hunter, 2020).

In fact, although they are opposite geographic regions, the Azores have very much in common with the inland regions of Portugal, both in terms of development constraints and the importance of tourism development and its model.

In a context of high competitiveness, low-density peripheral territories may be constituted as "counter-places" (Gössling & Wall, 2007) as opposed to mass destinations. The investment in authenticity, segmentation of demand and customized products and services are important strategies to ensure that tourism is able to assume a relevant and dynamic role in the economies of these regions and contribute to the settlement of the population and to the revitalization and enhancement of the natural, historical and cultural heritage.

However, a boost in the competitiveness of these destinations also depends on their capacity for innovation and the investment in creative practices supported by the enhancement of their attractiveness, qualification and competitiveness (Bucci, Sacco, & Segre, 2014). Along this line of thinking, the focus on training, qualifying and the settlement of human resources is key to the development of these territories and to the promotion of territorial cohesion (República Portuguesa, 2017).

REFERENCES

Adrianto, L., & Matsuda, Y. (2004). Study on Assessing Economic Vulnerability of Small Island Regions. *Environment, Development and Sustainability*, *6*(3), 317–336. doi:10.1023/B:ENVI.0000029902.39214.d0

Agência para o Desenvolvimento e Coesão. (2021). *Plano Global de Avaliação do Portugal 2020.* Agência para o Desenvolvimento e Coesão I.P.

Azevedo, F. (2017). *Estudo para a Comissão REGI - A situação económica, social e territorial dos Açores (Portugal).* European Commission.

Baldacchino, G. (2006). Warm versus cold water island tourism: A review of policy implications. *Island Studies Journal*, *1*(2), 183–200.

Bernardie-Tahir, N. (2011). *L'usage de l'île*. Petra.

Briguglio, L., & Kisanga, E. J. (Eds.). (2004). *Vulnerability and resilience of small states*. Commonwealth Secretariat and Islands and Small States Institute of the University of Malta.

Briguglio, L., Persaud, B., & Stern, R. (2006). *Toward an outward-oriented development strategy for small states: Issues, opportunities, and resilience building* (W. B. G. I. M. Fund, Ed.). World Bank Group / International Monetary Fund.

Bucci, A., Sacco, P. L., & Segre, G. (2014). Smart endogenous growth: Cultural capital and the creative use of skills. *International Journal of Manpower*, *35*(1/2), 33–55. doi:10.1108/IJM-08-2013-0193

Butler, R. W. (1980). The concept of a tourist area cycle of evolution: Implications for management of resources. *The Canadian Geographer. Geographe Canadien*, *24*(1), 5–12. doi:10.1111/j.1541-0064.1980.tb00970.x

Butler, R. W. (2006). Contrasting coldwater and warmwater island tourism destinations. In G. Baldacchino (Ed.), *Extreme tourism: Lessons from the world's cold water islands* (pp. 247–257). Elsevier.

Cooper, A. J. (2012). *Essentials of tourism*. Pearson.

European Commission. (2017). *My Region, My Europe, Our Future. Seventh report on economic, social and territorial cohesion*. European Commission.

European Commission. (2019). *Country Report Portugal 2019. Including an In-Depth Review on the prevention and correction of macroeconomic imbalances*. European Commission.

Gössling, S., & Wall, G. (2007). Island tourism. In G. Baldacchino (Ed.), *A world of islands* (pp. 429–454). Agenda Academic e Institute of Island Studies.

Grydehøj, A., & Kelman, I. (2020). Reflections on conspicuous sustainability: Creating Small Island Dependent States (SIDS) through Ostentatious Development Assistance (ODA)? *Geoforum*, *116*, 90–97. doi:10.1016/j.geoforum.2020.08.004

Hall, C. M. (2010). Island destinations: a natural laboratory for tourism: introduction. *Asia Pacific Journal of Tourism Research, 15*(3), 245–249. doi:10.1080/10941665.2010.503613

Hampton, M. P. (2002). Offshore pariahs? Small Island economies, tax havens, and the re-configuration of global finance. *Elsevier Science Ltd., 30*(9), 1657–1673.

INE. (2020). *Estatísticas do turismo 2019*. Lisboa: Instituto Nacional de Estatística, I. P.

IPDT. (2016). *Plano estratégico e de marketing do turismo dos Açores*. IPDT - Instituto de Turismo.

Lombard, N., & Labescat, G. (2010). Les îles, laboratoires du tourisme durable. *Espaces Tourisme & Loisirs, 278*, 8–37.

McElroy, J. L. (2003). Tourism development in small islands across the world. *Geografiska Annaler, 85*(4), 231–242. doi:10.1111/j.0435-3684.2003.00145.x

McElroy, J. L., & Parry, C. E. (2010). The characteristics of small island tourist economies. *Tourism and Hospitality Research, 10*(4), 315–328. doi:10.1057/thr.2010.11

OTA. (2021). *Satisfação do turista que visita os Açores 2019*. Ponta Delgada: Observatório do Turismo dos Açores.

Page, S. (2015). *Tourism management* (5th ed.). Routledge.

Pilati, M., & Hunter, A. (2020). *Research for REGI Committee – EU Lagging Regions: state of play and future challeges*. European Parliament.

República Portuguesa. (2017). *Programa Nacional para a Coesão Territorial*. Unidade de Missão para a valorização do Interior.

Richards, G. (2017). Creative tourism: Opportunities for smaller places? *Tourism & Management Studies, 15*(SI), 7–10. Advance online publication. doi:10.18089/tms.2019.15SI01

Saliba, C. (2007, Spring). Financial services and small island jurisdictions. *Bank of Valletta Review*, (35), 39–54.

Silva, F. (2013). *Turismo na natureza como base do desenvolvimento turístico responsável nos Açores* (Unpublished doctoral dissertation). Universidade de Lisboa, Lisboa.

Silva, F. (2017). Planeamento turístico nos espaços insulares. In F. Silva & J. Umbelino (Eds.), *Planeamento e desenvolvimento turístico* (pp. 467–482). LIDEL.

SREA. (2019). *Inquérito aos Residentes sobre o Turismo nos Açores 2018*. Angra do Heroísmo: Serviço Regional de Estatística dos Açores.

SREA. (2021). *Statistics Azores*. Retrieved Mai 12, 2021, from Serviço Regional de Estatística dos Açores https://srea.azores.gov.pt/default.aspx?lang_id=2

Terrasson, F. (2007). *La peur de la nature: Au plus profond de notre inconscient, les vraies causes de la destruction de la nature* (3rd ed.). Sang de la Terre.

Turismo de Portugal. (2017). *Estratégia Turismo 2027. Liderar o turismo do futuro*. Turismo de Portugal I.P.

Turismo de Portugal. (2021). *TravelBI. Taxa de sazonalidade*. Retrieved Mai 10, 2021, from Turismo de Portugal, https://travelbi.turismodeportugal.pt/pt-pt/Paginas/PowerBI/Sustentabilidade/taxa-de-sazonalidade.aspx

UN-DESA. (2010). *Trends in sustainable development small island developing states*. United Nations.

UN-OHRLLS. (2011). *Small island developing states - Small islands big(ger) takes*. United Nations - Office of the High Representative for the Least Developed Countries, Landlocked Developing Countries and Small Island Developing States.

Veron, S., Mouchet, M., Govaerts, R., Haevermans, T., & Pellens, R. (2019). Vulnerability to climate change of islands worldwide and its impact on the tree of life. *Scientific Reports*, *9*(1), 14471. doi:10.103841598-019-51107-x PMID:31597935

KEY TERMS AND DEFINITIONS

Cold-Water Island: Islands where the surrounding water is too cold for swimming most of the year, therefore without expression of the 3S (Sun, Sea, and Sand) tourist product.

Lagging Territories: Low-growth territories that differ from the average of countries or regions and/or with relatively low incomes.

Low-Density Territories: The classification of these territories is highly variable, but several indicators such as population density, GDP per capita, the population aging index, and population growth are generally taken into consideration.

Nature Tourism: Tourism activity that is supported by natural resources and the motivation to enjoy natural spaces.

Smart Tourism: A tourism model that brings together a set of paradigms geared towards a more sustainable, efficient, and inclusive organization, supported by technology, connectivity, innovation, governance, and creativity, with the aim of making the destinations more resilient, competitive, and better places to live.

Territorial Cohesion: Relatively balanced distribution of human activities in the territory with fair access for citizens and organizations to services of general interest.

Tourist Experience: Although a vague and comprehensive concept, it may be defined as one that aims to promote a more engaging and stimulating participation of visitors, to create rewarding and emotional experiences.

Warm-Water Island: Islands located in low latitudes, bathed by waters with a pleasant temperature for bathing and with great expression of 3 S tourism (Sun, Sea, and Sand).

Chapter 5
Creative Tourism as Sustainable Initiative of a Coastal Destination:
Creative Routes in Fethiye

Meltem Altınay Ozdemir
https://orcid.org/0000-0002-3002-6127
Mugla Sitki Kocman University, Turkey

Emre Ergun
Mugla Sitki Kocman University, Turkey

ABSTRACT

The study examines tourist attractions and creative experiences of Fethiye, a coastal destination in Mugla, Turkey, based on theoretical overlaps in creative tourism. Creative tourism is suggested as an initiative for sustainability of local lifestyle, unique culture, and authenticity of Fethiye in this study. Therefore, the study was maintained with an exploratory and conceptual viewpoint. The results are presented as Fethiye's tourist attractions (13 categories) and creative experiences (six themes). The prominent creative experiences are specified as house wine, squeeze of olive, cheesemaking, mushroom (morel) picking, agriculture cultivation (local fruits and vegetables), Dastar weaving (geotagging product), carpet weaving (carpet of Kaya), wool spinning, local festivals, traditional bazaars, and soap making. Finally, a creative tour case was offered, and two creative tour routes were proposed in the appendix. The study revealed that creative tourism has importance for sustainability of authentic culture, responsible consumption, and economic development of the local community.

INTRODUCTION

The "creativity" in tourism refers to tourists learning a destination not only by being physically present there but also by consuming local cultural products and engaging in local activities (Gordin & Matetskaya, 2012). Creativity in tourism increases originality and innovation that can affect competitive advantage

DOI: 10.4018/978-1-7998-7339-6.ch005

of companies and destinations (Bialk-wolf et al., 2013; Miles & Green, 2010; Richards, 2013; Richards & Wilson, 2006a). Creative tourism is considered as a new form of tourism that is separated from mass tourism and linked to development of urban tourism niche products (Rogerson & Visser, 2011; Waitt & Gibson, 2014) as well as rural tourism (Blapp & Mitas, 2018; Jarábková & Hamada, 2012; Mitchell & Fisher, 2010). Various concepts have been presented in the literature, such as creative tourism (Carvalho et al., 2016; Ohridska-Olson et al., 2010; Richards, 2011, 2014, 2017; Richards & Wilson, 2006), creative activities (Richards & Wilson, 2006), and creative industries (Booyens & Rogerson, 2015; Ohridska-Olson et al., 2010) that arose due to experience economy. The experience economy suggested by Pine II and Gilmore (1998) includes providing tourists with more opportunities for active involvement and environmental engagement, and more particularly, tourists are more interested in real experiences. "*Experiences are memorable*" is according to Pine II and Gilmore (1998). The more sensory an experience is, the more it can be effective and unforgettable (Hung et al., 2016; Pine & Gilmore, 1998; Winter, 2004).

The emergence of creative tourism has been defined as an extension or reaction to cultural tourism, because, unlike traditional cultural tourists, creative consumers seek more interactive experiences that assist them in their self-improvement and identity creation (Richards & Raymond, 2000; Tan et al., 2014; Tan et al., 2016). Therefore, creative experiences that correspond to a trend towards experience-based products in creative tourism attract attention (Alsos et al., 2014; Booyens & Rogerson, 2015; Prebensen, 2014). Creative tourism experiences combine different creative content elements and engage with creative lifestyles (Richards, 2019, p. 7). Thus, creative tourism involves the involvement of tourists in doing something experimentally to participate in the "real cultural life" of destinations authentically and memorably (Hung et al., 2016; UNESCO, 2006). In other words, tourists become co-producers of experiences they consume (Prebensen, 2014; Richards, 2010, 2012; Tan et al., 2014). Creative tourism, if well planned, brings diversity to tourists and preserves the cultural heritage of destinations. Indeed, creative tourism can regain and re-evaluate a place's identity. Meanwhile, it can also be a tool of economic development in these regions and attract tourists interested in the learning and interaction process (Carvalho et al., 2014; Marujo et al., 2020).

The present study examines current offers and future potential of creative tourism in Fethiye, a coastal destination in Mugla, Turkey, based on the aforementioned theoretical overlaps in creative tourism. The following research questions served as the study's guidelines:

- RQ_1: Besides natural tourism attractions, how creative tourism can be a tool for determining authentic tourist attractions in Fethiye?
- RQ_2: What kinds of creative experience are available around Fethiye?

THEORETICAL BACKGROUND

Creative Tourism: Concept and Context

Cultural tourism has been recognized as an element of tourism policy by national and regional governments of the world (Richards, 2016) and is generally shown as a major growth market in tourism worldwide (Richards, 2002). On the other hand, tourists participating in cultural tourism who wants to experience the local culture and the places they visit, expressing a shift from cultural tourism to creative tourism according to Richards (2002). The tourism industry and creativity have a long-term connection.

Destinations offering cultural tourism can develop their visitors' creative potential by offering destination-specific courses and educational activities (Királová, 2017, p. 72). Creative tourism, as a form of tourism, was first used by Pearce and Butler (1993) but they did not define the term (Richards, 2011, p. 1237). In the 1990s, with the work carried out in Europe's various regions, the concept has emerged more clearly and developed (Zoğal & Emekli, 2017, p. 24). Carvalho et al. (2014) and Tan et al. (2016) referred that creative tourism is accepted as a form of cultural tourism. Besides it aims to create engagement, authentic experience and participatory learning in terms of local identity, including arts, cultural heritage, handicraft, food and other activities that connect tourists, local and cultural heritage (Ohridska-Olson et al., 2010). Moreover, Raymond (2007) defines creative tourism as a form of sustainable tourism that offers authentic experiences of local culture using informal and practical seminars and creative experiences. According to Wurzburger et al. (2009), creative tourism aims at engagement and authentic experience from participatory learning about art, cultural heritage, or local identity.

Richards and Raymond (2000, p. 18) describe creative tourism as "*tourism that provides visitors opportunities to enhance their creative potential through active participation in learning experiences that are characteristic of their holiday destination*". Besides, it is a journey towards participatory learning and a true experience of a place's art, heritage, or unique character and provides a connection with residents of that place and those who create living culture (UNESCO, 2006). UNESCO defines creative tourism as "*travel directed toward an engaged and authentic experience, with participative learning in the arts, heritage, or special character of a place, and it provides a connection with those who reside in this place and creates this living culture*" (UNESCO, 2006, p. 3). According to (Raymond, 2007, p. 146), creative tourism is a type of sustainable tourism in which the culture is presented with creative experiences, and applied seminars are used when necessary.

Active participation, creative experiences, learning experiences, relation with local people is common keywords in all definitions of creative tourism. Meanwhile, (Királová, 2017, p. 73) has identified common elements in definitions of creative tourism as creative potential, active involvement, exceptional experiences, collaboration, acquisition of informal learning experiences and co-creation. Tourists gain knowledge and experience from the everyday lives of communities they are visiting within creative tourism. Also, communities can recognize creative tourism as a tool of (i) tourism innovation, and (ii) economic and sustainable growth benefiting communities. This also includes organizing, through direct experiences, learning activities such as engaging in activities and communicating with local people. The aim is that participants are not only tourists but also active residents of the society (Singsomboon, 2014). Tourists actively engage in activities to create unique works of their own and acquire individual experiences called "creative experiences" through this tourism form, by using simple materials and their imagination (Richards & Wilson, 2006b). Thus, one who learns by doing and finds joy and fulfilment in learning new abilities is a creative tourist; in other words, the creative tourist is interactive (Tan et al., 2014). Creative tourist aims to experience these values with active involvement rather than visiting unique cultural resources, unlike cultural tourists (Ohridska-Olson et al., 2010). The creative tourist is deeply involved and engaged in various activities in the cultural structure of the destination. This establishes a strong link between tourists and local people (Richards & Wilson, 2007).

Creative Tourism as a Sustainable Initiative

Creative tourism, which is of key value for tourism sustainability, involves more engagement with different destinations, cultures and individuals in scope, educational, emotional, social and participatory

(Altınay & İstanbullu Dinçer, 2017, p. 344). Tourists' active participation is involved in creative tourism although it is seen as the development of cultural tourism (Richards, 2011, p. 1245). An approach based on production rather than passive participation of tourists is encountered in this tourism (Akşit Aşık, 2014, p. 789). Besides, creative tourism has some advantages over cultural tourism (Richards & Wilson, 2006a, p. 1215). Being scarcity compared to traditional cultural products, allowing destinations a competitive advantage, being more sustainable and more mobile than cultural products, and also involving "the creation of values" were stated by (Richards & Wilson, 2006a, p. 1215) as the advantages of creative tourism. Richards and Marques (2012, p. 4) suggested new forms of creative tourism as follows but also these are accepted as sustainable initiatives of destinations in this study.

- A means for tourists to be involved in the creative life of the destination,
- A means to creatively use available resources,
- A means to make identity and distinctiveness stronger,
- A form of expressing and discovering oneself,
- "A form of "edutainment" meaning a form of education consisting of self-realization and entertainment,
- Provide an atmosphere for places and
- A resource for recreating and revitalizing places.

Creative tourists prefer to be active and experience the local lifestyle. They want to live traditional activities as it is and gain natural and authentic experiences. They aim to increase their personal experience and skills (Jelinčić & Ana Žuvela, 2012, p. 79). Creative tourists not only meet their social needs but also meet their needs to exceed themselves by improving their skills (Akşit, 2014, p. 789). This development goes hand in hand with changing the motivations of visitors in their search for originality (Királová, 2017).

According to Richards (2009), creative tourism is a recent form of tourism that can change existing patterns of tourism development and contribute to diversification and renewal of the tourist experience. It, in particular, is not a form of mass tourism; it can be an individualized, small-scale alternative to the development of mass cultural tourism. By doing this, creative tourism can help stimulate local economic, social, and cultural development (Richards, 2009). Small cities and rural areas need to build their creative values and activities that fit their nature and pace of life for sustainability, instead of trying to imitate larger places (Richards, 2016). If this occurs, as research (Blapp & Mitas, 2018; Brouder, 2012; Jarábková & Hamada, 2012) on rural destinations indicates, creative tourism can have a significant impact on local growth. Ensuring the sustainability of the regions' lifestyles and atmospheres, the emergence of business lines, the use of local people's creativity, and above all, the use of sustainable resources are among the most important benefits of creative tourism (Richards, 2009). Such that creative tourism is accepted the next form of cultural tourism that with the primary emphasis on active skill development, meets the need for a higher degree of self-realization. In this form of tourism, resources are produced with the collaboration of local people and tourists instead of the consumption of resources (Richards & Wilson, 2007). This reveals the sustainable nature of creative tourism and indicates that it contributes to the destination's development (Lindroth et al., 2007). Moreover, as expressed in previous studies (Blapp & Mitas, 2018; R. Carvalho et al., 2016; Királová, 2020), creative tourism is also closely related to community-based tourism. Referring to Richards (2011), creative tourism includes *active living consumption, intangible culture, the interaction between hosts and guests,* and *self-expression* by tourists. Creative tourism development is seen as a process where local communities and tourists collectively create culture by generating, sustaining and transforming values and meanings through interaction (Blapp & Mitas, 2018; García-Rosell et al., 2007). Creative tourism contributes to regional development through community participation. Therefore, creative tourism, which is considered to be tourism development strategy for rural areas or small residential areas, provides a sustainable interpretation of community-based tourism (Királová, 2020).

Referring to Singsomboon (2014) the success and sustainability of development using local resources to promote tourism in the creative tourism process depends on these factors:

- Have a strong community base, for example, a community with a production system;
- Have efficient natural resources;
- Have a talented leader who can create faith;
- Have a strong cultural foundation, such as having a unique culture;
- Have the participation of local people from the community;
- Operate continuously and
- Have innovative and creative tourism activities.

Creative tourism, in terms of participation, is tourism that is aware of the participation of "tourists-local people-tourism enterprises". Accordingly, it is aimed to use local resources in line with sustainability through creative tourism process to create sustainable tourism development (Singsomboon, 2014). If creative tourism is promoted, tangible benefits such as cultural capital, market expansion, innovation, preservation of cultural heritage, sustainability, brand visibility, job creation and export will be provided (Ohridska-Olson et al., 2010). Creative tourism relies on abstract processes (e.g. dancing, singing, cooking, painting, crafts) that are not tied to a specific place or time and have the potential to more evenly distribute demand across time and space (Richards & Marques, 2012). Because little infrastructure is needed for new products that can be launched quickly using existing resources (Richards & Marques, 2012). Moreover, creative tourism positively influences community cohesion and the self-confidence and pride of local people (Blapp & Mitas, 2018; Ivanova, 2013; Richards & Marques, 2012).

Creative Cities as Sustainable Destinations

Creative tourism can be also seen as part of the wider "creative turn" in society and social sciences which have included creative development strategies of major cities (Richards, 2019, p. 8). Creative Cities Network Program, created by UNESCO in 2004, is an initiative that brings cities together from various regions with different income levels, capacities, and populations to work in creative industries. The program aims to develop the creative, economic and social potential of cultural industries run by local actors and therefore supports UNESCO's cultural diversity ideals. The Creative Cities Network is structured around seven themes that can be selected by cities based on their creative industry sector preferences to direct their talent and energies. These themes have been identified as l*iterature, film, music, craft and folk arts, design, gastronomy* and *media arts*. Currently, Creative Cities Network has 246 members (UNESCO, 2021). The network plays an increasingly vital role in the economic and social development of cities. Richards (2010, p. 14) suggested the basic principles for development in a destination of creative tourism as follows:

- To know your place and who you are: Both tangible (cultural and historical monuments etc.) and intangible (atmosphere, environment, etc.) assets create a difference in location.
- To use local capacity: Rather than importing ideas and skills from elsewhere, the use of destination-specific features is effective in the development of creative tourism.
- To build existing facilities more creatively: It would be more interesting to renovate an existing structure (cave, factory) than to build a new five-star hotel.
- To improve quality rather than cheating: Instead of using gimmicks or following elements such as current trends and fashion, it is more important to emphasize the traditional values of the region by focusing on quality and originality.
- Using creative resources as a stimulator of future change: Creative use of local crafts will be the basis for innovative and authentic travel memories.

Researchers refer that each destination can be creative (Malachovský, 2020; Stipanovic & Rudan, 2014) in all regions of the world, including both urban and rural environments according to Richards (2011) (Blapp & Mitas, 2018, p. 1287). So much so that creative tourism has been studied in urban cities (i.e. Couret, 2012; Gordin & Matetskaya, 2012; Jelinčić & Ana Žuvela, 2012; Richards & Marques, 2012; Stipanovic & Rudan, 2014) as well as in rural destinations (i.e. (Blapp & Mitas, 2018; Brouder, 2012; Mitchell & Fisher, 2010). Local and specific places are highlighted and expressed through creative tourism initiatives. The small and interactive nature of activities incentivizes experimentation, flexibility, alteration of the products, collecting between residents and visitors, and networks of artisans and small entrepreneurs. It enables several experimental locations with "local design, implementation and control, content, models and approaches" (Duxbury et al., 2019). Passive consumption of cultural products is activated by creative tourism through communication, education and participation in the process, making creative tourism resources accessible. This ensures that cultural and local products are actively involved in the process. The increasing demand for creative tourism shows that international tourists want to participate in rural and cultural-oriented tourism activities rather than mass tourism activities (Altınay & İstanbullu Dinçer, 2017, p. 349). Previous projects and studies (i.e. Blapp & Mitas, 2018; Brouder, 2012; Jarábková & Hamada, 2012; Remoaldo & Cadima-Ribeiro, 2019) carried out to bring rural areas into tourism in recent years support this. Creative city projects have been initiated in some countries to support cities with creative products. For instance, Barcelona Creative Tourism (2005) in Spain; CREATOUR in Portugal (Remoaldo & Cadima-Ribeiro, 2019, p. 82).

METHODOLOGY

This study was conducted in Fethiye, one of the Aegean coastal destinations of Turkey, to carry out creative tourism with a sustainable approach. The climate, geographical location, virgin nature, and the implementation of tourism activities with mass tourism movements of Fethiye have led the city to come to the fore with holiday and adventure tourism. In its current condition, in a short period, a large number of visitors in the sea, sand, and sun triangle strains the city's carrying capacity and causes resources to decrease. However by preserving and improving socio-cultural and natural values and ensuring the continuity of attraction, that is, sustainability, it is possible to conserve resources and pass them to future generations. Contrary to the activities of holiday tourism that cause the cultural structure and lifestyle of the city to remain in the background, this study focuses on identifying the creative tourism activities appropriate for local creative activities of Fethiye, local development, rural and cultural life, and ensuring its sustainability via creative tourism. The continuity of traditional life in rural areas indicates that Fethiye has a high potential for creative tourism and the need for a sustainable perspective to carry these cultural values to tourism. Therefore creative tourism is suggested as an initiative for the sustainability of the local lifestyle, unique culture and authenticity of Fethiye in this study.

Research Objectives

- To research tourist attractions of Fethiye to add value
- To describe creative experiences of local lifestyle in creative tourism
- To create newer creative routes to develop creative tourism in Fethiye

Research Area-Where Is Fethiye?

Fethiye in the Aegean province of Mugla in south-western Turkey is one of the 13 districts (Fig. 1). Fethiye, located in the south of the province, has an area of 3.059 km^2. It is located where the Mediterranean and Aegean Seas meet. It is both a coastal town and a tourism center with its coast along Fethiye Bay. Fethiye, which is the largest district of Mugla, is a destination that still preserves its original lifestyle with its ecological, authentic and cultural values.

According to pre-pandemic statistics, 2 million 523 thousand tourists entered the province of Mugla in 2018 and 3 million 222 thousand in 2019. The majority of tourists to Britain, Russia, Germany, consisting of tourists from Ukraine and Poland, countries (Hotel Association of Turkey (TÜROB), 2020). According to Mugla's monthly foreign tourist arrival statistics, the most visited period of tourists is

Figure 1. Location of Fethiye
Map: Created by www.scribblemaps.com; Photo: The authors

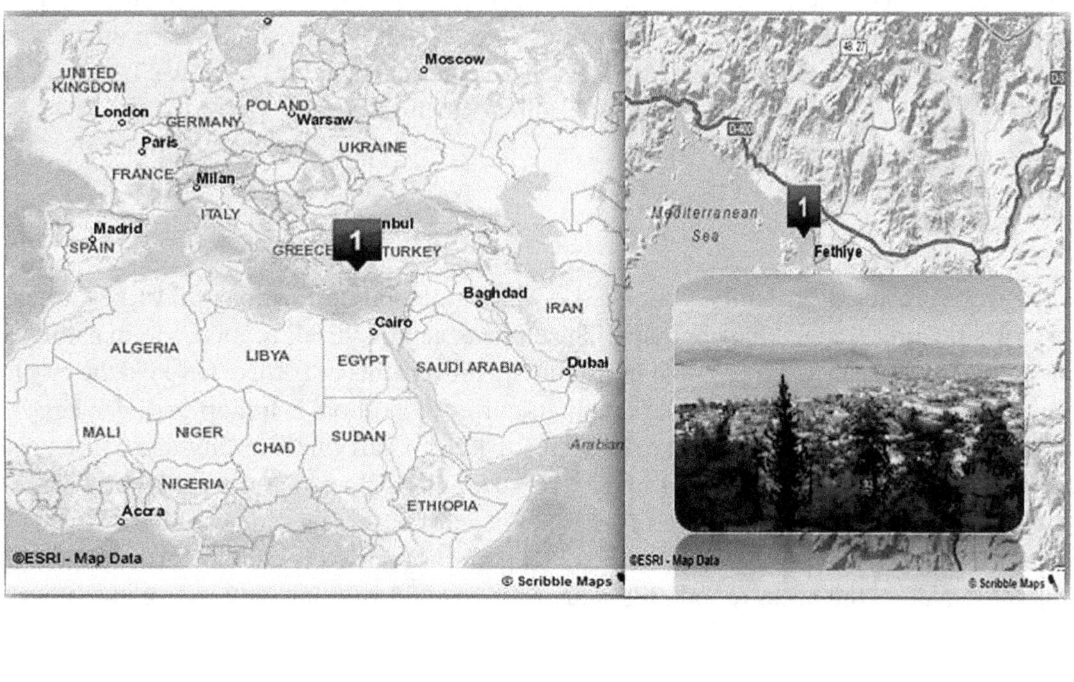

May-October. According to the data of 2019, 497 accommodation facilities certified by the Ministry of Culture and Tourism in Mugla have a capacity of 60,360 rooms and 130,611 beds. Although the total number of visitors to Fethiye cannot be reached, considering the facility statistics with tourism investment certificates, there are 76 accommodation businesses in Fethiye with a capacity of 7,635 rooms and 16,491 beds (Kartal, 2020).

Research Method

This research was carried out in particular for the Fethiye district of Mugla province located on the southwest coast of Turkey. The study investigates the description of Fethiye's creative attractions and activities from an exploratory and conceptual viewpoint. This region encompasses 41 neighborhoods (Fethiye District Governorship Building, 2020) and involves communities and tourism organizations with the potential to develop into creative tourist activities. This includes those with the ability to develop physical, economic, social, and cultural capital, as well as natural resources, as creative tourist attractions in rural zones. Basic social data on Fethiye's creative experiences was obtained from documents, texts, journals, and related online databases about cultural resources, lifestyles, customs, events, and natural resources.

RESULTS

Fethiye is a district located in the Southern part of Turkey. It is a coastal destination as well as rural. It was called "Makri" meaning far land in ancient. In Fethiye, where the Mediterranean climate is observed, summers are hot and dry, and winters are warm and rainy (Republic of Turkey Ministry of Culture and Tourism, 2021). Fethiye, whose name was Telmessos in ancient times, is the most important city in the west of Lycian civilization, one of the oldest Anatolian civilizations. Although there is no exact information about its establishment, it is estimated from the documents available that history of the city dates back to the 5th century BC. Famous as the city of oracles in ancient times, Telmessos is established in a wide area stretching from the slopes of the mountain on which Fethiye rests to the bay. When the prophets of Telmessos were going to prophesy, they were inspired by the blue waters of the Mediterranean. Perhaps that is why there are many legends of Fethiye. There are many sarcophagi and rock tombs in and around the city. The most important of monumental sarcophagus tombs belong to Lycian period in the city center. The sarcophagus is decorated with reliefs symbolizing the warriors. On Telmessos acropolis on hill rising to the south of the city are remains of a castle believed to have been built by Knights of St. John. The castle was also used in the Ottoman Period. Those who climb the hill can see the remains of the city walls, the cistern and small rock tombs on east side of the hill. As the result, Fethiye has a unique culture including various civilizations, e.g. meals, folks and traditions. These cultures can still influence local people, and they try to preserve all these cultures, although there are some changes from time to time (Republic of Turkey Ministry of Culture and Tourism, 2021). This section presents the results as Fethiye's current tourist attractions and potential creative experiences in two sub-sections.

Tourist Attractions and Creative Experiences in Fethiye

Fethiye's *tourism attractions* and *creative experiences* are examined under two sub-headings.

Figure 2. A photo of "Ölüdeniz"
Source: Retrieved from archive of Fethiye A.S.M.K. Vocational School

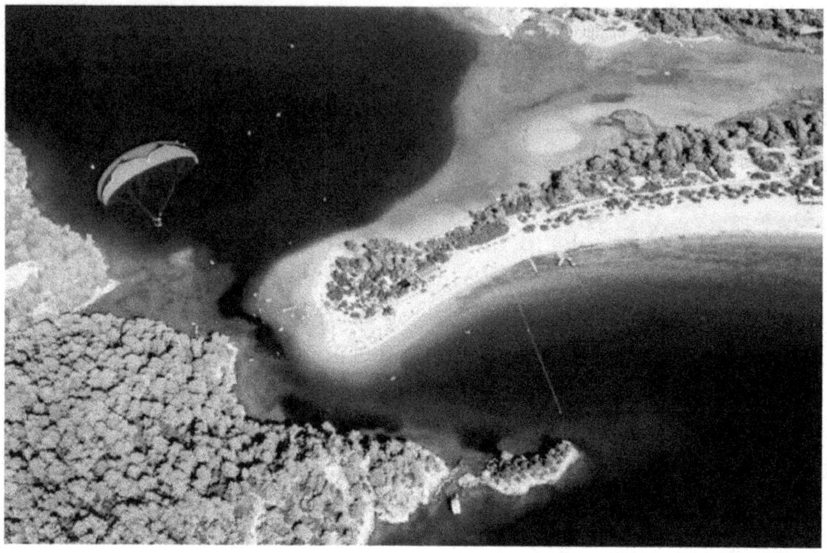

Tourist Attractions

Fethiye bays with turquoise clear waters offer tourists a unique experience. It is an important destination preferred with its coves and clean beaches. One of the most important points is Oludeniz, which hosts a natural beauty with its lagoon. Oludeniz is magical, it stands still. There is not a single moss at the bottom; it is covered with white sand. The light refracted by water and sand at its bottom gives a turquoise color. The shade of pines falls in Oludeniz (Fig. 2) and this enriches impressive turquoise (Republic of Turkey Ministry of Culture and Tourism, 2021).

With its natural structure, alternative sports such as paragliding, diving and hiking are available. There are very special regions with unspoiled nature such as Butterfly Valley and Kabak Bay. In addition to sea tourism, sports tourism and adventure tourism, Fethiye has a cultural tourism potential with its ancient cities (Telmessos, Cadianda, Tlos, Pinara). It welcomes its visitors with the annual Nif Traditional Cherry Festival, Oludeniz International Air Games Festival, and Uzumlu Mushroom Festival (Republic of Turkey Ministry of Culture and Tourism, 2021). Tourism attractions of Fethiye are given in 13 categories with specific examples. These categories are as follows (Table 1): *Bay and beaches, caves, archaeological and historical sites, shopping sites, farms, camping sites, natural parks, islands, festivals, villages, activities, trekking paths and geotagged products.*

As seen in Table 1, although Fethiye has many attractions in terms of tourism, it can be stated that tourists prefer the region mostly for sea-sand-sun purposes. The fact that most of the touristic businesses are open during the summer months means that the tourist demand to the region is more intense in the summer months. On the other hand, creative tourism is of great importance for Fethiye in terms of ensuring sustainability and spreading tourism to 12 months. In response to RQ_1, Fethiye's authentic tourist attractions, which are shopping sites, farms, camping sites, natural parks, festivals, villages, activities, trekking paths and geotagged products, can be improved by creative tourism. Moreover, authentic attractions unique to Fethiye can be discovered with creative tourism.

Table 1. Tourist attractions in Fethiye

Attractions	Descriptions
Bay and beaches	There are various bays and beaches in Fethiye. The most famous of them are Oludeniz, Belcekız Bay, Butterfly Valley, Kabak Bay, Gemiler Bay, Soguk su Bay, Katrancı Bay, Günlüklü Bay, Kıdrak Bay, Samanlık Bay etc.
Cave	There is only a cave that is located 2 km away from Oludeniz. It is called Fethiye Blue Cave since it is surrounded by deep blue water.
Archaeological and historical sites	Fethiye is one of the oldest cities of Lycian Civilization. For example, the Rock Tomb of the King Amintas of Lycian Civilization is here. Besides Fethiye has a historical value with "Kayaköy" (ghost town) ancient site, Telmessos ancient site, Lycian Sarcophagus, Cadianda ancient site, Tlos ancient site, Afkule monastery, etc.
Shopping sites	There are two famous markets in the center of Fethiye: Paspatur (historical) market and traditional fish market. Paspatur is the oldest historical one in the destination. It has narrow and colorful streets. It is a bazaar located by the sea and has both local and tourist shops. Although these stores are souvenirs, spices and clothes, the street of bars of Fethiye is also here, in other words, this is the entertainment center.
Farms	Pastoral Vadi ecological life farm[1], Tangala goat farm[2], Desperado Ranch[3], etc.
Camping sites	Katrancı forest camp, Butterfly valley, Kabak bay, Oludeniz Nature camping, Kayaköy jungle camping & tiny houses, Kayakoy art camp, Kıdrak camp, Inlice camping, etc.
Natural parks	Oludeniz nature park, Ömer Esen nature park, Katrancı nature park etc.
Islands	Fethiye is located in Fethiye gulf. Therefore many islands are available in this gulf as Sovalye island, Gemile island, Dikilitas island, Göcek island, Yassica island, Kizil island.
Festivals	Nif Traditional Cherry Festival, Oludeniz International Air Games Festival, Uzumlu Mushroom Festival, Milyonfest Fethiye, etc.
Villages	There are still rural villages in Fethiye where the local lifestyle continues. One of them, Uzumlu, is a village that is at the forefront for tourists, which applies to a slow city project. Other villages that have been brought into tourism are Yaniklar, Faralya, Yakacık, etc.
Activities	Babadağ paragliding, boat tour, diving, horse tour, jeep safari, photography, hiking
Trekking path	The Lycian way is a historical trekking route that runs along with Lycian cities
Geotagged products	Carpet of Kaya, Dastar of "Üzümlü"

Source: Republic of Turkey Ministry of Culture and Tourism (2021) and the authors

Creative Experiences

Creative experiences have implications for sustainable tourism as the sources of creativity are more sustainable than tangible products and creativity allows suppliers to renew their products relatively quickly (Prentice & Andersen, 2003; Richards & Wilson, 2006a). As a result, increasing numbers of tourism businesses have "transformed" into creative tourism in the context of the "experience economy" (Pine & Gilmore, 1998). This increased competition "forces" suppliers to add more experiential elements to differentiate their products; however, only those who can benefit from tourist demand will survive in this competitive market because unique experiences are invaluable. Creative experiences can be craft making, perfume making, porcelain painting, and dancing, as Richards and Wilson (2006a) say. As stated in the Barcelona Creative Tourism (2005) project, it can be defined as culinary arts, wine experiences, fashion, design, performing arts, music, photo, cinema, multimedia, traditional arts, folk, visual arts and crafts (DIY). So tourists can learn more about the local skills, specialties, traditions and unique qualities of the places they visit through creative experience (Richards & Wilson, 2006a). In this section, as the answer to RQ_2, the creative experiences of Fethiye are described and existing workshop practices for creative tourism management in the region are mentioned. Creative experiences are described in themes of *(1) culinary arts, wine experiences; (2) agriculture cultivation and farming (3) fashion, design; (4) traditional festivals; (5) visual arts, and crafts (DIY) and (6) local bazaars"*. "The main creative experiences of Fethiye are dastar weaving, house wine, squeezed olive oil, agriculture growing and farming, carpet and rug, crafts, folk, Turkish pancake "gözleme" and grape molasses.

Culinary Arts and Wine Experiences

House Wine

House wine is one of the creative tourism resources of Fethiye. Most of the wines unique to the region are produced in Uzumlu village of Fethiye by local people. Fethiye Municipality (2020) has been working to introduce and market the local wine to Turkey and all around the world and planning to construct a grape processing plant in Uzumlu to realize its purpose. On the other hand, the wines produced in Fethiye are known by the majority of local and foreign tourists. According to research conducted in the region, 56 percent of the participants stated that the word "Uzumlu" reminded them of the local wine (Bekar et al., 2017, p. 32).

Squeeze Olive Oil

Gastronomic products play an important role in transforming an area into a touristic destination. Olive oil, which is the main element of Mediterranean cuisine, is included in gastronomy tourism (Eskiyörük, 2016, p. 239). The amount of olives taken from a tree in Fethiye, which is located in the Mediterranean region, is about 2.6 kg and 1 kg of oil is obtained from 5 kg of olives. Fethiye is a popular destination in olive and olive oil production. It is stated that about 15 factories are processing olive products in Fethiye (Sönmez et al., 2015, p. 630).

Turkish Pancake "Gözleme"

Pancake is a kind of pastry prepared by filling the dough with potatoes, minced meat, cheese, or herbs and cooked on a pan with a wood fire. In Turkish cuisine, pancakes are of great importance in addition to bread (Ertaş & Gezmen-Karadağ, 2013, p. 122).

Cheesemaking

Turkey is a country rich in cheese country. Apart from white cheese, Kashar and Tulum cheeses, there are many local cheeses (i.e. Otlu, Mihaliç, Urfa, Malatya, Antep, Halloumi, Knitting, Ezine, Dil, Sürk, Cara) (Hayaloğlu, 2008, p. 21). To protect this rich local heritage Turkey has, the registration works are carried out by the Turkish Patent and Trademark Office today and geotagging is applied on the products (Turkish Patent and Trademark Office, 2020). From the application made for 32 different cheese varieties so far, Gaziantep squeezed cheese, Diyarbakır braided cheese, Edirne white cheese, Erzurum civil cheese, Erzurum moldy cheese, Erzincan tulum cheese, Ezine cheese, Karaman Divle sinkhole cheese, Kars cheese, Yozgat bowl cheese, Antakya Sürkü, Antakya moldy root, Malkara old cheddar cheese and Van herbal cheese were awarded geographical indications (Saygılı et al., 2020, p. 11). Although the cheese made from the milk of the goats fed on the slopes of the Taurus Mountains in Fethiye continues to be made by the villagers, they continue to be offered in the postmodern form under artisan production (for example; *Tangala Goat Farm).*

A Case: Tangala Goat Farm[4]

Tangala farm stands out especially with its cheese made from milk obtained from the goats and cows on its farm. The farm is open to all visitors in the form of breakfast, cheese tasting, cheese making and farm visits. Tangala Goat Farm, a member of TaTuTa, which is a good example of a creative experience, offers visitors an adventure of about 4 days, during which time visitors learn about milking and cheese making and enjoy the cheese they make with wine on the last day.

Mushroom Picking

Although Fethiye is rich in mushrooms, the village of Uzumlu is legendary with its true morel (Morchella) (Fig. 3). This mushroom (Pilz et al., 2007) which is consumed for its taste as well as its nutritiousness, is especially popular in European countries. Present in almost every region of Turkey, provided that the important export countries (Taşkın & Büyüklaca, 2012, p. 19). Turkey should be appropriate to the geographical morels are among the countries that export the largest. Pilz et al. (2007) stated that 50 million people in the world collect the true morels and India, Pakistan, Turkey, Nepal, Buttons, America, Canada and China are the biggest exporters of the dried true morels. The economic value of this type is so high that it increases even more when dried (Taşkın & Büyüklaca, 2012, p. 28).

Morel (Morchella) is such a unique local product for Fethiye that the "Dastar and Morel Mushroom Festival" is held every year when it starts to appear in spring. Most of the true morels grow in contact with living trees in undamaged forests. If the morel is dried, it is recommended to be eaten as its protein amount increases, but it is also consumed fresh. It is a sought-after mushroom as it has a unique flavor as well as nutritiousness (Iqbal, 1993). As a creative experience, mushroom search and gathering days are carried out in pine forests to collect various mushrooms that begin to mature since April. Then, this activity ends when the mushrooms collected are consumed in a restaurant in the village with house wine.

Viticulture and Grape Molasses

Turkey has suitable climatic conditions for viticulture. Almost all of the grapes grown in soil that is the most produced in the Aegean region of Turkey. So much so that it has 28% of the country's total vineyards and 45% of grape production. Grapes are consumed in Turkey in three ways: table, wine and

Figure 3. True morel in Uzumlu, Fethiye
Source: Retrieved from archive of Fethiye A.S.M.K. Vocational School

dried. In addition to this, it is also used as molasses, fruit pulp, and paste. It is an important Aegean destination where grape cultivation is carried out in Fethiye. In this respect, it is one of the regions where the wine and table grape production of Muğla is highest. Grape cultivation continues in Uzumlu village and locals are engaged in viticulture. Since ancient times, Uzumlu village has been doing this and is one of the rare settlements where home wine production continues. Molasses is also known as "Traditional Concentrated Fruit Product" (Karababa & Develi Isikli, 2005). It is reported that molasses can be produced from apples, mulberries, apricots, plums, watermelons, figs, sugar corn and sugar beets, and is most commonly produced from fresh grapes and raisins that cannot be exported. Each type of molasses is indicated by the name of the fruit from which it is produced (Yazıcıoğlu & Gökçen, 1976). Grape molasses is a product with a thick consistency obtained by clarifying the fresh and dried grape must with suitable enzymes without reducing the acidity, and thickening it under vacuum or in the open by its technique, or it is solidified with the addition of substances such as honey, gypsophila, milk and egg whites. Molasses is produced in every region of Turkey where viticulture (Kaya et al., 2005, p. 1483).

Agriculture Cultivation and Farming

Thanks to its fertile soil and favorable climatic conditions, Fethiye enables the cultivation of almost all fruit and vegetable varieties. In addition to beekeeping and honey production, Fethiye is also at the forefront of the country's economy in the production of tomato, citrus fruits, thyme, sage and sweet gum oil (Fethiye Chamber of Commerce and Industry, 2020). There are several businesses in the region to promote the agricultural products grown in the region to tourists by ensuring active participation and to ensure the sustainability of the products. In the enterprises, services such as ecological life education, using local seeds, planting and harvesting fruits and vegetables and growing herbs are provided (Üzümcü & Koç, 2017, pp. 16–17). For example, in *Pastoral Valley Ecological Life Farm* in Yaniklar village of Fethiye, visitors can visit the farm and collect the harvest. They can stay in the accommodation units of the farm and eat the vegetables collected from the field for their breakfasts. There are various horse farms in Fethiye. One of them is *Desperado Ranch* in Yaniklar village. Visitors can tour the farm here, grooming horses and riding by the sea. Short workshops are provided for horseback riding and visitors.

Fashion, Design

Dastar Weaving

Fabrics called dastar, which is obtained with a local weaving technique, are mostly produced in Uzumlu and İncirköy districts of Fethiye. A wide variety of clothing is produced from dastars, which are woven by using various embroideries on looms called "düven" (Fig. 4). Dastars are weaved in two ways which are thick and thin. Thin dastars hold the heads of local women as a traditional weaving used for covering. Thick dastars are used for making a jacket, inner skirt, and table cloth (Akpınarlı & Arslan, 2017, p. 19). Besides being authentic, dastar weavings made in the region reflect the unique understand and spirit of the local people. On the other hand, all kinds of industrial weaving to reduce costs and increase diversity causes weaving to lose its localness (Çürük, 2017, p. 182). Dastar, one of the elements that reflect the local culture of Fethiye, is trying to survive with the efforts of the people and the municipality.

Meanwhile, the traditional weaving "Üzümlü dastar", which is about to disappear, received a Geographical Indication Registration Certificate from the Turkish Patent and Trademark Office (Turkish Patent and Trademark Office, 2020).

Figure 4. A woman weaving dastar and "düven"
Source: Retrieved from archive of Fethiye A.S.M.K. Vocational School

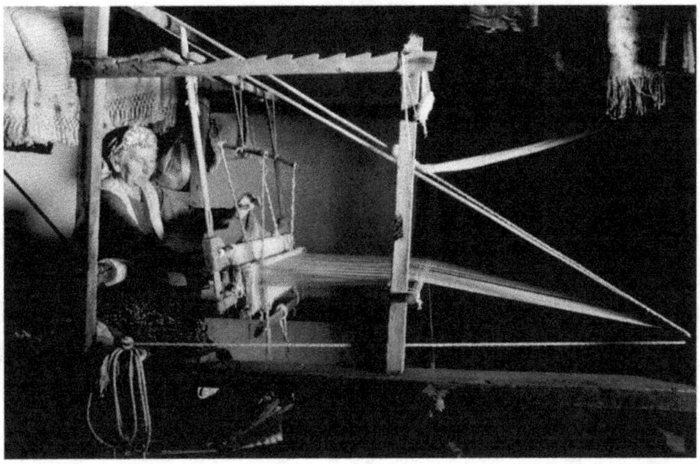

Figure 5. A "Yantırlı" Rug, Istar loom (rug tree) and "Tengerek" or "Kirman"
Source: Ölmez & Etikan (2013b)

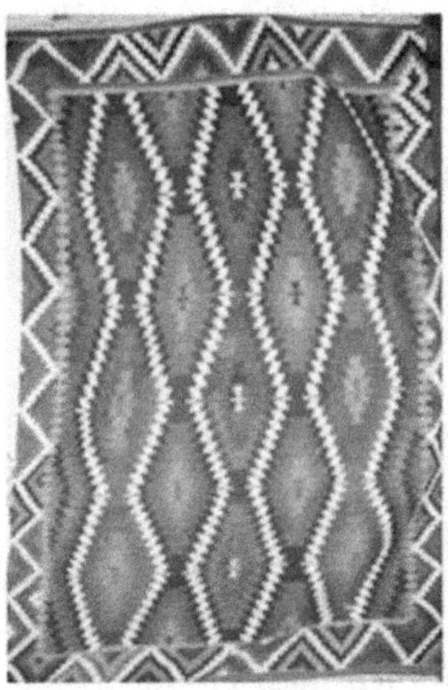

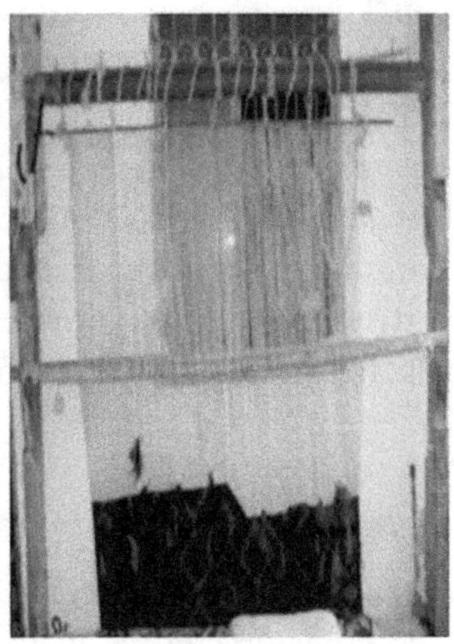

Carpets and Rugs

Fethiye is an important weaving center that has acquired a separate place with its unique color, motif, and pattern feature in weaving culture (Ölmez & Etikan, 2013a, p. 75) Plain weaving activities in Fethiye have been carried out for a long time. Today, plain weaving in Fethiye is not as common as it used to be. However, it continues as local handicraft. Fethiye plain weaving contributes to the home economy by meeting the needs of families and sometimes by providing them with income. Local people call plain weaving "rugs" (Ölmez & Etikan, 2013a, p. 75). Carpet and plain weavings (Fig.5), which are among the cultural riches of Fethiye, are important in terms of being produced and kept alive today. In the region where a traditional lifestyle is maintained, rug weaving has an important place both in daily use and in dowry. Seki, Kemer, Seydiler, Atlıdere, Eldirek, Boğalar, Esenköy, Kayabaşı,

Dodurga, Arsaköy, Kadıköy are some of the settlements where plain weaving is widely carried out, especially the plateau villages. Among the plain weavings in the weaving culture of the region, rug weavings, rugs and cicim techniques are used together, which are remarkable. In the region, plain weft-faced rug weaving technique is used to weave mats and prayer rugs in the size of the base, half base and runner (Ölmez & Etikan, 2013b, p. 63).

A Case: Carpet of Kaya

Kayaköy, one of the villages of Fethiye, is an important center for unique carpet weaving. Kayaköy carpets have their unique color, pattern and motif features (Etikan, 2007, p. 83). Local people obtained the weaving threads used in kayaköy rugs from sheep's wool and dyed the threads with natural dyes. "Turkish Knot Technique" has been used in all of the weavings (Oyman & Erdal Sideli, 2018, p. 241). Although the number of carpet weaving houses in Kayaköy has decreased, it is important to encourage the culture to be revitalized.

Sackcloth Weaving

Sackcloth weaving, which is unique to the nomadic and highland culture in Anatolia, is still seen in Fethiye, where the nomadic culture continues. In the region, warp and weft yarns made of black goat hair are known as "*karaçul*". These weavings are woven in the form of two wings and then stitched together in the middle. Used in the construction of tents in the beginning, the blackcloth weavings were later used as fate (mat), harar (big sack), saddlebag, or cover to protect animals from cold. In black weaving, mostly cream or brown colored goat hairs are used to create lines in the width of the weaving with these colors, and this type of sackcloth is also called bared sackcloth. "Alaçul" is a sackcloth on which a motif and pattern are created according to a certain technique using goat hair, motif yarn, or wool yarn dyed in the third yarn in warp and weft yarns. In Fethiye region, the motif is called "ala" (i.e. colorful). It is called by this name in the region because of the motifs in "Alaçul" (Etikan, 2016, p. 968).

Wool Spinning (Yarn Making) by "Tengerek" or "Kirman"

The process of turning sheep's wool into yarn is called "wool spinning". In rural areas, specific to the Anatolian culture and even nomadic culture, women turn their animal loads into threads with a tool called "Tengerek" or "Kirman". Kirman is rotated to ensure that the ropes are evenly thinned. These threads are generally used for knitting wool socks. Due to the coldness of the Fethiye plateaus, these wool socks are still knitted and used by the villagers (Oyman & Sideli, 2018, p.228). It is seen as a fun activity for someone who doesn't know. As a creative and authentic value, tourists will have a unique experience when they try rope making.

Traditional Festivals

Local festivals undertake many roles such as increasing the number of tourists, contributing to the social, cultural and economic development and strengthening the social peace and the image of the region (Özgürel, G., Alan, R. & Bingöl, 2015, p. 113). Nif Traditional Cherry Festival, Oludeniz International Air Games Festival and Uzumlu Mushroom Festival are held in Fethiye.

Traditional Nif Cherry Festival

A cherry festival is held in July every year in the Nif village of Fethiye, which has many cherry gardens. The festival features local stands featuring the best cherry-growing information, cherry eating contests and local dance performances.

Oludeniz International Air Games Festival

The International Fethiye Oludeniz Air Games Festival is organized by Fethiye Municipality in October every year In addition to the acrobatic shows with paragliders, skydiving and wingsuit shows are also performed at the festival (Fethiye District Governorship Building, 2020). Oludeniz and Babadag are the focus of attention of tourists from various countries who are interested in air sports.

Uzumlu Dastar and Morel Mushroom Festival

The mushroom festival has been held in April or May every year since 2008 in the Uzumlu village of Fethiye. Information on mushroom collection and consumption are provided during the festival (Bekar et al., 2017, p. 30). The festival, where the mushrooms and mushroom pancakes grew indigenous to the region are exhibited, also provides economic support to the local people (Bahar et al., 2020, p. 136).

Visual Arts and Crafts (DIY)

Natural Handmade Olive Oil Soap Making

Natural handmade olive oil soap is a type of soap produced in the village houses of Fethiye. For the production of olive oil soap, local people use the oil they collect from the olives in the region. In this way, superior quality soaps are obtained. On the other hand, the fact that industrially produced soaps are cheaper and more common than handmade soaps has reduced the production of handmade soaps (Şahin, 2019, p. 371).

Ceramic and Pottery

"Kayaköy Art Camp" in the Kayaköy region of Fethiye plays an important role in ceramic and pottery design. Candidates participating in the camp are taught how to make different ceramic objects using different techniques in workshops.

Glassblowing

Glass has been used as food containers, building materials and ornaments since ancient times (Yeşilay & Özsolak, 2017, p. 21). The technique used in glass production called as "glass blowing technique" dates back to 300 BC which was used by Syrian glass masters (Demirarslan & Demirarslan, 2010, p. 126). Gift items created by the glass blowing technique are exhibited in the center of Fethiye.

Local Bazaars

Fethiye has a lot of local bazaars. Local bazaars offer fresh vegetables and fruits growing in the gardens and farming of Fethiye. Most villagers grow local products; sell the milk of their animals and grape molasses made from grapes in their gardens. Especially the local bazaars on Tuesday and Friday are the most visited bazaars by tourists. Both bazaars are located near the coast in the city center. But Friday bazaar is not the same as on Tuesday since it is only sold the products grown by villagers. It is special for locals. There are villagers (i.e. producers) rather than sellers in this bazaar. Therefore it supports the villagers' economics. Apart from these local bazaars that are set up on various days of the week, "Paspatur" is the oldest bazaar in Fethiye. There are souvenirs, spice and clothing stores instead of fruit and vegetables. Some of the stores still sell local culture-specific products.

A Case Route: "Taste of Fethiye" Project

Travel Foundation, a charity that works for tourism to bring more benefits to people and the environment, has started a five-year project named "Taste of Fethiye" which can be considered as an example of creative tourism activities in Fethiye. The Project was supported by TUI Care Foundation, an independent foundation of TUI Group (The Travel Foundation, 2016). The project aimed to support local farmers in growing fruit and vegetables to supply regional hotels. Within the scope of the project, 750 farms were visited to improve agricultural practices, production and purchasing agreements with farmers and hotels were established and with the TUI Group, an agriculture excursion was developed during the excursion, a farm in Fethiye was visited and local farm products (such as olive, cucumbers, tomatoes, peppers, lentils) were introduced to the tourists. By promoting local products to tourists, the farmer has been provided with additional revenue. As a result of the project according to Travel Foundation data local "Taste of Fethiye" products were bought by 24 hotels in Fethiye, the revenue of farmers has increased by 24% and fruit and vegetable sales revenue has exceeded $1M (The Travel Foundation, 2016). The close relations of tourists with local producers, their active participation, and authentic experiences were common concepts in Richard & Raymond's and UNESCO's definitions of creative tourism (Richards & Raymond, 2000, p. 18; UNESCO, 2006, p. 3). Accordingly, this rural and community-based project is regarded as an application of creative tourism in Fethiye.

Creative Tourism Routes in Fethiye

Fethiye destination has great potential with various identities and different lifestyles. There are many routes in Fethiye where creative tourism can create value. For this reason, it is necessary to create a route for creative tourism to bring the above-mentioned creative values to tourism. These routes are based on the creative values that exist in the destination and live in rural areas. Fethiye offers creative activities for young, old, couples and families. For example; creative workshops for children are offered in *Pastoral Valley Ecological Life Farm*. Establishing more creative workshops in Fethiye, which includes creative activities suitable for all types of tourists, should be seen as a priority for managers. Routes are suggested to introduce, visit and experience the current creative workshops in Fethiye. These routes provide information about the local lifestyle. Thus, there are two creative routes ranked according to the popularity of creative attractions in Fethiye as Appendix 1.

DISCUSSION AND CONCLUSION

This study builds on theoretical overlaps that suggest that it can increase the potential of creative tourism as a sustainable initiative strategy. Creative activities were identified by examining the villages that host Fethiye's creative experiences. Findings have shown that Fethiye as a coastal destination has unique cultural and traditional creative experiences to build an authentic experience for tourists through sustainable initiatives. However, as the marine, nature and sports activities of Fethiye are at the forefront, unplanned misuse of resources can cause even creative experiences to disappear. Just as Binkhorst (2007) examined the creative tourism resources, which is a town of Sitges in the south of Barcelona, Spain, this study offers the creative resources of Fethiye which is an attractive destination with its marine, scenery and coastal resources on the southwest coast of Turkey. Therefore this viewpoint research described creative experiences of local lifestyle in Fethiye under six creative categories: (1) culinary arts, wine experiences; (2) agriculture cultivation and farming (3) fashion, design; (4) traditional festivals; (5) visual arts, and crafts (DIY) and (6) local bazaars. The prominent creative experiences of Fethiye according to these categories: house wine, squeeze of olive (make oil), cheesemaking, mushroom picking (especially morel), agriculture cultivation (local fruits and vegetables), Dastar weaving (geotagging product), carpet weaving (carpet of Kaya), wool spinning, local festivals, traditional bazaars and soap making. Some cases of the creative categories are given in detail. Besides, the project initiated by the world-renowned tour operator TUI for Fethiye in 2016 with the slogan "Taste of Fethiye" was mentioned. This project was started to support the production of local fruits and vegetables and small, local producers. Therefore the project includes a visit to a taste of Fethiye farm. The third aim of the research was to create newer creative routes to develop creative tourism in Fethiye. Meanwhile, as given in Appendix 1, two creative routes have been determined in Fethiye.

As Kiráľová (2020) recently stated, creative tourism protects the interests of the local community. In addition to being a coast-based holiday destination, Fethiye has a lifestyle that is suitable for creative experiences of local community. If the study is supported by operational, it emphasizes that it is possible to pass the authentic culture of Fethiye to tourists through creative experiences. As stated by Singsomboon (2014), Fethiye has all the factors (natural resources, productive community, unique culture, the participation of local people, innovative and creative tourism activities) that encourage creative tourism other than practical management. It is not late for Fethiye to embrace creative tourism as a sustainable initiative, given its potential. Destinations have recently begun to transform existing tourism opportunities into new products such as creative tourism (Malachovský, 2020). By designing an art street in Fethiye municipality, local community can teach traditional handicrafts to tourists with the "edutainment" role (Richards and Marques, 2012) of creative tourism. The "dastarhane" shown to tourists in the village of Uzumlu for Dastar weaving can be offered to tourists not as a sign but as a creative experience and so the length of stay can be extended. Fruits and vegetables can be gathered from the gardens of local community in the villages. Even, tourists can make local pickles of Fethiye as presents for their loved ones.

Theoretical Contributions and Managerial Implications

The study reveals that as a strategy of sustainable initiative, creative tourism has importance to the sustainability of authentic culture and economic development of local community. Sharing a daily lifestyle with tourists interested in interactive cultural experiences seems like a perfect experience at first glance. However, it was found that certain conditions were necessary. For example, for creative tourism to take place, it is necessary to establish workshops and create creative tourism management. It is a priority for municipalities, hotels, travel agencies and even universities to act together for the participation of local people in tourism. The creative initiatives live in some villages or rural areas of Fethiye, but not common and not well-known. Creative tourism can be a pattern of tourism for economic development of rural areas and local communities protecting local authentic lifestyles in the future. Creative tourism is interesting and is a trend, thus, it can be characterized by specific creative and sustainable initiatives of a local community. They are not government enterprises, service in specific farms, camps, and village houses. Thus, this is a deficiency of creative initiatives of Fethiye. On the contrary, it is necessary to establish creative workshops where training seminars will be given to tourists to be supported by state enterprises and to encourage local people. Creative tourism studies in Turkey are not common. Thus, the study done in this regard is thought to spread in Turkey in the future. Currently, the various destinations of Turkey or scientific studies on specific creative resources are not investigated in detail and inadequate. In this respect, this study is an outline specification of creative experiences of Fethiye into practice and a project subject that will turn into reality by sharing it with tourism stakeholders.

Limitations and Future Work

A lot of research needs to be done here on this subject. First, limited information is available on visitor profiles, their motivations, length of stay and the like, and the traditional tourism supply in Fethiye and the demand for creative experiences developed recently. It is necessary to continue researches to answer the questions of what is necessary for creative tourism management to be effective and how does it contribute to the local community and tourism as "a sustainable initiative". The findings obtained from the conceptual and exploratory viewpoint of this study show that there is not much awareness of the creative experiences of the region. On the other hand, it is doubtful whether the creators of tourism experiences have noticed that an important market is now available. Often, "not knowing" or "not understand" due to lack of publicity keeps tourists or visitors on the beach or in shops, bars, and restaurants. While this leaves the "other side of the region" unknown, Fethiye has a rich cultural and natural heritage that provides an unlimited resource for creative tourism experiences.

REFERENCES

Akpınarlı, H., & Arslan, P. (2017). Dokumacılığın Yerel Kalkınma Projeleri İle Yaşatılması Örneği: Üzümlü Dastar Dokumaları. In A. Aytaç (Ed.), *IX. Uluslararası Türk Sanatı, Tarihi ve Folklörü Kongresi*. Anka Basım.

Akşit, N. (2014). Yaratıcı (Kreatif) Turizm. *Uluslararası Sosyal Araştırmalar Dergisi*, *7*(31), 786–795.

Alsos, G. A., Eide, D., & Madsen, E. L. (2014). Innovation in tourism industries. In G. A. Alsos, D. Eide, & E. L. Madson (Eds.), *Handbook of research on innovation in tourism industries* (pp. 1–26). Edward Elgar.

Altınay, M., & İstanbullu Dinçer, F. (2017). Geleneksel El Sanatlarının Yaratıcı Turizm Kapsamında Değerlendirilmesi. *Journal of Recreation and Tourism Research*, *4*(1), 343–352.

Bahar, O., Özdemir, Ö., & Akyürek, S. (2020). Yeşilüzümlü Kuzugöbeği Mantar ve Dastar Festivalinin Kırsal Kalkınma ve Kültürel Değerler Üzerindeki Rolü. *Sosyal ve Beşeri Bilimler Araştırmaları Dergisi*, *21*(46), 131–159.

Barcelona Creative Tourism. (2005). *Barcelona Creative Tourism*. Retrieved from https://www.barcelonacreativa.info/

Bekar, A., Kocatürk, E., & Sürücü, Ç. (2017). Gastronomi Festivallerinin Algılanan Destinasyon İmajına ve Tekrar Ziyaret Etme Niyetine Etkisi. *Journal of Tourism and Gastronomy Studies*, *5*(2), 28–36. doi:10.21325/jotags.2017.109

Bialk-wolf, A., Pechlaner, H., & Nordhorn, C. (2013). The role of culture in building regional innovation systems and its impact on business tourism – The case of the Nuremberg Metropolitan Region. Poznan University of Economics Review, 13(4), 111–129.

Binkhorst, E. (2007). Creativity in tourism experiences: The case of Sitges. In G. Richards & J. Wilson (Eds.), *Tourism Creativity and Development* (pp. 125–144).

Blapp, M., & Mitas, O. (2018). Creative tourism in Balinese rural communities. *Current Issues in Tourism*, *21*(11), 1285–1311. doi:10.1080/13683500.2017.1358701

Booyens, I., & Rogerson, C. M. (2015). Creative Tourism in Cape Town: An Innovation Perspective. *Urban Forum*, *26*(4), 405–424.

Brouder, P. (2012). Creative outposts: Tourism's place in rural innovation. *Tourism Planning & Development*, *9*(4), 383–396. doi:10.1080/21568316.2012.726254

Carvalho, M. S., de Lima, J. I. S., & Kastenholz, E. (2014). Criatividade Cultural - que oportunidade para destinos rurais? *PASOS Revista de Turismo y Patrimonio Cultural*, *12*(3), 635–648. doi:10.25145/j.pasos.2014.12.047

Carvalho, R., Ferreira, A. M., & Figueira, L. M. (2016). Cultural and Creative tourism in Portugal. *PASOS. Revista de Turismo y Patrimonio Cultural*, *14*(5), 1075–1082. doi:10.25145/j.pasos.2016.14.071

Couret, C. (2012). Barcelona Creative Tourism. *Journal of Tourism Consumption and Practice*, *4*(2), 132–134.

Çürük, G. (2017). Yeşil Üzümlü Beldesinde Geleneksel Dastar Dokumacılığı. *Milli Folklor*, *29*(115), 170–183.

Demirarslan, S., & Demirarslan, O. (2010). İnşaat Sektöründe Cam ve Cam Kökenli Malzeme Kullanımı. *CAMGERAN 2010 Uluslararası Katılımlı Uygulamalı Cam Sempozyumu*, 125–130.

Ertaş, Y., & Gezmen-Karadağ, M. (2013). Sağlıklı beslenmede Türk mutfak kültürünün yeri. *Gümüşhane Üniversitesi Sağlık Bilimleri Dergisi*, *2*(1), 117–136.

Eskiyörük, D. (2016). Antik Çağda Zeytin ve Zeytinyağı: Kilikya Bölgesi Örneği [Olive and Olive Oil in the Ancient Age: Cilicia Region Example]. *Journal of Tourism and Gastronomy Studies*, *4*(1), 228–228.

Etikan, S. (2007). Kaya Köy Halıları. *Milli Folklor, Uluslararası Kültür Araştırmaları Dergisi*, *10*(19), 83–90.

Etikan, S. (2016). Fethiye Alaçul Dokumaları. *Journal of International Social Research*, *9*(42), 967–977. doi:10.17719/jisr.20164216206

Fethiye Chamber 0f Commerce and Industry. (2020). *Fethiye Farming*. Retrieved from https://www.fto.org.tr/index_eng.aspx

Fethiye District Governorship Building. (2020). *Fethiye*. Retrieved from http://www.fethiye.gov.tr/

Fethiye Municipality. (2020). *Fethiye*. Retrieved from https://fethiye.bel.tr/

García-Rosell, J. C., Haanpää, M., Kylänen, M., & Markuksela, V. (2007). From firms to extended markets: A cultural approach to tourism product development. *Tourism: An International Interdisciplinary Journal, 55*(4), 445–459.

Gordin, V., & Matetskaya, M. (2012). Creative tourism in Saint Petersburg: The state of the art. *Journal of Tourism Consumption and Practice, 4*(2), 55–77.

Hayaloğlu, A. A. (2008). Türkiye'nin peynirleri-Genel bir perspektif. *Türkiye, 10*, 21–23.

Hotel Association of Turkey (TÜROB). (2020). *2019 Muğla Tourist Statistics*. Access from http://www.turob.com/tr/istatistikler/mayis-2020-mugla-turizm-istatistikleri

Hung, W. L., Lee, Y. J., & Huang, P. H. (2016). Creative experiences, memorability and revisit intention in creative tourism. *Current Issues in Tourism, 19*(8), 763–770. doi:10.1080/13683500.2013.877422

Iqbal, M. (1993). *International Trade in NonWood Forest Products: An Overview. Working Paper Misc/93/11*. Rome, Italy: Food and Agriculture Organization of the United Nations. Sec. 7.1. http://www.fao.org /documents/show_cdr.asp?url_file=/docrep/ x5326e/x5326 e00.htm

Ivanova, P. (2013). Creativity and sustainable tourism development. *Икономика 21, 3*(2), 108–120.

Jarábková, J., & Hamada, M. (2012). Creativity and Rural Tourism. *Creative and Knowledge Society, 2*(2), 5–15. doi:10.2478/v10212-011-0022-4

Jelinčić, D. A., & Žuvela, A. (2012). Facing the Challenge? Creative Tourism in Croatia. *Journal of Tourism Consumption and Practice, 4*(2), 78–90.

Karababa, E., & Develi Isikli, N. (2005). Pekmez: A traditional concentrated fruit product. *Food Reviews International, 21*(4), 357–366. doi:10.1080/87559120500222714

Kartal, M. Y. (2020). *Muğla Alternatif Turizm Sektörü Yatırım Olanakları*. https://geka.gov.tr/tr

Kaya, C., Yıldız, M., Hayoğlu, İ., & Kola, O. (2005). Pekmez üretim teknikleri. *GAP VI. Tarım Kongresi*, 1482–1490.

Királová, A. (2017). Creativity as a Tool of Tourism Development. In A. Kiráľová (Ed.), *Driving Tourism through Creative Destinations and Activities* (pp. 67–93). IGI Global. doi:10.4018/978-1-5225-2016-0.ch004

Kiráľová, A. (2020). Creativity as a Tool of Tourism Development. In Destination Management and Marketing: Breakthroughs in Research and Practice (pp. 504-530). IGI Global.

Lindroth, K., Ritalahti, J., & Soisalon-Soininen, T. (2007). Creative tourism in destination development. *Tourism Review, 62*(3), 58–58.

Malachovský, A. (2020). Creative Tourism in Successful Destination Management as a Solution of Systemic Crisis in Slovak Tourism. In Destination Management and Marketing: Breakthroughs in Research and Practice (pp. 854-875). IGI Global. doi:10.4018/978-1-7998-2469-5.ch047

Marujo, N., Borges, M. do R., Serra, J., & Lourenço, P. (2020). Handicraft and Creative Tourism: The Project Mãos de Cá. In S. M. Cabeça, A. R. Gonçalves, J. F. Marques, & M. Tavares (Eds.), Creative Tourism Dynamics:Connecting Travellers, Communities, Cultures and Places (pp. 171–180). Grácio Editor.

Miles, I., & Green, L. (2010). Innovation in creative services. In F. Gallouj & F. Djellal (Eds.), *The handbook of innovation in services: A multi-disciplinary perspective* (pp. 178–196). Edward Elgar. doi:10.4337/9781849803304.00018

Mitchell, P., & Fisher, R. (2010). From passenger to driver: Creativity and culture in rural communities. *Tourism, Culture & Communication, 10*(3), 187–200. doi:10.3727/109830410X12910355180865

Ohridska-Olson, R. V., Ivanov, S. H., Ohridska-Olson, R., & Ivanov, S. (2010). Creative Tourism Business Model And Its Application In Bulgaria. *Proceedings of the Black Sea Tourism Forum'Cultural Tourism–The Future of Bulgaria.*

Ölmez, F. N., & Etikan, S. (2013a). Fethiye Alara Düz Dokumaları. *Süleyman Demirel Üniversitesi Güzel Sanatlar Fakültesi Hakemli Dergisi, 13*(11), 75–99.

Ölmez, F. N., & Etikan, S. (2013b). Muğla İli Fethiye İlçesi Yantırlı Kilimleri. *Art-e Sanat Dergisi, 13*(12), 62–82.

Oyman, N. R., & Erdal Sideli, E. (2018). Kayaköy (Muğla, Fethiye) Halıları. *The Journal of Academic Social Science Studies, 3*(67), 225–243.

Özgürel, G., Alan, R., & Bingöl, Z. (2015). Yerel Kalkınmada Bir Araç Olarak Festivaller: Munzur Kültür ve Doğa Festivali. *International Journal of Social and Economic Sciences, 5*(2), 113–121.

Pearce, D. G., & Butler, R. W. (1993). *Tourism research: Critiques and challenges (Issues in Tourism).* Routledge.

Pilz, D., Mclain, R., Alexander, S., Villarreal-Ruiz, S. B., Wurtz, Parks, C.G., McFarlane, E., Baker, B., Molina, R., & Smith, J. E. (2007). *Ecology and Management of Morels Harvested From the Forests of Western North America.* United States Department of Agriculture Forest Service Pacific Nortwest Research Station. General Technical Report, PNW-GTR-710, March 2007.

Pine, B. J., & Gilmore, J. H. (1998). Welcome to the experience economy. *Harvard Business Review, 76*(4), 97–105. https://www.ncbi.nlm.nih.gov/pubmed/10181589 PMID:10181589

Prebensen, N. K. (2014). Facilitating for enhanced experience value. In G. A. Alsos, D. Eide, & E. L. Madson (Eds.), *Handbook of research on innovation in tourism industries* (pp. 154–180). Edward Elgar.

Prentice, R., & Andersen, V. (2003). Festival as creative destination. *Annals of Tourism Research, 30*(1), 7–30. doi:10.1016/S0160-7383(02)00034-8

Raymond, C. (2007). Creative Tourism New Zealand: The Practical Challenges Of Developing Creative Tourism. In Tourism, creativity and development (pp. 145–157). London: Routledge.

Remoaldo, P., & Cadima-Ribeiro, J. (2019). Creative Tourism as a New Challenge to the Development of Destinations: The Portuguese Case Study. In M. Peris-Ortiz, M. R. Cabrera-Flores, & A. Serrano-Santoyo (Eds.), *Cultural and Creative Industries A Path to Entrepreneurship and Innovation* (pp. 81–99). Springer.

Republic of Turkey Ministry of Culture and Tourism. (2021). *Fethiye, Mugla*. https://mugla.ktb.gov.tr

Richards, G. (2002). Creative Tourism as a Factor Destination Development. In S. Flanagan, J. Ruddy, & N. Andrews (Eds.), *Innovation in Tourism Planning* (pp. 118–127). Dublin Institute of Technology.

Richards, G. (2009). Creative Tourism and Local Development: A global conversation. In Creative Tourism: A global conversation (pp. 78–90). Academic Press.

Richards, G. (2010). Tourism development trajectories: From culture to creativity? *Tourism & Management Studies*, *6*, 9–15.

Richards, G. (2011). Creativity and tourism. The state of the art. *Annals of Tourism Research*, *38*(4), 1225–1253. doi:10.1016/j.annals.2011.07.008

Richards, G. (2012). Tourism, creativity and creative industries. In Conference Creativity and Creative Industries in Challenging Times. NHTV Breda.

Richards, G. (2013). The Routledge Handbook of Cultural Tourism. In M. Smith & G. Richards (Eds.), *The Routledge Handbook of Cultural Tourism* (Vol. 6). Routledge.

Richards, G. (2014). Creativity and tourism in the city. *Current Issues in Tourism*, *17*(2), 119–144. doi:10.1080/13683500.2013.783794

Richards, G. (2016). The challenge of creative tourism. *Ethnologies (Québec)*, *38*(1–2), 31–42.

Richards, G. (2019). Creative tourism: opportunities for smaller places? *Tourism & Management Studies, 15*(SI), 7–10.

Richards, G., & Marques, L. (2012). Exploring Creative Tourism: Editors Introduction. *Journal of Tourism Consumption and Practice*, *4*(2), 1–11.

Richards, G., & Raymond, C. (2000). Creative tourism. *ATLAS News*, *23*, 16–20.

Richards, G., & Wilson, J. (2006a). Developing creativity in tourist experiences: A solution to the serial reproduction of culture? *Tourism Management*, *27*(6), 1209–1223. doi:10.1016/j.tourman.2005.06.002

Richards, G., & Wilson, J. (2006b). The creative turn in regeneration: Creative spaces, spectacles and tourism in cities. In M. K. Smith (Ed.), *Tourism, Culture and Regeneration* (pp. 12–24). CABI. doi:10.1079/9781845931308.0012

Richards, G., & Wilson, J. (2007). Tourism, creativity and development. In G. Richards & J. Wilson (Eds.), *Tourism, Creativity and Development*. Routledge. doi:10.4324/9780203933695

Rogerson, C. M., & Visser, G. (2011). Rethinking South African urban tourism research. *Tourism Review International*, *15*(1-2), 77–90. doi:10.3727/154427211X13139345020336

Şahin, G. (2019). Türk Kültür Coğrafyasından Özel Bir Örnek: Türk Sabunları. *Ağrı İbrahim Çeçen Üniversitesi Sosyal Bilimler Enstitüsü Dergisi*, *5*(2), 371–404.

Saygılı, D., Demirci, H., & Samav, U. (2020). Coğrafi işaretli Türkiye peynirleri. *Aydın Gastronomy, 4*(1), 11–21. doi:10.17932/IAU.GASTRONOMY.2017.016/2020.401/gas_v04i1002

Singsomboon, T. (2014). Tourism promotion and the use of local wisdom through creative tourism process. *IJBTS International Journal of Business Tourism and Applied Sciences, 2*(2), 32–37.

Sönmez, S., Efe, R., Cürebal, İ., & Soykan, A. (2015). Fethiye ve Çevresinde Zeytinin Yetişmesinde İklimin Rolü. *1.Teke Yöresi Sempozyum*, 622–630.

Stipanovic, C., & Rudan, E. (2014). Development concept and strategy for creative tourism of the Kvarner destination. *22nd Biennial International Congress Tourism & Hospitality Industry 2014*, 507–517.

Tan, S.-K., Luh, D.-B., & Kung, S.-F. (2014). A taxonomy of creative tourists in creative tourism. *Tourism Management, 42*, 248–259. doi:10.1016/j.tourman.2013.11.008

Tan, S. K., Tan, S. H., Luh, D. B., & Kung, S. F. (2016). Understanding tourist perspectives in creative tourism. *Current Issues in Tourism, 19*(10), 981–987. doi:10.1080/13683500.2015.1008427

Taşkın, H., & Büyüklaca, S. (2012). Kuzugöbeği mantarı (Morchella spp.)'nın Türkiye'deki bölgesel yayılımı. *Alatarım, 11*(1), 19–24.

The Travel Foundation. (2016). *Taste of Fethiye*. Retrieved from https://www.thetravelfoundation.org.uk/casestudy/taste-of-fethiye/

Turkish Patent and Trademark Office. (2020). *Patents*. Retrieved from https://www.turkpatent.gov.tr/

UNESCO. (2006). *Discussion Report of the Planning Meeting for 2006 International Conference on Creative Tourism*. UNESCO.

UNESCO. (2021). *Creative Cities Network*. Retrieved from https://en.unesco.org/creative-cities/

Üzümcü, O. O., & Koç, B. (2017). Ekolojik turizm kavramı: Fethiye örneği. *International Journal of Agricultural and Natural Sciences, 10*(1), 14–19.

Waitt, G., & Gibson, C. (2014). Tourism and Creative Economies. In A. A. Lew, C. M. Hall, & A. M. Williams (Eds.), *The Wiley Blackwell Companion to Tourism* (pp. 230–239). John Wiley & Sons, Ltd. doi:10.1002/9781118474648.ch18

Duxbury, N., Bakas, F. E., & Pato de Carvalho, C. (2019). Why is research–practice collaboration so challenging to achieve? A creative tourism experiment. *Tourism Geographies*, 1–26.

Winter, T. (2004). Landscape, memory and heritage: New Year celebrations at Angkor, Cambodia. *Current Issues in Tourism, 7*(4-5), 330–345. doi:10.1080/13683500408667989

Wurzburger, R., Aageson, T., Pattakos, A., & Pratt, S. (2009). *Creative tourism: A global conversation: How to provide unique creative experiences for travelers worldwide*. Sunstone Press.

Yazıcıoğlu, T., & Gökçen, J. (1976). *Kuru Üzümlerden Diffüzyon Yoluyla Pekmez (Konsantre) Elde Edilmesi İçin Geliştirilen Bir Yöntem TÜBİTAK Marmara Bilimsel ve Endüstriyel Araştırma Enstitüsü*. Gebze. Yayın No: 11.

Yeşilay, S., & Özsolak, İ. (2017). Sicak Cam Üfleme Tekniğinde Lüster Etkisinin Araştırılması. *TÜBAV Bilim Dergisi, 10*(4), 21–28.

Zoğal, V., & Emekli, G. (2017). Yaratıcı turizme kavramsal ve Coğrafi Bir Yaklaşım. *Eğe Coğrafya Dergisi, 26*(1), 21–34.

KEY TERMS AND DEFINITIONS

Active Participation: Direct experience of guests in creative tourism activities; tourists try to do the lifestyle activities of local people.

Creative Experience: This concept expresses that individuals actively engage in activities to create unique works of their own and acquire individual experiences through creative tourism by using simple materials and their imagination.

Creative Route: The route aims to provide creative experiences with the active participation of tourists.

Creative Tourism: A type of tourism that enables tourists to learn about the culture of the region they visit by communicating with the local people and by active participation.

Creativity: The ability to create something using original ideas and thinking out of the ordinary.

Sustainable Initiative: The sustainability of initiatives for future generations; a sustainable initiative is an approach aiming at the long-term success of an initiative strategy by focusing on the ethical, social, environmental, cultural, and economic dimensions.

Tourist Attraction: It is a place or activity including such as natural and cultural resources, historical buildings, natural beauty that tourists visit or experience for pleasure and interest while they are on vacation.

ENDNOTES

1 www.pastoralvadi.com

2 tangala.org

3 www.desperado-ranch.com

4 tangala.org

APPENDIX 1: PROPOSAL CREATIVE ROUTES FOR FETHIYE, MUGLA, TURKEY

Table 2.

First route
Visit Pastoral Valley Eco Farm
To ensure that guests taste fruits and vegetables produced by ecological methods
To engage guests in making noodles, jams and pickles.
Visit Dastar factory (Dastarhane) in Yeşilüzümlü
See how Dastar is produced in the Dastar factory.
Get information from residents about Dastar production and use
Visit the exhibit of products made by the local people from the Dastar
Visit a wine cellar in Yeşilüzümlü
Get information on wine production and types of wine produced in the region.
Taste the wine varieties produced in the region.
Visit a house that makes olive oil soap
Get information about the olive oil soap-making process.
Second route
Breakfast in Tangala Goat Farm
Taste the cheese produced on-farm.
Get information about different local cheese types.
Visit a carpet weaving workshop in Kayaköy
Show how the process works by weaving a carpet in front of visitors.
Get information about the color, pattern and motif features of carpets.
Have lunch at a local pancake restaurant
To show the guests how the dough is rolled and the pancakes are cooked on the pan.
Visit Kayaköy Art Camp
Provide detailed information about ceramic production.
To ensure that guests are also involved in the production.

Section 2

Sustainability and Challenges for the Inland Tourism Products

Chapter 6
Sustainable Tourism:
How Is This Issue Effectively Approached?

Fatima Lampreia Carvalho
https://orcid.org/0000-0002-1189-8843
FE, University of Algarve, Portugal

Silvia Fernandes
https://orcid.org/0000-0002-1699-5415
University of Algarve, Portugal

ABSTRACT

This work intends to verify if there is academic research that proposes innovative strategies for sustainable tourism. It analyses 70 valid documents including theses, dissertations, scientific papers, and reports. Main objectives to explore are which themes stand out most; the difference between academic production in Portugal and Brazil; and how the topics of sustainability, environment, governance, planning, and regulation relate to tourism. A software for qualitative analysis is used to enrich the discussion around sustainable tourism discerning preferential governance strategies. Few works use sustainability indexes, and, if used, most of them are descriptive. For real sustainable tourism, its governance has to be based on appropriate indicators. Measures can empower destinations, providing the information needed to decide accurate and creatively. A challenge for the Portuguese academy on tourism and hospitality is to have an ongoing role in implementing key indicators and in their validation and monitoring.

INTRODUCTION

Sustainable tourism is defined as the creation and promotion of a tourism industry that preserves or enhances a country's social, cultural or environmental capital. Data reveals that the environmental strength of a country is directly related to tourism revenue. Although there is no evidence of direct causality, the more pristine the natural environment of a local the more tourists are willing to access it. Consequently, as the natural capital deteriorates, destinations lose revenue. It is important to recognize that processes

DOI: 10.4018/978-1-7998-7339-6.ch006

and activities associated with tourism also damage the environment. Given the close relationship between natural resources and a very large segment of the tourism industry, a lack of progress on fostering sustainability will reduce tourism development opportunities. Besides environment, also cultural and socio-economic issues become involved in this equation of tourism sustainability. Many works and reports discuss this subject, revealing its increasing importance for today's global and local decisions about tourism development. Even the communication of sustainability of a destination influences the intention to visit it, as Melo and Farias (2018) acknowledged.

This work aims to establish what proportion of academic works, listed in the scientific repository of open access in Portugal (RCAAP), have proposed innovative governance strategies for sustainable tourism. And among these, which portion has used approaches grounded on international/national sustainability indexes. RCAAP was used as it promotes visibility, access and diffusion of the results of academic production and scientific research in Portugal and Brazil. Academic production on sustainable tourism not always refers to international indexes (such as the Travel & Tourism Environmental Sustainability Index and the Global Destination Sustainability Index). This might be due to different factors such as lack of funding, complexity of data collection and lack of statistical training in tourism schools. This analysis initially looked at 164 theses and other scientific works containing the required keywords: sustainable tourism, sustainability index, and governance. However, the final selection retained 70 stable documents which were well downloadable and according to the criteria considered.

Some Statistics

In 2017 the World Travel & Tourism Council (WTTC) estimated that travel and tourism contributed 10% of the world's GDP and accounted for just under 10% of its total employment. However, data from 2013 show that the contribution from tourism sector to global climate change was approximately 5% of carbon emissions (Gossling et al., 2013). A few national tourism economies can be considerably more carbon-intense than others, thus supporting the need to monitor their growth and sustainability in the coming years. According to the environmental footprint of nations, Portugal with a population of 10,291,027 (INE, 2011) was one of the countries that made a difference in setting a goal of achieving carbon neutrality in 2050. Electricity and transport are factors that account for this issue. The carbon footprint is now under the European average (of 8,7 tons CO2 emissions per capita).

Brazil, with a population of 208,862,818 (IBGE, 2017) has experienced success in slowing rain forest destruction, which has resulted in vast reductions of carbon emissions (Howard, 2014). The case of Brazil shows that it is possible to promote sustainability while still growing the economy.

The main objective of this work is to investigate in what ways academic works (especially doctoral theses, papers and reports) in Portugal and Brazil reacted to the projections on the growing pressure that tourism may be placing on the environment and society. Is the application of sustainability indexes an important concern in those academic works? And do they give practical recommendations from their appliance?

Thus, the overall research objective of present work is to analyze the academic production on sustainable tourism produced in Portuguese and Brazilian higher education institutions from 2004 to 2017. Key aspects to be analyzed are: 1) what contents are more salient; 2) what differences exist between academic production in Brazil and Portugal; and 3) how topics such as sustainability, environment, governance, planning and regulation relate to tourism sustainability and its indicators. Through these aspects this research intends to analyze: sustainable tourism indexes; visons of progress and general values about the relation between tourism and sustainability; goals of pro-sustainability actions; and preferential methods of governance for progresses in sustainable tourism.

To pursue these aspects this work is structured as follows: the first section analyzes the resulting production of academic works and their main themes and frameworks; the second section presents the research objectives and methods used to discuss them; the third focus on sustainability indicators and main approaches in theses and international reports; the fourth develops the content analysis on the collected documents using the software of qualitative analysis Nvivo; the fifth section discusses the obtained results, and is followed by the conclusion.

METHODOLOGY

Focusing on the collection of RCAAP, the present study inspected the academic production of Portuguese and Brazilian higher education, with the expression of research "sustainable tourism". A sample of 70 works was considered, including: 33 Master's dissertations, 9 theses of PhD, 18 scientific papers and 7 reports. The other three documents were undergraduate courses projects. In this sample, regarding the work in Portuguese institutions and addressing sustainability in tourism directly, one could observe a growing tendency to address sustainability and governance (for example, in recent works from 2015 to 2017). This academic production related these areas especially to the stakeholders and the environment. It turned out that there was a higher production of academic works in tourism in 2012, closely related to growth in tourism, local development, and economic evaluation. In particular, 6 works of the sample at hand addressed sustainability indicators (such as Stilwell, 2011).

Further academic works aimed to identify tourism's impact in the various dimensions of sustainable development and its connection to sustainable tourism. Carvalho (2012) approached sustainable tourism and ecotourism in the Portuguese reality analyzing the importance and efficiency of ecotourism certification as a tool to determine standards of quality via indicators and criteria.

The next section introduces our research objectives and methodological tools, with focus on the emerging topics in the previous academic background, to discern tendencies and lesson to be learnt.

Objectives and Tools

The main objective is to study academic production on sustainable tourism produced in Portuguese and Brazilian higher education institutions from 2004 to 2017. A reason for this period was the referred growing trend to address sustainability and governance. Also, this work resulted from an investigation done by the authors in 2018. Key aspects to be analyzed are what contents are more salient; what differences exist between academic production in Brazil and Portugal; and how topics such as sustainability, environment, governance, planning and regulation relate to tourism. One of the concerns to be addressed is the perception that academic outputs in tourism studies not always refer to indexes.

The main database used in this work is RCAAP, the Scientific Open Access Repository of Portugal (FCCN). It promotes, facilitates and supports the adoption of the open access movement in scientific knowledge across Portugal. Documents as theses, reports and academic articles were collected. The aim is to identify what kind of innovative governance strategies for sustainable tourism have been discussed in academic works submitted in Portugal's Higher Education institutions (HEIs). These works can be authored by Portuguese or Brazilian students. Another source used was the Brazilian portal of

Figure 1. Guiding framework for the proposed case study: Portugal and Brazil
Source: Own elaboration

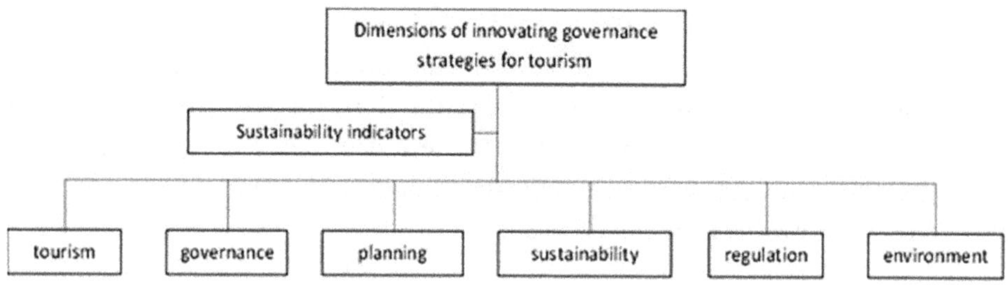

scientific publications (OASISBr) which is a multi-disciplinary search engine that allows free access to the scientific production from Brazilian universities and research institutes. After searching Master and PhD theses within RCAAP, by the keyword "sustainable tourism", a total of 1127 documents were listed whereas "sustainable tourism and success" returned with 66 documents and "tourism sustainability index" returned with 17 documents.

Due to the relevance of the subject, it seemed that more academic works listed under RCAAP, beyond doctoral theses, should be included in our case study. The aim is to search deeply on how the incorporation of sustainability indices in academic projects collaborates to the creation of new proposals for governance strategies for sustainable tourism. This analysis therefore also encompasses Master dissertations, Journal articles and international reports. The methodology used to analyze multiple documents is content analysis assisted by Nvivo12, a software used to produce a systematic qualitative analysis of documents' content. This analysis helped to establish what proportion of works discussed innovative governance strategies with basis on existent or new sustainability indices, and which ones are most referred or proposed. From the background identified and collected in the previous section, figure 1 illustrates the underlying areas emerged, whose content the software Nvivo will analyze.

These six thematic areas - tourism, governance, planning, sustainability, regulation, environment - which emerged via a first reading of the academic production selected, are now used as key works for a content analysis. Next section presents a theoretical approach to the central linking theme of this guiding map (or central node): Sustainability indicators. The reason for this is that sustainability indicators are the central tools used by policy makers, governments and non-governmental organizations to monitor and stir new and effective policies to foster sustainable tourism.

Sustainability Indicators

Since the Brundtland Report (1987), on the desirability of sustainable development, the academic community addressed what impact a revision of the model of progress would have on tourism. Studies on sustainable tourism tended to connect the management of needs to economic, social and environmental resource in tourist destinations. Using an NVIVO matrix of analysis on selected sources from the RCAAP database, it is apparent that major reports produced by non-governmental organizations, supra-national organizations, governments and their ministries reflected on the application of sustainability indicators. Greater focus on sustainability indicators could be found in reports produced by Regional Authorities, Municipal Authorities, Community members, the academic community, Tourism enterprises. In Portugal, the Ministry for Cities Territorial Ordering and Environment produced important reports. On March

2002, with the Resolution of the Council of Ministers, the Government began the process of elaboration of the National Strategy for Sustainable Development (ENDS) in Portugal. Following the Johannesburg Summit, it was concluded that there was a need to develop a Strategy Implementation Plan (PIENDS) setting indicators, concrete targets and deadlines. The indicators included general economic indicators, employment, education, innovation and investigation, economic reform, social cohesion and environment.

The World Tourism Organization (WTO) promoted the use of sustainable tourism metrics since the early 1990s as essential instruments for policy-making, planning and management processes at destinations. In 2004, a new guidebook was presented as the result of an extensive study on indicator initiatives worldwide involving 62 experts from more than 20 countries. This new guidebook described over 40 major sustainability issues, ranging from the management of natural resources (waste, water, energy, etc.), to development control, satisfaction of tourists and host communities, preservation of cultural heritage, seasonality, climate change, to mention just a few. For each issue, indicators and measurement techniques were suggested with practical information and examples. The publication also contained a procedure to develop destination-specific indexes, their use in tourism policy and planning processes, as well as applications in different destination types (e.g. coastal, urban, ecotourism, small communities). Numerous examples and 25 case studies provided a wide range of experiences at the company, destination, national and regional levels from all continents.

In 2005, Mota el al. carried a SWOT analysis to identify the Strengths, Weaknesses, Opportunities and Threats related to businesses competition showing that Portugal faced sustainability challenges. In its pattern predominate industries based on the intensity of low skills and services based on natural and low resources qualifications - namely tourism "Sun Beach". But there were opportunities with the intensification of tourism flows resulting from the search for homes in locations with environmental quality and landscaping, safety conditions and good health services (Mota et al., 2005).

One of the main tasks of the Mission Unit for Sustainable Development (UMDS) was to develop a System of Indicators of the National Strategy for Sustainable Development (ENDS) which would allow Portugal to evaluate the degree of success of its plan implementation, both in the Annual Progress Reports and in the preparation of the three-year cycles of its review. Mota et al. (2005) presented a first version of key indicators which were based on the critical areas addressed by the ENDS. This report recovered the important work on sustainability indicators that were developed over the course of the European Union, the OECD and various UN agencies. Key sustainability indicators in the ENDS referred to six objectives such as: (i) qualification of people in direction to the society of knowledge; (ii) sustainable, competitive and oriented economy for future; (iii) efficient management of the environment and natural resources; (iv) balanced organization of the territory valuing Portugal in European space and providing quality of life; (v) dynamics of social cohesion and individual responsibility; (vi) dynamics of social cohesion and individual responsibility.

International Reports

World tourism organizations and other international institutions are the most relevant bases of leadership on the subject for sustainability indicators whereas it is understandable that allusions to sustainability indicators are minor in PhD, MA theses and journal articles. Governmental, non-governmental and supra-national organizations' reports and guidebooks such as the WTO Indicators of Sustainable Development for Tourism Destinations refer to different types of indicators, each with different utility to decision-makers (WTO, 2004).

Table 1. Tourism sustainability indicators in main international reports

Report	Indicators	Objectives/Targets
National strategy for the Sustainable development ENDS 2005-2015	Sustainable Development Indicators (SDI) for monitoring the National Strategy for Sustainable Development	Qualification of Portuguese towards knowledge society; productivity growth with strong investment in sectors of marketable goods; model of development integrating environmental protection; better articulation of policies and territorial planning instruments; withstand the shock of ageing of population; international cooperation
Atlantic axis of the Northwest Peninsular (2011) Sustainability of the Euro-region Galicia-north of Portugal	Agenda 21's urban and sustainable transport indicators	Social demographic components; sustainable production and consumption; climate changes and clean energy; mobility; conservation and management of natural resources and planning of the territory
Global Destination Sustainability (GDS) Index Report (2017)	The Global Destination Sustainability (GDS) Index	Measure environmental performance; social performance; supplier performance; sustainability policy and communication
Sustainable tourism index: Enhancing the global travel environment (2017)	The Sustainable tourism index	Political and regulatory environment; environmental sustainability; socio-cultural sustainability; economic sustainability; travel and tourism industry
Travel and Tourism Competitiveness Report 2017 UNWTO and World Travel & Tourism Council	Travel and Tourism Competitiveness Index (TTCI)	Focus on ensuring the industry's sustained growth in an uncertain security environment while preserving the natural environment and local communities
Sustainable Development Goals (Portugal, Agenda 2030)	Sustainable Development Indicators for Portugal, Agenda 2030	Policies to promote sustainable tourism (jobs, local culture and products); achieve higher levels of economic productivity (diversification, innovation, high value-added); achieve universal and equitable access to safe water
Sustainability Report 2016 by Tourism Portugal	Tourism Sustainability Indicators for Portugal	Monitoring indicators of sustainability in tourism; constitution of an observatory of sustainability; definition of desired positioning of tourism sustainability of Portugal and underlying implementation strategy
Indicators of Sustainable Development for Brazil (IBGE 2015)	Indicators of Sustainable Development	Focus on the environmental dimension - pressure and impact factors related to the preservation of the environment, fundamental for quality of life of generations

Table 1 gives prevalence to 8 national and international reports containing debates and explanations of sustainability indexes methodology. The objective is to compare the sustainability indexes regarding their targets, proposals and custodian agency.

The most directly useful indices may be those that help to predict problems. Several other types of indicators are early warning indicators (e.g., decline in number of tourists who intend to return); indicators of stresses (e.g. water shortages, or crime indices); measures of the current state of industry (e.g. occupancy rate, tourist satisfaction); measures of the impact of tourism development on the biophysical and socio-economic environments (e.g. level of deforestation, changes in consumption patterns, income levels in local communities); measures of management effort (e.g. clean-up cost for coastal contamination); and measures of management results or performance (e.g. changed pollution levels, greater number of returning tourists) (WTO, 2004).

Studies on Indexes

A review of theses and articles in RCAAP on sustainable tourism revealed that the variability of contexts and points of views can give a contribution to an academic debate on the topic of indexes. About the preparation by European Union of an Agenda 21 for the Tourism Sector, Albuquerque (2004) produced a master dissertation highlighting the need to monitor, collect and process statistical information on the Coastal zones of Portugal to allow the periodic preparation of a sustainable development strategy evaluation report. The dissertation initiated a dialogue with Portugal's National Strategy for Sustainable Development – ENDS, 2005-2015. The municipal coastal zone of Mira, in central Portugal, was selected as a case study to reflect on the development of an ENDS Indicator System to assess the success of its implementation plan preparation of three-year cycle. For the author accommodation capacity, sea indentation indicator, physiological balance, local quality of life and environmental impact should compose periodic sustainability indicators to implement new management strategies responsive to the load capacities of a territory. At the time, tourism suffered from a markedly seasonal and mass character. But there was also evidence that tourists no longer only pursued sun and beach tourism, instead seeking multiple areas of interest. Betting on territory's specificities and singularities, both at the environmental and cultural levels, this work advised experts to develop new tourism products so that the seasonality would be mitigated.

On another case study, of a destination in Brazil, Andrade (2006) proposed new methods to monitor sustainability in the island of Florianopolis (Santa Catarina) whilst capturing the ecological footprint (EF). The underlying argument was that the urban space comprehends a set of social and economic activities that generate negative and positive impacts on the natural environment. It would be therefore useful to measure the impacts generated by tourism development via capability indicators to establish the environment's capability to support a specific activity. According to evidence, EF would be an alternative indicator to screen the connections between the tourism activity and the ecological sustainability level in a specific destination.

Coelho (2009), studying Porto Seguro (Bahia, Brazil), scarcely mentioned sustainability indexes. This work limited the argument to indicators of population growth rates, increase in trade and flow of tourists, as well as job creation, public and urban changes resulting from tourism. It focused on the Bahia's Tourism Development Program (PRODETUR-BA) and the importance of it to foster private investment and leverage the region's economic development. This work raised doubts whether Bahia's Tourism Development Program interventions were really in the name of sustainable tourism. The author hoped that PRODETUR could modify the profile of mass tourism in the North of Brazil, establishing quality tourism to guarantee the physical integrity of natural resources via a diversification within the tourist activity with incentive to business tourism, ecological tourism, adventure and cultural tourism.

Amazonas (2014) carried out an interdisciplinary analysis of business administration, tourism, and law to analyze environmental management in lodging facilities in the city of João Pessoa (Paraíba, Brazil). The objective was to produce evidence on the incorporation of sustainable practices, the extent to which guests were held to environmental and social responsibility standards, and the overall level of sustainability in hospitality management. These metrics were assessed based on a sustainability index. The choice of hotels was based on the hotel guide of the Brazilian Association of Hotel Industry (ABIH/PB) which totaled 14 hotels (categories of 3 and 4 stars) according to the old classification of EMBRATUR (Brazilian Tourist Board) and of medium size, with a reasonable organizational structure. The author developed indexes such as the ISGH (Hotel Management Sustainability Index) and IEHQA (Tourist Demand on Environmental Issues Index).

From the data collected, a diagnosis was made to evaluate the application of sustainable practices in hotel management at João Pessoa, which could serve as a basis for future studies in the area and as a reference for new ventures. It was found that the ISGH was higher for hotels that belonged to hotel chains. Besides evaluating reasons that lead an entrepreneur to seek sustainability, the research focused on understanding the importance given by consumers to hotels that adopt sustainable practices. Through the Guest Requirement Index with the Environmental Questionnaire (IEHQA), the interviewed guests presented a high level of demand for adopting sustainable practices. This shows that investments in hotels at João Pessoa should go beyond compliance with environmental legislation.

Content Analysis of the Sample

This work's objectives referred before, according to the related literature, require a qualitative approach. Qualitative research intends to evaluate, interpret or explain social, cultural or economic phenomena. Related works analyze data like interviews, surveys, web pages, audio visual material, theses and journal articles. The software Nvivo, for qualitative analyses, can help to explore and find patterns in data. It does not favor a particular methodology as it is designed to facilitate common qualitative techniques, no matter what method is used. The Nvivo12 was used to organize the 70 documents on sustainable tourism, collected from RCAAP, amongst the categories of theses, journal articles, master dissertations and international reports.

Having in mind our objectives, which are 1) highlight what contents are more salient; 2) explain what differences exist between academic production in Brazil and Portugal; and 3) discuss how topics such as sustainability, environment, governance, planning, and regulation relate to sustainability indicators, we have produced outputs that can answer these research issues. In this way, we need to describe the cases and nodes firstly derived and used.

Results From Coding Cases

According to Nvivo's functioning and to the content obtained, used in the development of our guiding framework, the cases here correspond to the categories or types of academic works collected from our search in RCAAP. Table 2 illustrates those types and respective works that make up the sample used in this research (with a total of 70 documents).

Hence, these cases will be gathered into nodes which are the topics/themes derived, according to the guiding framework of this research (figure 1). Nodes are the structures that allow gather related material so that we can look for emerging patterns and ideas.

Results From Coding Nodes

In Nvivo, coding is one of the several techniques for making sense of our data. The way to handle coding depends on our methodology and research design. Coding the sample's content at topics or themes will constitute the nodes. This is fundamental in most qualitative projects as it organizes our material into broad topic areas. This tends to be a cumulative rather than a one-stage process, as the node for each topic can be explored toward more detailed coding.

When we open a node in Nvivo, it shows all its content which can be text, audio, video, pictures (that have been coded at that node). The node can be linked to a memo to describe the analytical insights or coded on to other topics. Regarding the guiding framework for this research, the nodes derived were: tourism, governance, planning, sustainability, regulation, environment, and as a central node the topic of Sustainability indicators.

Table 2. Types of academic works collected (cases)

Types	Cases (Works)
PhDs	1: PhD_Teixeira_Fernanda(2010) Gestao_Conhec_Tur_Parq_Nac_BR 2: PhD_Silva_Francisca(2005)EDUC_Sup_Sust_BR 3: PhD_Rabinovici_Andrea_D (2009)Org_não_Gov_conflitos_BR 4: PhD_Panzeri_Viviane(2012) TurSust_ Sao Simao SP_BR 5: PhD_Oliveira_Ermelinda(2015) Tur_Sust_Beiras_S_Estrela_PT 6: PhD_Farsani_Neda(2012) Tur_Sust_Geoparques_PT 7: PhD_BOVO_Raquel(2006) Sub_Pol_Pub_BR 8: PhD_Borges (2016) Gov_Sust_Alentejo_PT 9: PhD_Albuquerque Helena(2013) Des_Sust_ Ria_Aveiro_PT
Master dissertations	1: DM_Veloso_CHRISTIANE(2015) Tur_sust_Teoria dos Stakeh_PT 2: DM_Varela de Oliveira(2006) Cidades_Tur_Sust_Rio G_Norte BR 3: DM_Torres(2007) GESTÃO_Patr_Hist_Curitiba_BR 4: DM_Tomasi(2011)DESEN_REG_Sust_BR 5: DM_Stillwell_Diogo(2011) Indicadores de Sustentabilidade 6: DM_SOUZA_Karoline(2013) Dimensao etica_Tur_Sust_BR 7: DM_SILVA_Ana_Filipa(2017) Prop_des_sust_centro_Porto_PT 8: DM_SILVA Michel(2014) Percepcao ambiental BR 9: DM_SANTA CRUZ_Iratan(2007) Tur_alternat_sust_Alagoas_BR 10: DM_Sampaio(2005)Estratégias para a superação_BR 11: DM_Rodrigues_Sonia(2012)Tur_sust_rurais_ residentes_PT 12: DM_Rodrigues_ Ana(2012) Estrat_Tur_Sust_aldeias de Xisto_PT 13: DM_Oliveira_Huascar(2009)Aval_indicadores_ pot_tur_Ceara_BR 14: DM_Moutinho_Diana(2011) Des_Sust_ Mata_Sesimbra_PT 15: DM_Milton Hoppen(2004) A Aplicacao do ICMS._BR 16: DM_Martins_Catia(2012) Tur_rural_Des_sust_PT 17: DM_Libombo(2010) Resp_Social das Empresas Turísticas_PT 18: DM_GUAMBE_Cecilia(2012) Tur_Sustentavel_Dest_Natur _PT 19: DM_Garcia_Ana(2011) Tur_Sust_Ilhas Acorianas PT 20: DM_Fernandes_Sandra(2011)Tur_sus_Comun_Piscatorias_PT 21: DM_Faria Telma(2009) Aldeia _sust_COVILHÃ_PT 22: DM_Dinkhuysen (2017) Guias_Tur_PT 23: DM_CARVALHO_JConrado(2005) Des_sust_Lencois_ Maranh_BR 24: DM_Brandao (2012) Tur_Indig_Local _BR 25: DM_Branco Soares (2015) Relatório_Carta Europeia_Tur_PT 26: DM_BATHKE_MARIA ELIZA (2002) Tur_Sust_Rural_BR 27: DM_Araujo_Paula(2016) Tur_Surf_Matosinhos _PT 28: DM_Antunes_Luiza(2016) Governanca_Territ_PT 29: DM_ANDRADE_BEATRIZ (2006) TUR_Pegada_Eco_BR 30: DM_Amazonas_Iuri(2014)Gest_Amb_Hotel_BR 31: DM_ALBUQUERQUE Helena(2004) Tur Sustent zonas costeiras
Reports	1: REPOR8_IBGE_2015_ Desenvolvimento_SUST_BR 2: REPOR7_Turismo de Portugal (2016) Relat_sust_PT 3: REPOR6_INE_SUSTANABILILTY_SDGOALS_STATS_2018 4: REPOR5_World Economic Forum (2017)_TTCR_2017 5: REPOR4_Sustainable_Tourism_Index_The Economist 6: REPOR3_GDSI (2017)Road_Circular_Economy 7: REPOR2_Eixo Atlantico Noroeste Peninsular _Norte PT 8: REPOR1_ENDS_2005-2015_PT

continued on following page

Questioning the data/content from the works collected, a first query was the word frequency. This first exploratory technique (word cloud) shows that, when taking all sources into account, the central term in the selected academic works is 'tourism'. The word cloud also illustrates a more close connection of tourism to the term 'development' than to the terms 'sustainable' and 'sustainability'. There is an even lower relevance of the term 'indicators', due to its low frequency in the cloud. This corroborates

Table 2. Continued

Types	Cases (Works)
Journal articles	1: JA_Oliveira&MAnso(2010) Turismo Sust_Utop_Real_PT 2: JA_Neves& Mateus(2015) Edu_amb_Tur_sust_PT 3: JA_Nascimento&Silva(2009) TUR_PÓS_MODERNO_Dilemas GestSust_BR 4: JA_Marujo&Carvalho(2010)Tur_Plan_Desenv_sustentável_PT 5: JA_LESSA_Lucio(2008) Educ_ambiental_desenv_tur_Sust_ Maceio_BR 6: JA_Freitas_Souza_Zambra (2014) Gest_Sust_Tur_BR 7: JA_Franca e Araujo(2010) DES_SUST REDEF POD_LOC_Tur_PORT 8: JA_Ferreira et al(2009) Des_Sust_Sudo_Contin_PT 9: JA_Castro & Martins (2011) Tur_Sust_NAT_PT 10: JA_ Bernardes_Carlos(2013) Instr_ ordenam_Territorio_PT 11: JA_Pocas_Isabel et al(2011) An evaluation of changes rural landscape 12: JA_PORT_Ferreira et al(2009) Turismo e Desenv Sustent_Sudoeste PT 13: JA_Wimmler et al(2015) Multicriteria support methods renewable energy 14: JA_Dias_Jose_G(2016) Environmental sustainability measurement 15: JA_Ferreira et al (2014) Resource evaluation Costa Caparica 16: JA_Gil_Luis(2011)Environmental, sustainability and ecological 17: JA_Guerra et al(2017) Smart cities smart tourism_city of Porto 18: JA_Leitao et al(2013)The iGPI collaborative project moving IAM
Bachelor dissertations	1: BCHAP_ENG_Matos & Batista(2013)Allotment-gardens_urban-sust 2: BA_Carvalho_Nuno(2012) Ecoturismo-Zmar PT 3: BA_Soldi(2007) Diagnóstico Parque florestal Rio_BR 4: DB_Coelho_Ivan(2009) Crescimento Turismo Porto Seguro_BR

the limited number of works with focus on indexes of sustainability. Regarding the two sides of the term 'tourism', the top side ('development') appears more related with terms such as environment, economy, impacts, policies, sector, strategy, tourists, growth. The underside ('sustainable') appears more related with terms such as destinations, community, resources, governance, quality, planning, residents, indicators.

Handling qualitative data is not usually a step-by-step process. Instead, it tends to be an iterative process to explore, code, memo, code some more, query, etc. According to a Nvivo12 project map involving files and nodes (figure 2), the themes 'Tourism', 'Regulation' and 'Environment' appear at the center of it. Nodes such as 'Planning', 'Sustainable tourism', 'Indicators' and 'Governance' appear at the middle left of the map.

For more detail, from a query based on matrix analysis, with focus on 32 master theses in Brazil and Portugal, Albuquerque (2004) has 377 mentions of the term tourism, 77 mentions of term planning, 74 mentions of term sustainable, and 15 mentions of regulation. The thesis of Andrade (2006) on ecological footprint (written in a Brazilian high education institute) was the only academic work to include more than 40 mentions to sustainability indexes. Another work made a single mention to sustainability indexes (Amazonas, 2014). The thesis of Antunes (2016), on territorial governance in Portugal, was the only with strong reference to governance, with 75 mentions to the topic.

Looking at five major international reports, published between 2004 and 2017, a report by the World Economic Forum (2017) stressed themes such as indicators and sustainability index together 311 times, with the highest mentions to governance (197), environment (359), regulation (154), and tourism (1741).

Analyzing the journal articles of the sample, Dias (2016) with a work related to environmental sustainability measurement, mentioned sustainability indicators 22 times and sustainability index 16 times (a greater extent than other articles in the sample). Wimmler et al. (2015), whose work has a multi-criteria method applied to renewable energy, also highlighted the topic sustainability index. Ferreira et al. (2014), whose work is about resource evaluation, was the one that stressed more the topic of governance.

Figure 2. Project map - Academic production on sustainable tourism (Portugal and Brazil)
Source: Own elaboration

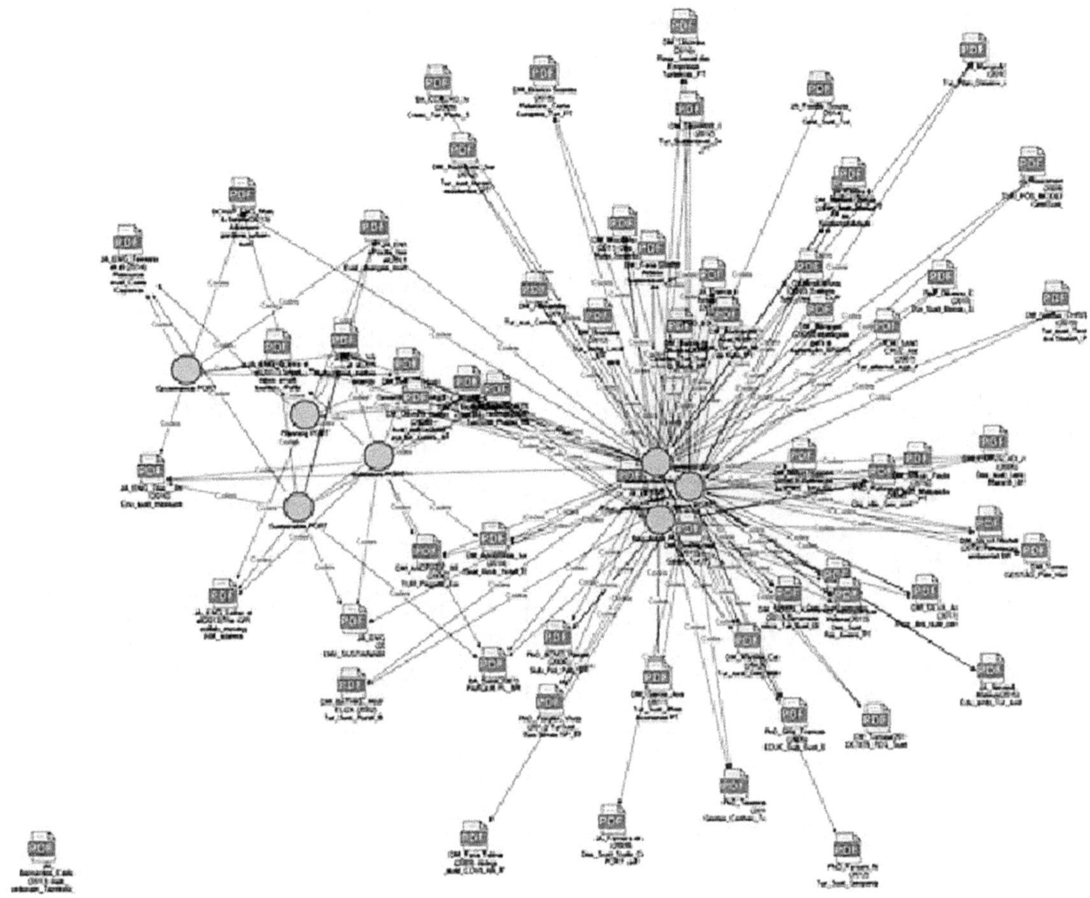

DISCUSSION OF RESULTS

Researchers working in methodologies such as phenomenology and discourse analysis may get straight into detailed coding and then, later on, combine and group the nodes into related categories. The process of coding brings us closer to data in order to ask questions like 'how do the data help answer the research question?', etc. Coding facilitates the use of queries and visualizations allowing us to look for connections between topics and test our understanding. Queries and maps support the iterative nature of qualitative research and help to investigate hunches as we progress through a project. They can help us focus on the questions we want to ask and prompt us to code accordingly. They also help to determine whether we need to gather additional files of data or re-frame the questions asked before.

In Nvivo, the techniques of re-coding and querying contribute to integrate results to better compare and discuss them. Thus, from a query based on cluster analysis of the nodes derived, figure 3 illustrates two main clusters or groups: 1) one that reflects the use of sustainability indicators for regulation matters; and the other 2) where governance relates to sustainable tourism and planning.

This scenario suggests an increasing concern with governance and regulation, and that sustainability indicators are addressed especially for regulatory issues. Thus, there is a growing awareness of the

Figure 3. Cluster analysis of the nodes derived
Source: Own elaboration

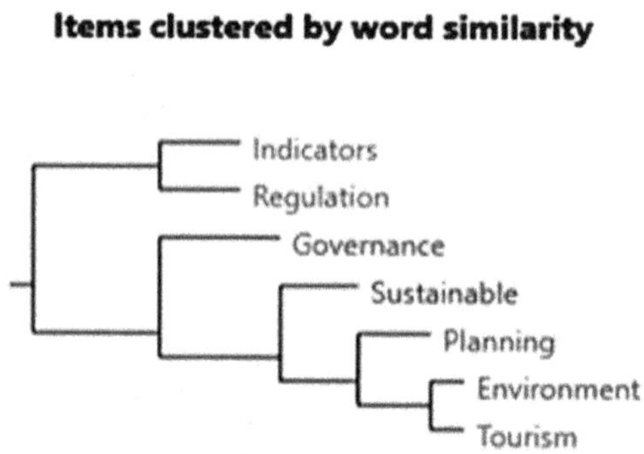

Figure 4. Academic production clustered by country (Portugal right vs. Brazil left)
Source: Own elaboration

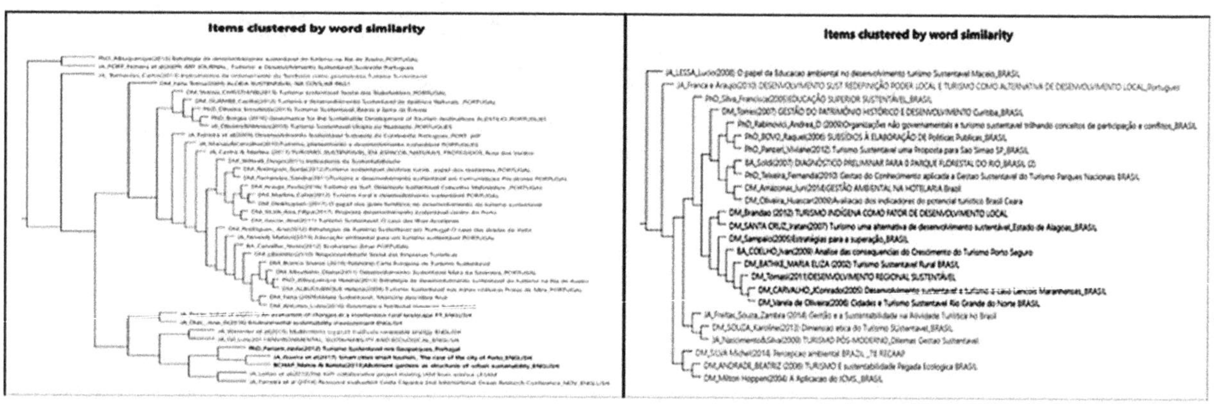

importance of regulation and the need for effective and lasting policy measures. However, from this research, we acknowledge that quantitative indexes for governance and planning are still minimally approached. There is only one international report and two scientific articles that play appropriately on the theme of using sustainability indices to support tourism governance. The other works that mention these issues are more descriptive.

Comparing the works from Portugal and Brazil (figure 4), the former focus on themes more related to territorial planning, stakeholders involved, natural destinations, governance, rural tourism, eco-tourism and resources evaluation. The latter focus on themes more related to environmental education, redefinition of local power, participation, tourism consequences, ecological footprint and ethical issues.

Thus, Portuguese investigation on sustainable tourism has been focusing on resources/ stakeholders and Brazilian research on local power/community. This latter is due to a focus of the Brazilian academic works on local policies, of the regions where the universities are placed, which facilitate data collection. Getting resources in Brazil is a more complex issue, passing through the scrutiny of economists and technocrats.

This is in line with aspects previously referred about the Brazilian context, such as the hope that the public power, local community and companies can join forces to structure a tourism development plan, considering the balance ratio of the dimensions of sustainability. Also, we acknowledge that in the Brazilian case there is more diversity of themes, which is in line with the educational policies in Brazil that have sought to provide the conditions for new initiatives for the diversity of educational projects in higher education, in order to meet the diversified demands in the different regions of the country.

CONCLUSION

Sustainable tourism development requires good information and integrated approaches to management, supported by the right indicators. As WTO (2013) suggested, indicators are tools that provide accurate information for decision-making, leading to the implementation of solutions, better planning and management. They are not an end in themselves, but rather signals of important trends and changes; catalysts for discussion on plans, risks and impacts to the destinations. They can also serve as performance measures for progress towards sustainability. Indicators are not a one-time procedure, they are most useful when measured repeatedly and consistently through long-term monitoring programs. Only when information is available over time, on a regular basis, can the most effective use of indicators occur, providing context for the understanding of changes and their importance to policy priorities.

In the present work, based on selected academic research from RCAAP, quantitative indexes for governance and planning are still weakly approached. Only two scientific articles and one international report appropriately approach the application of sustainability indices to support tourism governance and regulation. The other works that touch these issues are more descriptive.

The aim of indicators development is not solely the measurement of factors, production of tables or publication of reports, but better decision-making for the sustainable development of the destination. Indicators, therefore, should be an integral part of planning and monitoring processes. The information generated by government authorities, at different levels, by different private and civil sector organizations or through research, can be all important for sustainable tourism indicators. Also, indicators at local levels can be building blocks for applications at regional and national levels. Making indexes accessible and sharable is the only route to informed decision making and involvement of all stakeholders in the decision process. Measures can empower destinations, providing the information needed to negotiate future investments, development standards, joint ventures, and the sharing of benefits. They also support approaches to consensus on what is important to sustain and how to achieve it, such as the limits to acceptable change.

As WTO (2004) also proposed, many indicators begin as measurements done through scientific research. The academic community has been important in initiating indicator studies, developing methodological tools and generating information. Too often, however, academic studies have been one time initiatives. There is the initial study of a problem and the establishment of baseline information, but without follow up through linear studies and monitoring. A challenge for the academic community is to obtain sufficient resources to support an ongoing role in implementation of key indicators, as well as applied research on their validation and evaluation of their impact.

Another need is for academic institutions to transform information generated through scientific research into forms where it can be easily understood and used in policy making by practitioners. Finally, in order for future tourism professionals to be familiar with the need for indicators and practices of monitoring, academic and training institutions should include, in tourism studies curricula, cases and methodologies for development and use of indicators.

ACKNOWLEDGMENT

This chapter is financed by National Funds provided by FCT- Foundation for Science and Technology through project UIDB/04020/2020.

REFERENCES

Albuquerque, H. (2004). *Turismo sustentável nas zonas costeiras: O caso das praias de Mira*. Dissertação de Mestrado, Universidade de Aveiro, Portugal.

Amazonas, I. (2014). *Gestão ambiental na hotelaria: Tecnologias e práticas sustentáveis aplicadas nos hotéis de João Pessoa*. Dissertação de Mestrado, Universidade Federal da Paraíba, Brazil.

Andrade, B. (2006). *Turismo e sustentabilidade no município de Florianópolis: Uma aplicação do método da pegada ecológica*. Dissertação de Mestrado, Universidade Federal de Santa Catarina, Brazil.

Antunes, L. (2016). *Governança territorial, inovação e turismo sustentável: Uma análise comparative das redes Casas Brancas e Aldeias de Xisto*. Dissertação de Mestrado, Universidade de Coimbra, Portugal.

Carvalho, N. (2012). *Ecoturismo - Estudo de caso: Zmar-Eco camping resort. Projeto final de licenciatura*. Universidade Atlântica.

Coelho, I. (2009). *Análise das consequências do crescimento do turismo no município de Porto Seguro a partir da década de 90 sob o ponto de vista do desenvolvimento sustentável*. Projeto de Bacharelato, Universidade Federal da Bahia.

Dias, J. (2016). Environmental sustainability measurement in the Travel & Tourism Competitiveness Index: An empirical analysis of its reliability. *Ecological Indicators*, *73*, 589–596. doi:10.1016/j.ecolind.2016.10.008

Ferreira, J., Silva, S., Seixas, L., Lopes, A., Fortes, C., Reis, M., Poseiro, P., Capitão, R., Malvarez, G., & Tenedório, J. (2014). *The resource valuation and the challenge of coastal ecosystem-based management implementation*. Retrieved from http://repositorio.lnec.pt:8080/handle/123456789/1007022

Gössling, S., Scott, D., & Hall, C. (2013). Challenges of tourism in a low-carbon economy. *Wiley Interdisciplinary Reviews: Climate Change*, *4*(6), 525–538. doi:10.1002/wcc.243

Howard, B. (2014). Brazil leads world in reducing carbon emissions by slashing deforestation. *National Geographic*. Retrieved from https://www.nationalgeographic.com/science/article/140605-brazil-deforestation-carbon-emissions-environment

IBGE – Instituto Brasileiro de Geografia e Estatística. (2017). *Estatísticas – população*. Retrieved from https://www.ibge.gov.br/estatisticas-novoportal/sociais/populacao.html

Melo, F., & Farias, S. (2018). Sustainability communication and its effect in consumer intention to visit a tourist destination. *Tourism & Management Studies*, *14*(2), 36–44. doi:10.18089/tms.2018.14204

Mota, I., Pinto, M., Sá, J., Marques, V., & Ribeiro, J. (2005). *Estratégia Nacional para o Desenvolvimento Sustentável ENDS 2005-2015*. Retrieved from https://infoeuropa.eurocid.pt/files/database/000015001-000020000/000019537.pdf

Report, B. (1987). *Our Common Future*. United Nations World Commission on Environment and Development. Retrieved from https://en.wikisource.org/wiki/Brundtland_Report

Stilwell, D. (2011). *Indicadores de sustentabilidade aplicados ao contexto do desenvolvimento turístico português*. Dissertação de Mestrado, Faculty of Sciences, University of Lisbon.

Sustainability Index, G. D. S. (2017). *Sustainable destination management trends and insights: A path to a brighter future*. Retrieved from https://www.100carbonneutral.com/assets/sustainable-destination-management-trends-and-insights--a-path-to-a-brighter-future.pdf

Wimmler, C., Hejazi, G., Fernandes, E., Moreira, C., & Connors, S. (2015). Multi-criteria decision support methods for renewable energy systems on islands. *Journal of Clean Energy Technologies*, *3*(3), 185–195. doi:10.7763/JOCET.2015.V3.193

World Economic Forum. (2017). *Travel & Tourism Competitiveness Report 2017*. World Economic Forum. Retrieved from http://reports.weforum.org/travel-and-tourism-competitiveness-report-2017/

WTO - World Tourism Organization. (2004). Indicators of sustainable development for tourism destinations. A guidebook. WTO.

WTO - World Tourism Organization. (2013). *Sustainable Tourism for Development*. European Commission. Retrieved from https://www.e-unwto.org/doi/pdf/10.18111/9789284415496

KEY TERMS AND DEFINITIONS

Governance: Effective governance practices must reflect the changing business and policy environment, and the evolving roles and competencies of government tourism organizations.

Indexes: Metrics that serve to measure the performance of a group of assets. An example is a ratio derived from a series of observations and used as an indicator.

Monitoring: It is the systematic process of collecting and analyzing information to track an issue's progress toward reaching its objectives. It usually focuses on processes (when/where activities occur, who delivers them, etc.).

Regulation: It is the management of complex systems according to a set of rules and trends. In tourism, it includes legislation affecting this industry.

Sustainability: It focuses on meeting the needs of the present without compromising the ability of future generations to meet their needs. It involves three pillars: economic, environmental, and social.

Sustainability Indexes: Are instruments to measure the responsibility in social and environmental areas. The analysis of this issue in tourism is reflected in the publication of specific metrics and reports.

Sustainable Tourism: A balance must be found between limits and usage so that continuous changing, monitoring, and planning ensure that tourism can be managed.

Chapter 7
Slow Tourism in Low-Density Territories: NUTS III – Beiras and Serra da Estrela

Mafalda Marques
Estoril Higher Institute for Tourism and Hotel Studies, Portugal

João Reis
Estoril Higher Institute for Tourism and Hotel Studies, Portugal

ABSTRACT

This research focuses on slow tourism in low-density areas of Beiras and Serra da Estrela, Portugal, and explores how these genuine and authentic spaces can become attractive for tourism development. Compared to visitors who travel to a destination to visit as many attractions as possible in the shortest time, slow tourists refuse to be treated as a number and seek to receive from and give to their destination in the same proportion. Their involvement in a host community, practiced in a personalized, deep, and informed way, contributes to their well-being. The community's inhabitants recognize that their cultural identity is valued, feel motivated to serve as ambassadors, and note that tourist activities can lead to improvements in their quality of life. This research includes the results of interviews, a benchmarking in Italy, and an analysis of a village tourism. The findings reveal the opportunities they hold for slow tourism, with very positive effects for local development.

INTRODUCTION

Describe the general perspective of the chapter. End by specifically stating the objectives of the chapter. Access to information creates numerous advantages for travelers in search of tourist activities. However, such research sometimes only allows the visitor to obtain superficial knowledge of tourist destinations. It also can prevent visitors from fully engaging in their target locale so that they can enjoy personalized experiences that meet their preferences.

DOI: 10.4018/978-1-7998-7339-6.ch007

Furthermore, although many rural areas require development, they have unique characteristics and are places of rare beauty that are favorable to the development of slow tourism, as is the case of Beiras and Serra da Estrela (Figure 1). The present study analyses the adequacy of the tourist offer of low-density territories to the growing demand for slow tourism.

Figure 1. Beiras and Serra da Estrela territory.
Source: own elaboration

Potential slow destinations are often rural, people live in accordance with the time and with the seasons, the true speed of life. Remote places are untouched places, unknown destinations, and unstepped routes, out of season scenarios, far from the madding crowd (Guiver & McGrath, 2016).

The population of this territory, mostly aged, feels forgotten and abandoned and, for that reason, much of its knowledge is also being depreciated, forgotten, and, consequently, lost. Thus, the purpose of this chapter is to capture the value of the rural areas of the territory under analysis and provide them with a renewed image, taking advantage of a seemingly new concept - slow tourism.

In short, it analyzes how a territory consisting of depopulated, degraded, and poorly developed areas, repulsive to many of its inhabitants but attractive to a certain type of tourist (slow tourists) due to its unique characteristics, can be transmuted into a pleasant space for all stakeholders through the development of tourist activities.

Although those who live there view the region as unattractive, it has all the attributes that the slow tourist seeks and values. In this sense, it is classified as a territory that, despite being one of the richest in cultural and natural heritage, also has many associated weaknesses inherent to low-density areas, which must be seriously evaluated. Thus, in this investigation, from the research question: How slow tourism contributions to the revitalization of low-density territories? some opportunities for the development of this territory will be presented, based on the tourist demand of slow tourists.

SLOW TOURISM

Slow tourism is a concept that values participating in experiences using the time necessary to deepen the reality of the target destination. It prioritizes personalizing experiences according to the rhythms and preferences of each visitor. The slow tourist, being unique, is integrated into the community that welcomes them and gives to and receives from them in the same measure. This visitor is more conscientious than the average traveler and ends up contributing to the well-being of the local population. By showing interest in cultural and natural heritage, they are easily guided by those who truly know the region and are proud to make known the same to those who can value it, even if these same inhabitants do not always appreciate their surroundings. In this way, slow tourists do what they like, taking the time they need to experience and appreciate what surrounds them. As a result, the traveler achieves holistic experiences that will last in their memories forever, and that complete and enrich them as a human being.

It is important to take time to observe, to be in the moment and take pleasure from simple things, taking time to wander around, chat to people, enjoy a drink at a local café or just watch the world go by, sipping coffee on a sidewalk terrace, and taking leisurely trips to neighbouring villages (Guiver & McGrath, 2016).

The experiential dimension is an integral part of slow tourism and the most relevant aspect at the core of slow tourism (Serdane et al., 2020).

Mass tourists tend to consume everything they can in a short time. They are unable to truly appreciate what surrounds them, so they visit and only superficially witness the iconic monuments of the region. When they return home, they know little of the reality of the place they visited. They observed only what was shown to them by the tour guides and interpreters, contributing little to the well-being of the local community. Generally overlooked, the host population is increasingly impoverished with this type of tourist activity, and they are the main ones harmed by the negative impacts of tourism.

The slow travel philosophy focuses the importance of getting to know one small area well, rather than seeing "only a little bit of many different areas" (Schlichter, 2007, quoted in Guiver & McGrath, 2016:).

The fast pace of life of contemporary people living in developed countries results in a search for peace and relaxation, which is reflected in new forms of tourism (Zielińska-Szczepkowska, 2020). One of these manifestations is the Slow phenomenon that had as its origins in food and later extended to other activities such as cities and tourism. The aims of slow tourism, food and cities is the search for the "good life", for better lifestyle quality for both residents and visitors (Osorio, 2020).

Slow tourism, in contrast to packages aimed at those who like to rush through and move on quickly, allows tourists with a preference for a slower pace to have the time to make connections with the destination, people and local culture (Caffyn, 2012, quoted in Özdemir & Çelebi, 2018).

Caffyn (2012) mentioned five key dimensions of slow tourism (place, people, time, travel and individual) and summarizes the basic elements of this form of tourism (minimizing travel distances; maximizing the length of stay; psycho-physical relaxation; meeting local culture and customs; emphasis on local products including local restaurants and shops; creating authentic experience). A destination is considered "slow" if it encompasses a larger number of the elements mentioned above (Caffyn, 2012, quoted in Marić et al., 2021).

Guiver & McGrath (2016) explore five themes emerging from the literature: time; conscious decision-making; engaging the senses; holiday duration and location.

A new type of tourist is emerging – one who is conscious, responsible, critical, respectful of nature, open to learning about many things, as customs, culture, and religion of local populations (Zielińska-Szczepkowska, 2020).

In the late 1980s, the "slow philosophy" emerged from the Italian Slow Food movement, then quickly triggered the worldwide Slow Movement in daily life, such as Slow City and Slow

Tourism to promote a slower pace and way of life. Under the guidance of "slow philosophy", it is suggested that slow tourism can be a different tourism choice that contributes to contrast with the negative consequences of mass tourism (Shang et al., 2020).

The slow movement (slow food and slow city) emerged in Italy in the 1980s and 1990s, started in 1986 by Carlo Petrini as a response to the opening of a Mc Donald's restaurant in a place of cultural importance in Rome. Slow food was a form of collective retaliation against the phenomenon of increasing global consumption of fast food. The slow movement was formed officially in 1989 and was later expanded to 132 countries (Fullagar et al., 2012).

Carlo Petrini opposed the widespread McDonaldization of life, especially in the context of food. He believed that everybody was entitled to tasty, healthy, local food. The small local Slow Food Movement Association (created by him) transformed itself into the International Non- Governmental Organization – Slow Food International, within three years after its birth (Bac, 2013).

The slow tourism movement has grown rapidly at the beginning of the 21st century, contradicting the vulgar phenomenon of Western societies that encourages consumption, excessive growth, and frequent travel (Honoré, 2004, quoted in Ferreira, 2014).

Slow cities emphasize the distinctive features and sustainable development of the place. The slow city movement has the purpose to find the cities' own identity and soul. As the core feature of a slow city, "Slower for a better life" not only realizes the wishes of residents to pursue quality of life but also conforms to the reality that tourists seek to have a slow rhythm of life (Shang et al., 2020).

The practice of slow tourism remains a reality for millions of people who annually choose to camp, travel in caravans, rent accommodations with family and friends, or use a secondary residence. These tourists are fully immersed in the local culture, favoring the enogastronomy of the region and walking or cycling through the surrounding area. Many tourists practice slow tourism involuntarily, without realizing that they are engaging in the trend (Guiver et al., 2016).

Bac (2012) summarizes that the slow components of tourism can be observed in different fields, including transportation, accommodation (small units of local accommodation and rural tourism), gastronomy (respecting the principles of slow food), and culture (valuing local culture and traditions). Heitmann et al. (2011, in Bac, 2012) believe that slow tourism is characterized by the pleasure of discovery, learning, and sharing. This slowdown in travel creates the opportunity to interact and establish a connection with the local population at a deeper level.

The immersion in a culture, the understanding of and interaction with hosts and locals in their everyday lives, the depth of the meaning of their exchange without having to be limited by a time-based schedule but managing the pace of the events according to an individual perception of time, are what make the experience the predominant component of slow tourism (Serdane et al., 2020). Experience is a central concept in this type of tourism. Tourists try to create valuable memories, awaken positive emotional responses and inspire minds, seeking unforgettable experiences in tourist destinations (Shang et al., 2020).

The essence of slow travel is having time to engage with people, and to enjoy both the trip and the destination, rejecting quick flights and getaways and opting instead for bike rides or even walking. Relaxation begins at the beginning of the trip; traveling between the area of residence and the destination is no longer stressful, and the visitor enjoys the trip and observes how the landscape changes along the way. Slow travelers explore communities along their journey, stop according to their preferences, and look for places that are recommended by the local population. It is a cultural change that aims to slow the pace of travel and do everything at the right time (Dickinson & Lumsdon, 2010: 148).

Gastronomy reflects the territories in which it is created, giving meaning to flavors and knowledge (Gimenes, 2008, in Costa et al., 2014: 64). The quality and authenticity of local products, combined with the guarantee of being produced in a more responsible way, is reinforced when tourists have the opportunity to learn about the production methods and the characteristics of the products they are consuming, for example, through animation activities (Simões & Cristóvão, 2003: 182). Slow-food fans, when traveling, like to interact with the local population - talking to shop owners, producers, fishermen, and waiters - as a way of integrating and connecting with the destination they are visiting (Pecsek, 2016). Slow tourism represents an "antidote" to mass tourism. In such a sense, a feverishly organized trip "to see as much as possible" is replaced by a slow journey of relaxation (Balletto et al., 2020:4).

The goal of this type of tourism is to know the reality of a tourist destination, not just what is sold for tourism purposes. Also important is to learn about the location's heritage that is not found in the brochures or promoted on the Internet, but that is highly appreciated by the local population, to visit and experience what surrounds the region, calmly and without being on a schedule. Slow tourists seek to truly explore what they like, assimilating what surrounds them, and exchanging ideas with those who know the space very well and who can suggest the best experiences.

This research emphasizes the new tourist who, being more conscientious, intends to live memorable experiences and feel a complete and unique sense of being in the world. Thus, the tourist is offered the possibility to participate in the day-to-day life of the place they visit, giving them the opportunity to make a difference and contribute to a better world.

Figure 2. Slow tourism and its concepts framed on the supply and demand side.
Source: own elaboration

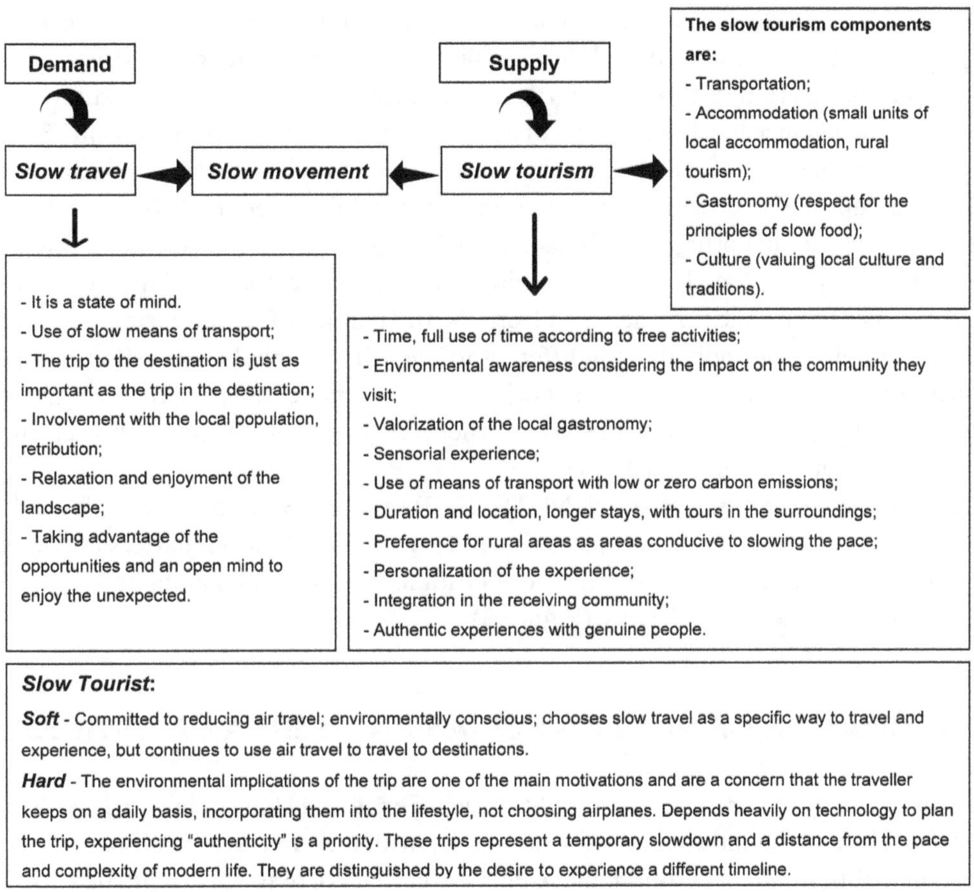

As a result, slow travel is a concept that translates demand, and slow tourism portrays supply (Figure 2).

LOW-DENSITY TERRITORIES

Low-density territories are generally associated with the inland of regions of Portugal and present insufficiencies in the following dimensions (Martins & Figueiredo, 2008, in Olímpio et al., 2013; Silva et al., 2010):

- Urban: highlighting the reduced dimension and expression of a large part of its urban agglomerations;
- Economic: a reduced number of economic activities and job opportunities, and small local markets;
- Institutional: insufficient number of entities with attributions and competences of proximity that support the local development;
- Social: aged population and/or residents with few qualifications and a reduced entrepreneurial spirit and reduced capacity for initiative; and
- Relational: reduced effectiveness in establishing partnerships and networks, and deficient participation rates of the population

However, low-density territories are also endowed with unparalleled wealth (Marques & Silva, 2009, in Silva et al., 2010), which is manifested through the existence of heritage elements of high value (geological resources, water, forestry, environmental, landscape, cultural, historical) and that are, most of the time, unexplored. Wealth is also produced through factors very favorable to investment in competitive sectors in terms of production (e.g., wine, dairy, olive oil, fresh and dried fruits, and forestry), and in a set of conditions that enhance the development of activities related to tourism and leisure. Such a heritage represents a peculiar character of the area, a hard one and apparently in contrast with a beautiful and still little-known landscape and endowment of natural resources (Balletto et al., 2020).

Low-density territories in Portugal are characterized by problems that are a cross-cutting issue to other sectors; they are closely associated with specializing in traditional sectors that are not very competitive, resulting from exhausted business models and from the difficulties of restructuring and modernization (Silva et al., 2010). However, nobody is indifferent to the villages that bring visitors closer to nature, heritage, and traditions, for a reunion with their origins. The inland regions still have the ability to dazzle visitors (Godinho, 2017: 4). Villages are the spaces where the past meets the future, preserving memories that are important to retain (Santos, 2017: 4).

With regard to the territory under study, NUTS III - Beiras and Serra da Estrela, it can be said that, except for some population centers concentrated mainly in the county headquarters and in neighboring regions, it is a predominantly low-density territory (with fewer than 50 inhabitants/km2), concentrated in numerous villages, some of rare beauty (Figure 3).

CHARACTERIZATION OF NUTS III - BEIRAS AND SERRA DA ESTRELA

The territory under analysis corresponds to the Statistical Territorial Units of level III (NUTS III), Beiras and Serra da Estrela, and the Intermunicipal Community with the same designation, and consists of the following municipalities: Almeida, Belmonte, Celorico da Beira, Covilhã, Figueira de Castelo Rodrigo, Fornos de Algodres, Fundão, Gouveia, Guarda, Manteigas, Mêda, Pinhel, Sabugal, Seia, and Trancoso (CIMBSE, 2021).

The territory of Beiras and Serra da Estrela is strategically located in central Portugal, halfway between Lisbon and Madrid, being Vilar Formoso, the main border of the country. It occupies an area of 6,305 km2, corresponding to 22% of the central region, with 211,705 inhabitants (2019 census). The most representative sector is the tertiary sector, representing around 68%, followed by the secondary sector, representing around 27%. The number of companies installed in this region in 2011 was 21,611 (CIMBSE, 2021).

The main advantages of this territory are: its strategic location in the Iberian Peninsula; its preserved natural and cultural resources; great tourist potential; the notoriety of the "brand" Serra da Estrela; the existence of small units of tourist accommodations and catering; high-quality endogenous products DOP like cheeses, sausages, liqueurs, olive oil, wines; there are also high-altitude wines - rye bread, Loriga cake, mountain honey, traditional infusions, and characteristic woolen and design pieces; local research centers and units; local/regional development agencies; support services for investors; watersheds; higher education institutions; and the presence of cultural and leisure facilities (CIMBSE, 2021). The DOP products like: the Beira Interior olive oils, the Bravo Esmolfe Apple, Terrincho lamb, Serra da Estrela lamb, Serra da Estrela cheese, Beira Baixa cheese, Serra da Estrela curd cheese, Beira Baixa curd cheese. At the same time, we have the IGP products, also valued in this territory, such as: The Cherry

Figure 3. Population density of Beiras and Serra da Estrela.
Source: own elaboration

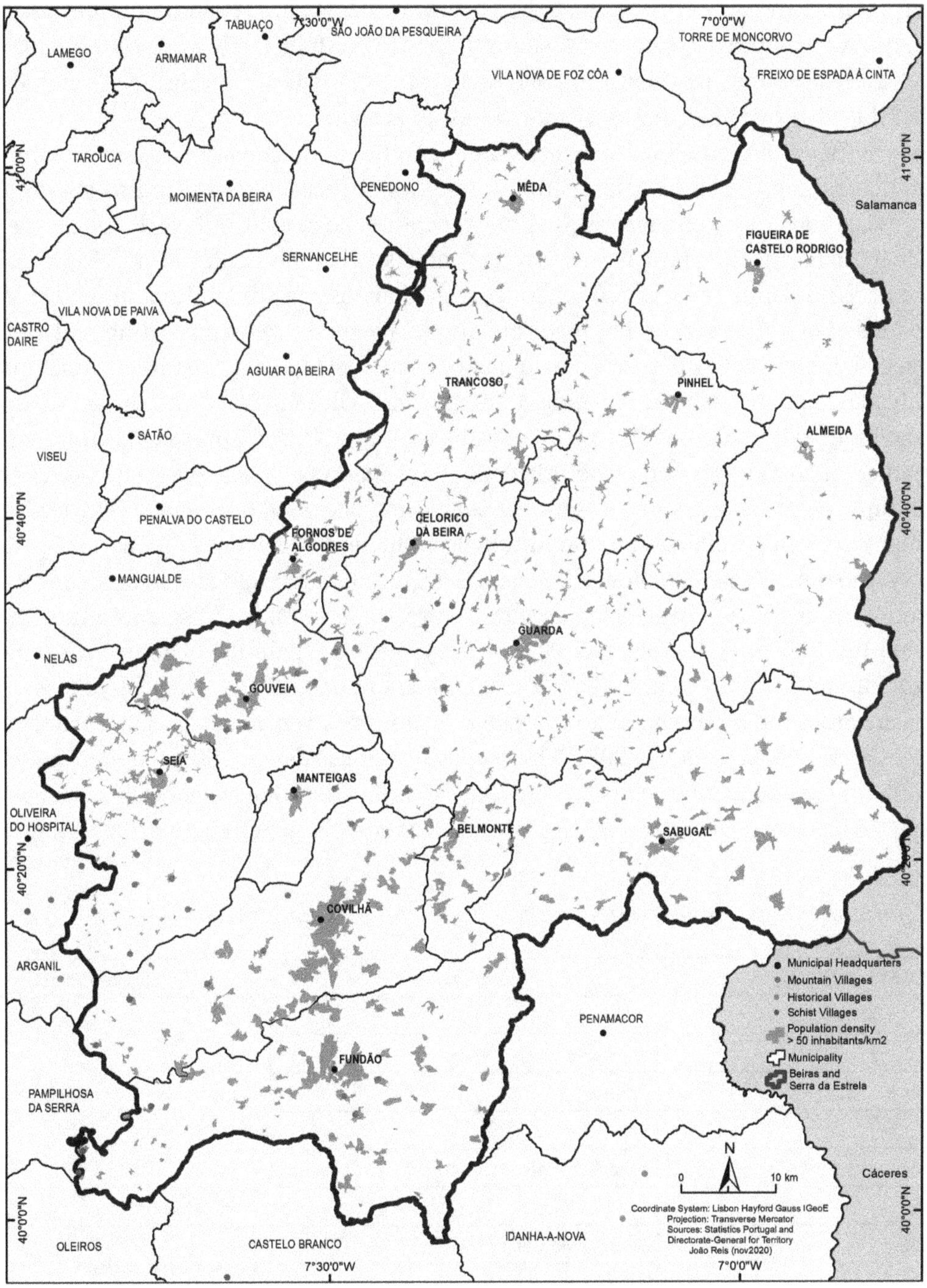

from Cova da Beira, the Peach from Cova da Beira, the kid from Beira, the Serra da Estrela Lamb, the Beira Lamb, the Kalega Olive from Beira Baixa, among others. These products are highly appreciated by this type of tourist and this certification, besides valuing and authenticating the high quality of the products, generates more profitability for the producers.

Also noteworthy is the friendliness of the population, the dogs of the Serra da Estrela, the river beaches, the Bread Museum (Seia), the Toy Museum (Seia), the Natural Electricity Museum (Seia), the Serra da Estrela Interpretation Center - CISE (Seia), the Discoveries Museum (Belmonte), the Jewish Museum (Belmonte), the Agriculture and Cheese Museum (Celorico da Beira), the Wool Museum (Covilhã), the Automobile Miniature Museum (Gouveia), and the Ecological Park (Gouveia). All are examples of the potential that this territory holds for tourism and leisure activities.

Numerous activities are possible, such as flying, paragliding in Linhares da Beira, being a shepherd for a day and corralling the herds, and visiting a traditional cheese factory and following the cheese-making process, among many others. It is possible to explore the glacial valleys, lagoons, waterfalls, and dams, which are magnificent landscapes. Travelers can also visit the Center for Ecology, Recovery and Surveillance of Wild Animals - CERVAS (Gouveia) and sponsor a wild animal in recovery, where the animals are treated and returned to their natural habitat. There are several sporting and cultural events that take place, namely, the Festival da Transumância, medieval festivals, historical reenactments of the Battle of Salgadela and the siege of Almeida, fairs, theaters, film festivals, and music, among others.

Before the industrial revolution, the people of this region produced almost everything necessary for their subsistence, including scissors, wool blankets, jams, baskets, shoes, pots, cheese, clothing, tools, agricultural implements, and other necessary utensils, using ancestral techniques. The tradition of masters, craftsmanship, and specialization in traditional production is evident in many pieces that can still be seen today. In some cases, the knowledge can be acquired, resulting from skills being passed down from generation to generation over many centuries (Pró Raia, n.d.). In 2008, more than 100 artisanal activities persisted in Serra da Estrela, including pottery, weaving, painting, woodwork, cooperage, toys, tapestries, traditional costumes, embroidery, linen, basketry or handicrafts, sausages, cheese, and sweets. The pieces were produced by artisans who worked in their homes or in small artisanal production units of family businesses (Guia da Cidade, 2008). There are still some artisans who keep their traditions intact.

On the other hand, there are also historical villages, mountain villages, and schist villages, although not all belong to the territory under analysis. The associations in which these villages are included are

Table 1. Historical villages of Portugal and their municipalities and NUTS III

Historical Villages	Municipalities	NUTS III
Almeida	Almeida	Beiras e Serra da Estrela
Belmonte	Belmonte	
Castelo Mendo	Almeida	
Castelo Novo	Fundão	
Castelo Rodrigo	Figueira Castelo Rodrigo	
Idanha-a-Velha	Idanha-a-Nova	Beira Baixa
Linhares da Beira	Celorico da Beira	Beiras e Serra da Estrela
Marialva	Mêda	
Monsanto	Idanha-a-Nova	Beira Baixa
Piódão	Arganil	Região de Coimbra
Sortelha	Sabugal	Beiras e Serra da Estrela
Trancoso	Trancoso	

(Aldeias Históricas de Portugal,2021)

distinguished because they are integrated in a network. There are 12 locations that are part of the network of the Historical Villages of Portugal. Only three are outside the geographical area that integrates NUTS III - Beiras and Serra da Estrela: Idanha-a-Velha, Monsanto, and Piódão (Table 1).

The mountain villages network consists of 41 villages located in the following municipalities: Celorico da Beira (5), Covilhã (5), Fornos de Algodres (2), Fundão (3), Gouveia (5), Guarda (5), Manteigas (4), Oliveira do Hospital (2), and Seia (10). Only the villages of São Gião and Alvoco das Várzeas, in Oliveira do Hospital, do not belong to the territory under analysis (Table 2).

Table 2. Mountain villages and their municipalities and NUTS III

Mountain Villages	Municipalities	NUTS III
Alcaide	Fundão	Beiras e Serra da Estrela
Alcangosta	Fundão	
Aldeias	Gouveia	
Algodres	Fornos de Algodres	
Alpedrinha	Fundão	
Alvoco da Serra	Seia	
Alvoco das Várzeas	Oliveira do Hospital	Região de Coimbra
Cabeça	Seia	Beiras e Serra da Estrela
Cadafaz	Celorico da Beira	
Cortes do Meio	Covilhã	
Erada	Covilhã	
Famalicão da Serra	Guarda	
Fernão Joanes	Guarda	
Figueiró da Granja	Fornos de Algodres	Beiras e Serra da Estrela
Figueiró da Serra	Gouveia	
Folgosinho	Gouveia	
Lapa dos Dinheiros	Seia	
Loriga	Seia	
Mangualde da Serra	Gouveia	
Melo	Gouveia	
Penhas da Saúde	Covilhã	
Penhas Douradas	Manteigas	
Póvoa Velha	Seia	
Prados	Celorico da Beira	
Rapa	Celorico da Beira	
Sabugueiro	Seia	
Salgueirais	Celorico da Beira	
Sameiro	Manteigas	
São Gião	Oliveira do Hospital	Região de Coimbra

continued on following page

Table 2. Continued

Mountain Villages	Municipalities	NUTS III
São Jorge da Beira	Covilhã	Beiras e Serra da Estrela
Sazes da Beira	Seia	
Teixeira	Seia	
Trinta	Guarda	
Vale da Amoreira	Manteigas	
Valezim	Seia	
Valhelhas	Guarda	
Verdelhos	Covilhã	
Vide	Seia	
Vide entre Vinhas	Celorico da Beira	
Videmonte	Guarda	
Vila de Manteigas	Manteigas	

(Aldeias de Montanha, 2021)

The schist villages network is formed by 27 villages distributed throughout the central region of Portugal. Only the villages of Barroca, Sobral de São Miguel and Janeiro de Baixo are integrated into the territory under study (Table 3).

The associations, each on a smaller or larger scale, were able to help to affirm these villages as tourist destinations since they knew how to organize what supports the concept. There are many other villages with great tourist potential and unique beauty. The rich history of the interior gives each town and village a valuable historical heritage.

The Historical Villages of Portugal are integrated into the granite or schist surroundings. These villages that once played a prominent role in the first line of defense of the country are now trying to find their place in the 21st century, asserting themselves as tourist destinations. The network that connects them is based on an irreplaceable historical heritage that allows visitors to take part in unique experiences in villages of medieval origin. The Historical Villages of Portugal have managed to create very interesting events that valorize the towns as a tourist product that, in addition to offering recipes, reinforces the identity and feelings of belonging of these populations, who believe that they live in special places sought by visitors who appreciate their location.

Rural transport is the primary link between villages and cities (usually the county seat) that are usually within 15 to 20 km, though they can be more distant, especially in low-density territories. The most common form of public transport is the bus, with more travel times offered during the school period; however, there are also taxis and, in some cases, a rail network. The taxi offers greater flexibility, suiting the specific needs of travelers; however, in rural areas, this type of service is scarce, given the low demand and frequently higher fares compared to cities (White, 2011: 12). As rural areas are devoid of an efficient public transportation network, it is important to emphasize the importance of taxis. Taxis serve both tourists and residents and can be important for visitors partaking in a gateway in a location without buses or trains. The territory under study can be accessed via several routes. From Lisbon, there is easy access via the A23 to Castelo Branco or to the EN18. From the Aveiro-Porto region, travelers can take the A25 to Celorico da Beira and Fornos de Algodres, or to the N1 and N16. The N221 is also a

Table 3. Schist villages and their municipalities and NUTS III

Schist Villages	Municipalities	NUTS III
Água Formosa	Vila de Rei	Médio Tejo
Aigra Nova	Góis	Região de Coimbra
Aigra Velha	Góis	
Aldeia das Dez	Oliveira do Hospital	
Álvaro	Oleiros	Beira Baixa
Barroca	Fundão	Beiras e Serra da Estrela
Benfeita	Arganil	Região de Coimbra
Candal	Lousã	
Casal de São Simão	Figueiró dos Vinhos	Região de Leiria
Casal Novo	Lousã	Região de Coimbra
Cerdeira	Lousã	Região de Coimbra
Chiqueiro	Lousã	
Comareira	Góis	
Fajão	Pampilhosa da Serra	
Ferraria de São João	Penela	
Figueira	Proença-a-Nova	Beira Baixa
Gondramaz	Miranda do Corvo	Região de Coimbra
Janeiro de Baixo	Pampilhosa da Serra	
Janeiro de Cima	Fundão	Beiras e Serra da Estrela
Martim Branco	Castelo Branco	Beira Baixa
Mosteiro	Pedrógão Grande	Região de Leiria
Pedrógão Pequeno	Sertã	Médio Tejo
Pena	Góis	Região de Coimbra
Sarzedas	Castelo Branco	Beira Baixa
Sobral de São Miguel	Covilhã	Beiras e Serra da Estrela
Talasnal	Lousã	Região de Coimbra
Vila Cova de Alva	Arganil	

(Aldeias do Xisto, 2018)

potential route and has excellent road conditions. However, for tourism with a low ecological footprint, it is also possible to access the region via rail transport and bicycle circulation. The slopes are smooth, and the roads have little traffic, good pavement, and excellent scenery. Local routes that start and end in historic villages or in a Bikotel accommodation are also available and can be completed in a single day or divided into several stages. There are also several mountain-biking centers and several proposals for mountain-biking routes. Each center has a reception area with parking, changing rooms, self-service bicycle stations, and a network of trails that include several levels of difficulty (Guia de Viagem Foge Comigo, 2014). Some urban centers within this NUTS III area are also accessible through the Expressos Network.

METHODOLOGY

In the present study, it was chosen a qualitative approach and the following stakeholders were interviewed through structured and semi-structured in-depth interviews. The interview results were collected and treated using simplified content analysis.

An exhaustive fieldwork was carried out in this territory in order to identify, if there was adequate offer for the slow tourism product and if the residents would be willing to play their role as ambassadors in their villages. The stakeholders interviewed were:

- Organizations: the Association of Transhumance and Nature, the Agency for the Tourist Development of Schist Villages (ADXTUR), the Association for the Development of Historical Villages (AHP), and the Association for Integrated Development of Mountain Villages Network (ADIRAM);
- Accommodation units that were carefully chosen according to their relevance to the subject under study (Quinta de Pêro Martins, Natura Glamping, and A Queijaria);
- Restaurants in the municipalities in this NUTS III region (O Brasão, Fio de Azeite Taberna, Taberna do Alcaide, A Cerca, A Pedra do Lagar, and the restaurants of the Palace Hotel and SPA Termas de São Miguel, Inatel Manteigas, and Longroiva Hotel Rural and Termal SPA);
- A tourist animation company (Beltour);
- Some thermal resorts (the Palace Hotel and SPA Termas de São Miguel, Termas Caldas de Manteigas, Termas Longroiva Hotel Rural and Termal SPA, and Termas de Almeida - Fonte Santa); and
- Residents distributed in the counties of Almeida, Belmonte, Celorico da Beira, Fornos de Algodres, Figueira de Castelo Rodrigo, Gouveia, Fundão, Guarda, Manteigas, Mêda, Pinhel, Sabugal, and Seia. The goal was to analyze the existence of tourist offerings appropriate to this niche market and to find out how the local population would play a role in welcoming visitors.

The questions to the local population intended to evaluate the disadvantages of the territory, if they were aware of the importance of tourism and if they would be receptive in welcoming tourists. The remaining interviewees were approached in order to understand if there was adequate offer to the slow tourist demand.

A case study was performed in Chão do Rio - Turismo de Aldeia, and a benchmarking study was carried out to assess the effectiveness of the solution discovered by villages in Italy that face similar problems. These villages also seek their place in the 21st century, remaining alive and authentic. As such, they have identified consumers within a market niche who value that same authenticity and who seek to live life in the village as it is. These villages have made efforts to attract the slow tourist because they know that they are a tourist destination of choice for this group, and they do not have to change the original essence of the village to be successful.

SLOW TOURISM IN BEIRAS AND SERRA DA ESTRELA

Quinta Chão do Rio, located in the village of Travancinha, 12 km from the Serra da Estrela Natural Park in the municipality of Seia, is the materialization of the concept of slow accommodations in village tourism. It has offered a sustainable accommodation experience since 2010. With 8 acres and six houses, Quinta Chão do Rio is harmoniously integrated into nature; guests can enjoy authentic experiences that reflect the innocence of their childhood and make memories that last and bring them joy.

All houses have a dining area, private bathroom, and a balcony with a porch, and pets are allowed. These stone houses, with roofs covered with brush, were strategically placed around the biological pool (swimming pool in a natural environment with the presence of aquatic flora and fauna), which was inspired by the lagoons of Serra da Estrela. Guests are encouraged to explore the nearby trails on foot or on bicycles that are provided free of charge. Children can have fun with hand-held cars, pick up eggs from the removable chicken coop, harvest seasonal vegetables from the garden, pick cherries, feed the sheeps, touch the rattles of the sheep that are hanging on the trees, swim in the biological pool, or play freely and safely outdoors. This is a place where time slows down and it is possible to enjoy each season of the year as it is (Chão do Rio, 2018).

A Queijaria, a local accommodation in Barreira, a village in the municipality of Mêda that was once one of the largest pottery centers in the country, produces rustic goat cheese, called Faias do Massueime, and constitutes a slow accommodation with slow activities. A Queijaria is a typical stone house in which the artisanal cheese-making is conducted on the ground floor, and a small museum of pottery and cheese is located elsewhere in the building. It has three suites, as well as a living room and a kitchen that are combined in open space, and a terrace that includes a colorful carousel for children (CM Mêda, 2018). It is a house that pays homage to pottery, cheese, and shepherds throughout the premises. It is possible to complement the accommodation with the experience of herding and milking the goats that are found on the owner's farm. It is also possible to make artisanal goat cheese, an organic cheese with an intense flavor, with just three ingredients (salt, thistle, and goat's milk).

The rattles event - Festival dos Caminhos da Transumância - held in Alpedrinha in the municipality of Fundão, celebrates the descent of the mountain herds in transhumance (the seasonal movement of livestock between mountains and lowland pastures) from Estrela to the meadows of Idanha. It is a festival that attracts many visitors. It was chosen as the featured slow event because it is possible to travel by train to Alpedrinha, and the city council provides buses to and from Fundão, ensuring that the village is not invaded by cars. Guided tours are provided in Alpedrinha, beginning from the tourist office, for those who want to learn about the heritage of the village, including the famous furniture of Alpedrinha. The event includes a market for biological products, and several stages for nightlife entertainment. There are also many daytime entertainment features circulating on the streets of the village, and there are many small taverns where visitors can purchase traditional Portuguese food and drink. Finally, spaces are designated for small producers who sell their products directly to the consumer, including honey, cheese, craft beer, liqueurs, and small clay mugs. Traditional porridge blankets, rattles of various sizes and shapes, and other indigenous products are sold near the "lounge," where drinks and other snacks were served.

Several residents open the doors of their homes to receive visitors and tell stories. The program includes several workshops given by artisans (inlaid materials of Alpedrinha, basketry, musical instruments, cutlery, construction of drums and ceramics) and an atelier, as well as a place dedicated to the exhibition of several sheep, goats, and canine breeds (herd dogs). Children are not forgotten, and several kid-friendly activities were offered. On Sunday morning, the Pedestrian Walk of Gardunha takes place with a herd and with the folk group "Os Chocalheiros de Vila Verde de Ficalho". The walk is over 9 km, starting in Fundão and ending in Alpedrinha. The rattle kit, at a cost of 10€, includes a bracelet with rattle, t-shirt, water, password for the bus, and shepherd soup at the end of the walk.

In 2018, there were 12 events that animated the 12 historical villages, and participants were invited to be temporary residents. The events were organized to emphasize the distinctive products of the villages - that is, something that belongs to the collective memory of each village, perhaps from a character, a legend, or a tale. The purpose was to identify the distinctive and unique symbols that are linked to each village and to turn them into programs capable of creating unforgettable memories among the inhabitants and tourists (DN, 2018). The entertainment program was very extensive, and the identity characteristics of the village were featured in terms of gastronomy, traditions, music, theater, and guided tours with the motto, "Experience, Interpret, Revive, Feel" (CM Sabugal, 2018). The efforts resulted in lively plays, concerts, workshops, showcooking, and recreation.

From the fieldwork, it was noted that the population is friendly and welcoming, and many residents like to talk to people who express interest in listening to them. They answered questions with patience and the calm characteristics of the countryside. Tourists are a novelty; given the low demand, locals still do not perceive tourism as an activity that generates wealth. Residents demonstrate a great community spirit that allows them to be helpful when asked about some services, revealing contentment by helping to direct visitors to the people or places they are looking for, and filling in the information gaps on services available in the territory. They know how to receive and give guidance and easily include visitors in the life of the village. They are clearly the best hosts that a slow tourist can find.

Except for the localities with the highest population density, there are few restaurants, which is in line with low demand; if there was a significant increase in tourist demand, the villages would not be able to respond. The meals available are mostly traditional Portuguese cuisine. Beira cuisine is rich and very tasty and retains its simplicity and flavor.

The three associations surveyed (representing historical villages, mountain villages, and schist villages) admit that the increase in tourists in the villages is beneficial. However, they consider that there must be good management of population flows to minimize the negative impacts of tourism. They do not intend to become the target of much demand, nor do they have an interest in being a mass tourism destination; rather, they seek to attract quality tourists, who are responsible and understand the difference between slow and mass tourism. Some villages are already facing problems resulting from excess capacity during peak demand due to excursionism, precisely because they are small villages.

BENCHMARKING IN ITALY

The benchmarking study was carried out in Italy, the country where the slow tourism concept was born. The intention was to investigate at least one successful case of adapting low-density territories to attract slow tourism. Italy was also chosen because some of its southern territories are very similar to the Portuguese interior under study. The Associazione Borghi Autentici d'Italia supports and represents a significant part of a hidden Italy, which every day finds motivations to immerse itself in the rediscovery and development of its identity, an identity that manifests itself in several ways - in history, in the traditions of places, in landscape, in artisanal production - that is, it translates the way of life of the population. To perform this benchmarking, the information was collected from the respective websites and analyzed.

This association promotes articulated local development by considering existing resources as starting points for the design of concrete and feasible strategies to improve the local social, environmental, and economic context. The main purpose is to improve the living conditions of the population and make the location attractive for living and working in these spaces (Associazione Borghi Autentici d´Italia, 2018). There are several completed projects and others in progress, with the most interesting being:

- The Little Festival of the Roots (PFR) brings together traditions, stories, and magic. The villages hosting the festival organize a program that includes conferences, shows, musical events, and gastronomic proposals; it is a fun way to promote the discovery of its roots, from a learning and sharing perspective. The aim of the PFR is to make known the hospitable communities rich in history and biodiversity; they are authentic villages that express their culture and that are committed to fighting for a better, inclusive, and sustainable Italy.
- The Observatory of Authentic Villages is an organization that aims to understand current trends and future prospects, studying measures that can be carried out and shared. Their motto is, "Watch to make it happen. Together".
- The National Network of Hospitable Communities is a project created and promoted by the Association of Authentic Villages of Italy and co-financed by the Ministry of Cultural Heritage and Tourism Activities. It aims to create an unprecedented experience in the 27 municipalities that are part of the National Network of Hospitable Communities project, and to bring to life a network of tourist, cultural, environmental, and productive destinations that share a common goal: hospitality based on the inclusion of the guest, accepted as a temporary citizen at the pace of life of the local community. They promote a different form of tourism, which intends to modify the concept of "vacation," transforming it into a travel experience designed for those looking for authentic traditions and seeking an experience to be shared. Tourists and residents are the protagonists of these experiences. Associated with this initiative is a Product Portal for Authentic Villages, where it is possible to purchase some characteristic products (Associazione Borghi Autentici d´Italia, 2018). One of the actions developed in this project was the community of good and authentic food, involving 90 Italian villages, which aimed to highlight the value of local food and wines, giving adequate recognition to villages committed to preserving and conserving agri-food biodiversity, promoting production, and making recipes that respect the natural balance of ecosystems. These villages have implemented specific initiatives to promote and propose authentic products and recipes that reveal local culture.

The Italy Slow Food Foundation was founded in Florence in 2003 through the special contribution of the region of Tuscany. This foundation coordinates and promotes slow food projects to protect food biodiversity worldwide. Some of the most interesting projects are conducted by: Slow Food Fortress, Ark of Taste, Slow Food Cooks Alliance, and Earth Markets. These tools promote a model of agriculture based on biodiversity, respect for the territory, and local culture. An agriculture in balance with the environment has the main objectives of food sovereignty and access by all communities to good, clean, and fair food. Working in more than 100 countries, the foundation supports thousands of small producers in its projects, guaranteeing technical assistance, training, communication, and interaction between producers. Slow Food Fortresses support small traditional productions that are at risk of disappearing. They also aim to assist territories in various ways, including protecting a threatened rural landscape or ecosystem, recovering traditional products and old manufacturing techniques, and saving from extinction indigenous breeds and varieties of fruits and vegetables.

There are over 500 Slow Food Fortresses involving over 13,000 producers. The Ark of Taste is a project that consists of identifying traditional products throughout the planet that symbolize the culture and traditions of a people, including some that are at risk of being lost. In Portugal, 24 authentic and endangered products have been identified and flagged. The Slow Food Cooks Alliance has more than 700 chefs from bars, restaurants, and street food vendors, scattered around the globe. They have very different profiles, histories, and cuisines; however, they share a commitment to protecting food biodiversity through the protection of gastronomic knowledge and local cultures. They undertake to use daily the ingredients of the Fortress, the Ark of Taste, and locally produced products, and to indicate in the menus the names of the producers who supply the ingredients, as well as to participate in events where they cook together.

Similarly, the Earth Markets are farmers' markets that follow the guidelines of the slow-food philosophy. These markets are collectively managed meeting places where local producers present their quality seasonal products directly to consumers at fair prices (for those who buy and for those who produce). They also guarantee environmentally sustainable production methods. In addition, they preserve the food culture of local communities and help defend biodiversity. These markets include only select producers, especially small farmers and artisanal producers who have more difficulty with the large-scale distribution circuit but nevertheless have high-quality products (Fondazione Slow Food, 2018).

CONCLUSION

The results of the study suggest that the territory of As Beiras and Serra da Estrela presents weaknesses inherent to low-density regions, and that the few residents, who stubbornly resist the urban call, admit that the territory lacks many necessary improvements. They do not see tourism as the immediate answer to all the problems that affect the area, however, they are excellent at welcoming visitors and, when questioned by them, they prove to be excellent hosts.

On the other hand, the uniqueness of the territory and all the associated tourist resources make it a pleasant destination for slow tourists who appreciate it exactly because of its unique characteristics. They understand and value differences, becoming temporary inhabitants of the places where they integrate. Their travel stories are a sum of unforgettable moments, some of them resulting from obstacles shared and overcome together with the residents. It is a wonderful territory that attracts the slow tourist, since these travelers are open to new and different experiences.

Slow Tourism and the Environment

Slow tourism, in addition to contributing to a better world, has in its genesis the reduction of the ecological footprint and concerns not only environmental factors but also the inhabitants of the territory. It views tourism not only as a commercial transaction because the slow tourist respects and wants to understand the differentiating characteristics of the territories; they are aware of the environmental, social, and economic impacts of their activities and their own responsibility. The slow tourist is educated, has the capacity for discernment, and seeks quality and participation. They want to contribute to a better world and leave a little of themselves in the territory. As a result, they seek experiences with meaning that they will remember forever.

This territory has all the attributes to become an excellent slow-tourism destination, provided that the weaknesses identified (i.e., the characteristics of the low-density territory) are minimized. There is also huge potential for the development of voluntary/solidarity tourism, not only in the Faia Brava Reserve but also in the organization of events and activities promoted by ADIRAM, as well as in helping the elderly by reducing their isolation and offering a variety of support.

Financial Benefits of Slow Tourism

Tax benefits, for those who want to invest in the interior of the country and for students attending educational establishments located in these territories, were contemplated in the State Budget 2019 (Dinheiro Vivo, 2018). The current state budget has continued to provide benefits to stimulate demand and promote domestic migration to these inland territories. Currently, there is a published guide that summarizes all the benefits available to families and businesses that are permanently transferred to these territories (XII Government of the Portuguese Republic, 2020). With regard to the territory under analysis, with the exception of some population agglomerations concentrated mainly in the municipalities and neighboring regions, it is a predominantly low-density territory (with fewer than 50 inhabitants/km2). Thus, these measures can be an opportunity to minimize some of the weaknesses inherent in low-density areas; that is, they can increase the number of residents and businesses in the territory; expand the multiplier effect; generate more production, services, and employment; and produce a more interesting ambience not only for those who work there, but also for those who visit the interior of the country.

Although many authors focus on the environmental motivations and benefits of slow tourism, most of these are from reducing long-haul flights rather than activities at the destination. However, destinations adopting principles of slow tourism may reduce the financial leakage to foreign suppliers and so generate more income and employment for local people (Caffyn, 2012, quoted in Guiver & McGrath, 2016).

The situation caused by the COVID-19 pandemic led to telework being adopted as a measure to contain the spread of the virus. Many companies have reduced operating costs by abdicating workspaces in urban areas, allowing their employees to perform professional activities remotely. In some cases, this situation may prevail in the post-COVID era, allowing digital nomads to choose to develop their activities in places of rare beauty that offer good living conditions, far from the hustle and bustle of the city, with lower living expenses. This could create opportunities for low-density territories if they know how to position themselves in the interests of this market niche.

Slow Tourism and Slow Food

It is also necessary to relate tourism with other sectors of activity based on sustainable development. Tourism is multidisciplinary and needs to be analyzed in all its aspects, in all its development strategies designed for in-depth analysis, and in line with all sectors. Tourism can help to develop the territory economically, together with other sectors, and thereby benefit the resident community. Given that slow tourism is directly related to slow food, it is important to value the mosaics of the territory and invest in agriculture, not only with regard to new rural areas, but also in terms of people who work the land for their subsistence and who have surpluses. Cooperatives and local markets work well in this case, and there are already initiatives in place to promote this. It cannot be ignored that fresh food produced locally, which is sown and harvested at the right time, has many advantages but exhibits a less attractive appearance; however, these products are rich in nutrients and high in taste and quality. It is understood, therefore, that the development of tourism in the region must necessarily involve the development of agriculture, with the valorization of indigenous products. Schist villages, with their chains of shops, are a good example of stimulating the economy that is particularly strengthened during the summer months, with tourists and emigrants loading their vehicles with local produce and goods.

The territory of As Beiras and Serra da Estrela has very attractive food and wine. In fact, there are residents who produce wine (especially in Beira Alta), raise pigs and make homemade sausages, bake traditional breads, and make high-quality artisanal cheese.

However, the lack of a strong network and a structured, cohesive, and publicized tourism product are limitations identified in some villages. In many cases, it is only possible to know the existence of certain services locally, by consulting the information available on the spot or by questioning the local population. That is, if you plan to book in advance, it may not be possible to do so in many cases, and tourists may not be aware that this service exists. Thus, it is recommended that the appropriate entities draw up an inventory of primary and secondary resources, and articulate the offerings to tourists, creating various tools that disseminate the information to those who intend to travel to these territories (for example, through applications for smartphones, in this case, adapting to the device used and to offline operation).

The Value of Infrastructure

On the other hand, infrastructure should also be created for the benefit of residents and not just tourists, as the population will use the infrastructure throughout the year. A good public transport network is utopian in a low-density territory, however, good accessibility is essential, not only to reduce the isolation of the population in accessing services not offered in the villages, but also to allow daily commuting for workers who want to live permanently in the territory.

In light of this, the Figueira de Castelo Rodrigo Tourist Office offers electric bicycles free of charge. This excellent initiative should be extended, even at a symbolic cost, to other tourist offices and allow riders, for example, to collect the bicycle at one office and leave it at another (note that because younger residents are interested in the use of these bicycles, this would be a long-term investment). The tourist office should be chosen strategically in relation to the proximity of public transportation and can be defined as a gateway to and from the territory. Moreover, the ecooltra, a scooter-sharing service, is being widely disseminated in Lisbon and neighboring counties, and could be extended to the interior in several strategic locations.

The Serra da Estrela Artisans Association has organized some financed training courses (Aasestrela, 2021), however, with more than 100 remaining artisanal activities of pottery, weaving, painting, woodwork, cooperage, toys, tapestries, and embroidery, among others, there are many artisanal activities for which training can be provided so that new artisans can emerge and continue the manufactured work, ensuring the valorization of these arts and crafts, which are still seen as marginal and poorly remunerated. With more state support, it will be beneficial for the region to create a cohesive network where entrepreneurs can create and innovate, resulting in the presentation of projects that celebrate artisanal products. There are some good examples carried out by AHP that demonstrate that the market is receptive to innovative products from local artisans. Although some innovations are introduced, it is a question of not letting ancient techniques die when they have, for many centuries, supported families and continue to support the artisans who still exist and who will be the masters of the new generation. It is a way of valuing the art of the craftsman and demonstrating the importance of this knowledge. The initiatives created by AHP have shown that the market is also receptive to "know-how" workshops.

Slow Tourism Outdoors

It is important to invest in initiatives that encourage outdoor leisure, namely, nature walks and cycling, taking advantage of the territory's strengths. The increase in consumer demand in this segment requires it, and this territory has excellent conditions for the practice of these activities on large and small routes, the maintenance of which must be insured. Indeed, synergies should be fostered in favor of the dissemination of the three Great Routes (GR): GR45 - Great Route of the Vale do Côa; GR22 - Great Route of the Historical Villages of Portugal; and GR33 - Great Route of the Zêzere. These are located relatively close to one another and are part of the territory that has been identified as excellent for hiking trails and geocaching.

In short, tourism is not the solution to all the problems of low-density territories, but it can be an excellent ally against depopulation if its development is properly planned throughout all sectors of the economy. In this case, it can prevent traditions from dying, reduce the isolation of the population, and recover the recipes of the region's grandmothers, delicacies that are slowly made and that will please tourists' palates.

The Uniqueness of Rural Areas

It is understood that slow tourism combines exceptionally well with rural areas, since, given their characteristics, they are the most suitable for this niche market that wants to enjoy these spaces with the time necessary to discover them. Rural areas are endowed with natural and cultural heritage of high interest, and they are places where people are remembered by name, the cuisine is rich, and if properly organized they can become a destination dedicated to a market segment that privileges travel to smaller regions where tranquility prevails. Due to the characteristics of the territory under analysis, it is understood that it has huge potential to become a renowned slow-tourist destination. The slow tourist can enjoy authentic and quality experiences in this region because they visit it without the mass excursionism that sometimes occurs in the villages. One of the fundamental characteristics of rural tourism is that it is associated, in general, with shorter stays. The way to counter this trend and increase the average stay will be through this niche market, which tends to stay longer in the territory when it is not traveling to reach its destination.

If, in mass tourism, the traveler is impelled to carry out an endless list of tasks that force them to discover the largest number of attractions in a destination, slow tourism does not prompt visitors to engage in all the activities existing in the rural world. Rather, it is only intended that the slow tourist can find something that meets their needs. In other words, these visitors can have time for themselves, without having a designated purpose. Thus, there are activities in the territory under analysis that meet the interests of each visitor, including dolce far niente. In short, if tourists intend to complement their stay and enrich their experience at the destination, there are a variety of activities with which they can engage but that are not imposed on the visitor.

The associations promote their calendars of events that include various activities that feature the essence of each village, promote their traditions, reenact historical moments, or offer festivals highlighting a product created in the village. In many cases, these were local events, but they have been promoted by the associations and have become the target of national demand. The events allow tourists to get to know the village, participate in various activities, taste enogastronomy, engage in cultural events, and interact with the local population.

The villages have no interest in becoming the target of mass tourism demand, as some of them struggle when the load capacity is exceeded. In other words, for these villages, niche tourism - namely, slow tourism - considers the characteristics of this segment of demand and is the most suitable for integrated and sustainable development. Everyone can benefit from this more intimate and personalized contact, in which residents have a lot to teach, and slow tourists a lot to discover and learn, as long as they are open to new experiences, including authentic activities that are not programmed and that can arise spontaneously.

Although excursionism is not very beneficial for this territory, mass tourism did arise in some historic and schist villages in the summer of 2020. This was the result of visitors escaping the hustle and bustle of the overcrowded coastline and the spread of COVID-19. Some villages were not prepared for this activity, nor are they intended to be, since the excess load capacity prevents the enjoyment of these quiet spaces and the ability of slow tourists to maximize the time spent there.

Instead, the accommodation that best reflects the essence of this territory is tourism in rural space and local accommodation (AL), with A Queijaria being one of these examples. Just like AL, it accommodates the traditional characteristics associated with the authenticity of the houses of Beiras, although in many cases it is the only existing accommodation. To enjoy a meal in local restaurants, to attend local pastry shops, and to stay overnight in an AL is to be part of the territory; it is what distinguishes a traveler from a tourist, and it is the total immersion in the local culture, feeling the thrill of the place, with its people and their way of life. Due to the low demand that still exists in low-density territories, tourists mainly find welcoming residents who greet visitors and help them feel less isolated.

Moreover, the seasonality of tourism can be mitigated in the territory because activities related to cheese production are more concentrated in the winter, when the animals are born, thus having a greater abundance of milk. This feature allows the organization of events related to cheese and the shepherd, which will make some places in this territory more attractive, encouraging an increase in demand at this time of year, alongside the demand for snow in the Serra da Estrela. On the other hand, respect for the seasonality of products that are sown and harvested at the right time is the motto for the organization of fairs and events, some of which have already been held, namely, the Cherry Festival in Alcongosta, the Míscaro mushrooms Festival in Alcaide, and the Cheese Fair in Seia.

In this territory there are several thermal spas that enrich the stay of the slow tourist on a seasonal schedule, especially between April and November, the most emblematic being the thermal baths of Unhais da Serra, which are part of the H2otel Congress and Medical Spa Hotel, and the medicinal waters of the village of Longroiva, which are explored by Longroiva Hotel & Termal SPA, and the Palace Hotel & SPA Termas de São Miguel in Fornos de Algodres.

The Comparison With Italy

The benchmarking analysis carried out in Italy allows us to consider some good practices that could be implemented in the territory of Beiras and Serra da Estrela. Indeed, the observatory of authentic villages is a place of study that seeks to understand current trends and future perspectives to establish measures that can be implemented and shared. Such an observatory could be created in the Portuguese territory under study, as a subdivision of the Tourism Observatory of Central Portugal, to delineate strategies and concrete measures for the development of slow tourism.

The Slow Food Fortresses and the Ark of Taste support small traditional productions that are in danger of disappearing. They seek to protect a threatened rural landscape or ecosystem and to engage in the recovery of traditional products and ancient manufacturing techniques, preserving flora and fauna that can be adapted to the reality of Beiras and Serra da Estrela. In this sense, local development associations will be able to make efforts to prevent these artisanal productions from disappearing, creating ways to monetize and adapt them to the present day, endowing them with new utility, through innovation, with the involvement of tourism, and using financial support from the programs of valorization and development of the interior.

The National Network of Hospitable Communities, as mentioned above, aims to create a unique experience and give life to a network of tourist, cultural, environmental, and productive destinations that share a common goal. Within this network, the community of good and authentic food was an excellent project that created a feeling of pride and belonging, which can be adapted to the villages of Beiras and Serra da Estrela so that local people believe in their potential based on traditional knowledge and can share it with slow tourists in an environment of unforgettable cultural immersion. On the other hand, the Alliance of Slow Food Cooks has a lot of similarities with Xisto Week, which is developed by ADXTUR.

In short, tourism is not the solution to all the problems of low-density territories, but it can be an excellent ally in mitigating depopulation if the development of these interior territories is properly planned in conjunction with other sectors of the economy. From this perspective, it can prevent traditions from disappearing, reduce the isolation of the population, and recover ancient knowledge of artisanal techniques to value the ways of life of local communities and make the experiences of tourists unique.

For these reasons, it is essential that territories and low-density regions be analyzed and viewed in a different way - that is, by exploring the added value and strengths of each region to transform them into strategic forces through collaborative network processes that involve local communities. The articulation of the interior with the coastal urban centers can help create a better spatial distribution of the population and reduce the regional disparities in the country.

Local development associations, even though they have few resources, have performed remarkably well with local communities, stimulating pride in living in these territories, raising self-esteem, and creating a sense of belonging. Sometimes these are determinants that prevent the total abandonment and forgetting of these low-density territories.

REFERENCES

Aasestrela (2021). *UFCD 0244 – Execução de Peças de Olaria.* https://www.aasestrela.com/site/index.php?option=com_content&view=category&layout=blog&id=1&Itemid=46

Aldeias de Montanha. (2021). *Aldeias de Montanha.* https://www.aldeiasdemontanha.pt/pt/aldeias/

Aldeias do Xisto. (2018). *Aldeias.* https://aldeiasdoxisto.pt

Aldeias Históricas de Portugal. (2021). *Planear viagem.* https://aldeiashistoricasdeportugal.com/planear-viagem

Bac, D. P. (2012). From Slow Food to Slow Tourism. Department of Economics, University of Oradea, Faculty of Economics.

Balletto, G., Milesi, A., Ladu, M., & Borruso, G. (2020). A dashboard for supporting slow tourism in green infrastructures. A methodological proposal in Sardinia (Italy). *Sustainability (Switzerland), 12*(9), 3579. Advance online publication. doi:10.3390u12093579

Chão do Rio. (2018). *Chão do Rio.* https://www.chaodorio.pt/pt-pt

CIMBSE-Comunidade Intermunicipal das Beiras e Serra da Estrela. (2021). *Quem Somos.* http://cimbse.pt/apresentacao/quem-somos

Costa, R., Costa, C., Brandão, M., & Breda, Z. (2014). *Turismo nos Países Lusófonos: Conhecimento, Estratégia e Territórios* (Vol. 1). Escolar Editora.

d'Italia, A. B. A. (2018). *L'ASSOCIAZIONE*. https://www.borghiautenticiditalia.it/lassociazione

Dickinson, J., & Lumsdon, L. (2010). *Slow Travel and Tourism*. Earthscan. doi:10.4324/9781849776493

DN – Diário de Notícias. (2018). *Ciclo "12 em Rede - Aldeias em Festa" anima aldeias históricas até dezembro*. https://www.dn.pt/portugal/ciclo-12-em-rede---aldeias-em-festa-anima-aldeias-historias-ate-dezembro-9256131.html

Ferreira, P. F. (2014). *Slow cities (Cittaslow): Os espaços urbanos do movimento slow*. Academic Press.

Fondazione Slow Food. (2018). *Fondazione Slow Food - Salva la biodiversità, salva il pianeta*. https://www.fondazioneslowfood.com/it/

Fullagar, S., Markwell, K., & Wilson, E. (2012). *Slow Tourism. Experiences and Mobilities*. Channel View Publications. doi:10.21832/9781845412821

Godinho, A. M. (2017). Prefácio. In A Viagem que Ainda Ninguém Fez. Por 49 Aldeias – Maravilha de Portugal. Edição 7 Maravilhas.

Governo da República Portuguesa, X. I. I. (2020). *Guia Fiscal do Interior*. https://www.portugal.gov.pt/pt/gc22/comunicacao/documento?i=guia-fiscal-do-interior

Guia da Cidade. (2008). *Na Rota do Artesanato da Serra*. https://www.guiadacidade.pt/pt/art/na-rota-do-artesanato-da-serra-19679-05

Guia de Viagem Foge Comigo. (2014). *Aldeias Históricas de Portugal*. Foge Comigo.

Guiver, J., & McGrath, P. (2016). Slow tourism: Exploring the discourses. *Dos Algarves: A Multidisciplinary e-Journal, 27*, 11-34. DOI: doi:10.18089/DAMeJ.2016.27.1

Guiver, J., McGrath, P., & Torkington, K. (2016). Special Issue Slow Tourism [Editorial]. *Dos Algarves: A Multidisciplinary e-Journal, 27*, 5-10. . doi:10.18089/DAMeJ.2016.27.0

INE – Statistics Portugal. (2019). *Annual Provisional Resident Population Estimates*. https://www.ine.pt/xportal/xmain?xpid=INE&xpgid=ine_indicadores&contecto=pi&indOcorrCod=0008273&selTab=tab0&xlang=pt

Marić, D., Leković, K., & Tomić, S. (2021). Slow Tourism Consumers´s Recommendations. *Teme*, 1429-1440.

MêdaC. M. (2018). *A Queijaria*. https://cm-meda.pt/diretorio/a-queijaria

Olímpio, M., Baltazar, M., & Grosso, L. (2013). Empreendedorismo em territórios de baixa densidade. In *Futuro do Mundo Rural em Questão: Atas do I Congresso de Estudos Rurais do Norte Alentejano*. Estudo Rurais.

Osorio, S. (2020). Review of slow tourism, food and cities slow tourism, food and cities: Pace and the search for the "Good Life". *Tourism Geographies, 22*(4-5), 4–5, 874–876. doi:10.1080/14616688.2019.1618907

Özdemir, G., & Çelebi, D. (2018). Exploring dimensions of slow tourism motivation. *Anatolia, 29*(4), 540–552. doi:10.1080/13032917.2018.1460854

Pecsek, B. (2016). Revitalizing tourism in small regional towns through folklore-driven slow tourism: The example of Matyó land, Hungary. *Dos Algarves: A Multidisciplinary e-Journal, 27*, 94-119. . doi:10.18089/DAMeJ.2016.27.6

Pró Raia – Associação de Desenvolvimento Integrado da Raia Centro Norte (n.d.). *Guarda*. Pró-Raia / Câmara Municipal da Guarda – Sec. Turismo.

Sabugal, C. M. (2018). *12 em Rede Aldeias em Festa, AHP_Marialva_Castelo Novo_Linhares da Beira*. https://www.cm-sabugal.pt/12-rede-aldeias-festa-ahp_marialva_castelo-novo_linhares-da-beira_/#more-16690

Santos, L. (2017). Prefácio. In A Viagem que Ainda Ninguém Fez. Por 49 Aldeias – Maravilha de Portugal. Edição 7 Maravilhas.

Serdane, Z., Maccarrone-Eaglen, A., & Sharifi, S. (2020). Conceptualising slow tourism: A perspective from Latvia. *Tourism Recreation Research, 45*(3), 337–350. doi:10.1080/02508281.2020.1726614

Shang, W., Yuan, Q., & Chen, N. (2020). Examining structural relationships among brand experience, existential authenticity, and place attachment in slow tourism destinations. *Sustainability (Switzerland), 12*(7), 1–23. doi:10.3390u12072784

Silva, A., Lima, F., & Chamusca, P. (2010). Estratégias de eficiência colectiva em territórios de baixa densidade: reflexões a propósito do Minho-Lima e do Tâmega. XII Colóquio Ibérico de Geografia. Porto: Faculdade de Letras da Universidade do Porto.

Simões, O., & Cristóvão, A. (2003). *Turismo em Espaços Rurais e Naturais*. Edições IPC.

Vivo, D. (2018). *Viver no interior vai aliviar IRS*. https://www.dinheirovivo.pt/economia/viver-no-interior-vai-aliviar-irs-12795759.html

White, P. (2011). *Equitable Access: Remote and Rural Communities "Transport Needs"*. International Transport Forum. Discussion Paper No. 2011-19.

Zielińska-Szczepkowska, J. (2020). Slow Tourism – The Fashion for Slowness or a Conscious Choice? *Folia Oeconomica Stetinensia, 20*(2), 468–483. doi:10.2478/foli-2020-0060

KEY TERMS AND DEFINITIONS

Beiras and Serra da Estrela Territory: It is a predominantly low-density territory, made up of numerous villages, some of rare beauty. It consists of the following municipalities: Almeida, Belmonte, Celorico da Beira, Covilhã, Figueira de Castelo Rodrigo, Fornos de Algodres, Fundão, Gouveia, Guarda, Manteigas, Mêda, Pinhel, Sabugal, Seia, and Trancoso.

Low-Density Territories: Are usually associated with the inland territory (with fewer than 50 inhabitants/km2). They are characterized by the existence of heritage elements of high value, that are, most of the time, unexplored.

Rural Areas: Have unique characteristics and are places of rare beauty that combines exceptionally with the development of slow tourism.

Slow Tourism: The joy of discovering the place at the right pace, valuing its uniqueness. It is a different tourism choice that contributes to a better world, also characterized by the pleasure of discovery, learning, and sharing.

Slow Travel: At its core means relaxation, traveling between the area of residence and the destination is no longer stressful, and the visitor enjoys the trip and observes how the landscape changes along the way, is having time to engage with people, and to enjoy both the trip and the destination, rejecting quick flights and getaways and opting instead for bike rides or even walking.

Chapter 8
Slow Tourism in the Sustainability of Local Culture

Zeynep Çokal
Nevsehir Haci Bektas Veli University, Turkey

Nilgün Demirel
Igdır University, Turkey

ABSTRACT

The essence of slow tourism is the sustainability of the locals. Slow tourism includes tourism activities by visiting slow cities, tasting slow food, and enjoying the moment, avoiding the worry of uploading photos to social media accounts while experiencing all these. Slow tourism is based on the need to experience travel within the framework of sustainability and to prioritize the locals during these travels. What is important in this tourism activity is not how long a destination is spent or how many destinations are visited, but how productive time is spent at the destination and the emotions that emerge afterward. In addition to these, many practices are carried out to ensure the sustainability of local culture, especially in slow cities, within the scope of slow tourism. Practices such as the continuation of local handicrafts, ensuring the continuity of local life under the influence of ancient civilizations, and sustainability of the local architectural style are included in slow city destinations. The essence of all this is to ensure the continuity of the local culture.

INTRODUCTION

Local culture expresses the essence of societies and all their values from the past. Carrying the traces of the past into the future and ensuring the continuity of local cultures has become a very important issue today. Local cultures create their specific regions around the world to reflect the identity of the group to which they belong. Local culture reflects all their experiences from the past to the present, which affects the lifestyle of local people. The disappearance of local culture under the effect of globalization reduces local differentiation in the world. However, local distinction is one of the main attraction of tourism. Also tourism is the key point of intercultural communication. Recognition of different cultures, intercultural

DOI: 10.4018/978-1-7998-7339-6.ch008

transfer and interaction are the main goals of tourism. People set trips to get to know and experience the local culture in different regions. Local culture encompasses everything local communities have, from tangible and intangible heritage elements to the natural environment. All these also constitute the sources of tourism supply. Many alternative types of tourism, especially cultural tourism, focus on experiencing authentic experiences. Individuals who want to escape crowded cities tend to prefer destinations where calm is greater and locality is maintained. For this reason, the preservation of local culture is also very important for tourism.

The sustainability of tourism resources has become one of the main problems of today. Being sustainable in the world is one of the most important criteria for buying or traveling, while in tourism, this is reflected. Many issues, such as the sustainability of destinations, sources of supply, and tourist attractions, are on the agenda of tourism researchers. As solutions and ideas arise on these issues, sustainability philosophies adopted in different areas are reflected in tourism and adopted in this area. One of these sustainability philosophies, which is reflected in tourism, is the slow tourism movement.

The slow movement began with the slow food movement in order to ensure the sustainability of eating habits and to respond to the issue of fast food. The Slow Food philosophy of time then the slow-motion slow city (CittaSlow), slow living, slow trade, slow money, slow art, slow design, slow like reading the lower branches spread (Markwell, Fullagar and Wilson, 2012: 228). Its slow movement is a movement that emphasizes the need to devote more time to achieving results rather than aiming to slow people's pace and actions. In this movement, the purpose is to ensure that time is lived to the fullest, that everything consumed is sustainable, and that everything original remains original and is passed on to future generations (Jang and Jung, 2015). After the philosophy that emerged in the form of the slow movement, it also became widely used in tourism and was a very important factor, especially for the sustainability of destinations. The preservation and sustainability of the local culture, which is very important for the sustainability of tourism, is supported by this philosophy. As a reflection of this, slow tourism has emerged as an alternative type of tourism. The essence of slow tourism is local sustainability. Slow tourism covers tourism activities by visiting slow cities, tasting slow food, and enjoying the moment by moving away from the anxiety of uploading photos to social media accounts while experiencing all this. Like all other formations that adopt the slow philosophy, slow tourism emphasizes the need to stay in the moment and enjoy travel, overcoming the anxiety of time. The slow tourism movement, on which sustainability is based, is supported by practices such as slow city and slow food. Slow food is based on the sustainability of local flavors. As with the slow food movement, slow cities focus on good food, healthy environments, sustainable economies, preserving local and traditional culture, and argue that life should be spent comfortably, slowly, enjoyable, and enjoyed (Yilmaz and Çokal, 2019). Slow food and slow city philosophies constitute the supply sources of slow tourism.

Slow tourism has six sub-dimensions: interaction, authenticity, sustainability, slowness, time, and emotion, which are also associated with these experiences. In order for a tourism activity to be considered as slow tourism, it should cover these six dimensions (Dall'Aglio, 2011).These dimensions of slow tourism represent the quality of the time spent in the destination rather than the quantity, which is the essence of the slow tourism movement.

Slow tourism, which is based on environmental and economic sustainability and adopts calm at its core, includes elements such as slow travel, slow tour, slow transport and slow tourist (Fullagar, Wilson and Markwell, 2012; Dickinson and Lumsdon, 2010). Considering these elements, slow travel includes travel to destinations where local cultures live, slow transport involves the use of means of transport that do the least damage to nature. Slow tours qualify for tours to destinations where the local is protected.

With the development of slow tourism day by day, travel businesses have started to organize tours in order to serve people who have adopted the slow philosophy and want to travel slowly. Slow tours organized by travel businesses are alternative tourist tours and programs organized with the aim of getting to know, taste and participate in local life and local products (Slowmotionporto, 2018). In addition, many applications are made within the scope of slow tourism in order to ensure the sustainability of local culture, especially in slow cities. Practices such as the effort to maintain local crafts, ensuring the continuity of local life where the influence of ancient civilizations continues, and the sustainability of local architectural style are taking place in slow destinations. At the core of all this, it is aimed to ensure the continuity of local culture.

The sustainability principle, which is one of the basic principles of slow tourism, includes ensuring the sustainability of local culture and transferring elements of culture to future generations. In this context, slow tourism is an alternative type of tourism that has a very important place in the sustainability of local culture. Therefore, in this section, the relationship and sustainability of slow tourism with the local culture will be discussed, first of all, the relationship between local culture and tourism and slow tourism and sustainability will be discussed. The study was prepared by taking into account the concept map, which is a step of the qualitative approach. Phenomenologically, it is aimed to compile the literature of slow tourism in the sustainability of local culture and to create the main titles according to the most mentioned themes. For this reason, the main concepts were determined from a critical and analytical point of view. Technically, literature review is preferred. All sources in the literature on the subject were reviewed and the main titles in which slow tourism ensures the sustainability of local culture were compiled and detailed thematically. Also, the sustainability of the local; folklore/authenticity, local gastronomy/ food identity, local festivals, local handcrafts, sustainable environment, sustainable infrastructure, sustainable architecture in terms of slow tourism title are included. Finally, in terms of the sustainability of local culture in Turkey, applications made in slow tourism destinations are discussed.

LOCAL CULTURE

Culture; is a concept that includes many components such as art, history, language, religion, lifestyle, and architecture. There are factors such as recognizing a culture, gaining different experiences, learning the culture of the country in destinations as the travel motivation of tourists. Conservation of cultural heritage, which is considered based on cultural tourism and brings local cultural elements to the fore, produces many positive results. This stands against globalization, especially economic, and creates a unique value so that not all destinations are alike (Meydan Uygur and Baykan, 2007). In the globalizing and resembling world, tourists move with the desire to experience different cultures and daily life forms. They are interested in intangible local cultural assets and are no longer interested in the built cultural heritage of the past, but in the local culture that brings the past culture to the daily life experiences. With the protection of tangible and intangible cultural heritage and the emergence of new elements specific to the regions, tourism types based on intangible local cultural assets exhibit a rising trend in tourism today. Many countries mobilize abstract local cultural assets in developing their underdeveloped regions (Çağlı, 2012). The slow tourism movement is one of them. It is based on the necessity of experiencing a travel experience within the framework of sustainability and keeping the locals at the forefront during these travels. Tourists participating in the slow tourism activity establish meaningful and sincere relationships with local people and nature, enjoy their lives, and gain small joys and experiences. What is important in this tourism activity is not how long a destination is spent or how many destinations are visited, but how productive time is spent at the destination and the feelings about the local culture that emerge afterward.

The concept of local means to have common features within a community. The identity of a place and its relationship with society is important. As to the identity of a community and how it can be defined is identified with bridges between its past and present. Since healthy and happy people living in the desired environment are more likely to contribute to their communities, healthy environment must please and elevate the spirit of the people by developing local identity (Sassi, 2006). Commonly the term local culture is used to characterize the daily life experience in specific regions. It reflects the qualities that define the idiosyncratic emotions, personal preferences, and changing tastes of ordinary people. Given the strength of local cultures, accepting the existence of an inclusive global culture is nowadays difficult (Watson, 2020). As with the contradictions in many sectors in general, the global integration paradox suggests that two opposing perspectives need to be addressed. In the 1970s, it is recommended that we think globally and act locally on environmental trends. Many industries prefer the ease of thinking globally over the challenge of thinking locally. It is much more difficult to act locally; doing so requires local insight (Bruner, 2004).

If tourism cannot be developed with plans based on sustainability principles in the context of the continuity of cultural heritage, it also causes negative effects on the cultural assets of the regions. As a result, local cultural assets deteriorate, become alienated, and lose their originality. Because today's tourist wants to be in the natural environment and to know the local cultural assets. Countries showing this sensitivity will have the opportunity to direct this touristic demand to their own countries. In this context, natural and cultural assets can be used as a marketing tool as it is an important factor that causes today's tourists to participate in tourism activities. Each area inhabited has a unique cultural view and cultural structure. In other words, culture can differ in each country, and even within the same country, it is possible to talk about its differences and originality among themselves (Meydan Uygur and Baykan, 2007).

THE RELATIONSHIP OF TOURISM WITH LOCAL CULTURE

Elements such as natural environment, climate, cultural heritage, and local culture, whose value is understood after fast life and mass tourism movements, have taken their places among the subjects discussed in tourism. In fact, items specific to sustainability that come to the fore with grobalization have become the product of tourism indexed with sales and profit. One of them is the marketing of local culture. Rather than being a marketing target, local culture has enabled it to be made an element of originality within slow cities, and even to revive old traditions and customs. With the guiding philosophy of slow tourism, interest in destinations continues to develop within the framework of sustainable development, ecology, and locality (Dickinson and Lumsdon, 2010).

As the level of communication between tourists and local people increases, local people can be influenced by the cultures they encounter. Changes occur in lifestyle, eating and drinking habits, dressing of the society, and they can shape their life according to the culture they are influenced. However, they can adopt the lifestyle of the tourists by moving away from their religious beliefs and moral values. This situation may lead to the destruction of traditional art, customs, values, namely to traditional culture (Meydan Uygur and Baykan, 2007). The tourism industry has been accused of commodifying local culture and heritage as a product of attractive holidays. In addition to being held responsible for the negative effects of mass tourism on local communities and strengthening the social distinction between tourists and locals, the word 'tourist' has negative connotations (Heitmann, 2011). While tourism often has similar negative examples in local communities, there are ways in which potential problems can be recovered. With the participation of local people and the right policies, tourism has the potential to act as a tool for sustainable community development and poverty reduction (Beeton, 2006). Slow tourism can create positive results in local communities, such as new experiences, quality time. Unlike mass travel, slow travel and individual movements can make the experience of both the host community and the tourist more different and special (Wearing, Wearing, and McDonald, 2012).

Local culture is seen not only as a tourist attraction but also as a resource from which community-based tourism development can be built (Giampiccoli and Kalis, 2012). The development of community-based tourism in a destination depends primarily on the high and active participation of local people in tourism planning, development, and management. However, it depends on local people's cooperation with other stakeholders, preserving natural and cultural touristic attractions and transferring them to future generations (Tasci et al, 2013). Globally, in the general tourism sector, there is an increasing focus on community-based tourism as a means of enhancing community development, poverty reduction, cultural heritage, and conservation. Community-based tourism's commitment to local culture stems from its respect for local culture, minimizing the effects of tourism, and evaluating attractions in the context of sustainability and cultural heritage (Giampiccoli and Kalis, 2012). It is a managerial issue that needs to be emphasized that the local culture, which is the subject of tourism, is planned correctly and based on sustainable principles in the places where the communities live, and thus preserved and transferred to the future. As part of this process, microsocial factors need to be analyzed in the conceptualization of touristic destinations. It is important to correctly evaluate the social value, tourist, and local cultural elements that arise in line with the interactions and to identify the negativities. It should not be forgotten that; local value increases depending on tourist-local people interaction, communication, and relationships (Wearing, Wearing, and McDonald, 2012).

SLOW TOURISM AND SUSTAINABILITY

Faster lives have begun to be lived for development to be rapid. To keep up with the pace of development, the acceleration of life has also had negative effects on humanity and nature. Speed-oriented life has also brought with them rapid consumption. Fast life and fast consumption, which is one of the biggest benefit of globalization and modernization, have become a problem for people and societies over time. As people move faster, their living spaces have also started to be designed by this speed, and everything designed by speed has led to the need to replace it with a new one over time (Yılmaz and Çokal, 2019: 204). If consumption is fast, the lack of production at the same speed, the inability of natural resources to renew themselves at the same speed has caused the danger of depletion of resources. All this has caused significant problems with sustainability.

Sustainability is an approach in which the environment is featured when it is defined as controlling the negative effects of man on the ecosystem so that diversity and non-renewable resources can be transferred to future generations (Keskin, 2012:94). But the sustainability of everything involved in human life, not just the environment, is a major problem. The sustainability of food, cities, culture, and, most importantly locality have become very important. With the emergence of these problems, solution-oriented approaches have started to come into question. One of these approaches has been slow movement.

The slow movement is a movement that arose in response to the fast life that emerged as a result of globalization. While its slow movement has its influence in all areas of society, the tourism area has also come under this influence. The concept of slow tourism is relatively new, but its origin is based on the slow food movement like other slow movements. The main goal of slow tourism is sustainability, as in the slow food and slow city movements that began in Italy in the 1980s. The understanding that slow tourism is much more than slow mobility or transport between destinations over time has begun to prevail (Dickinson and Lumsdon 2010: 70). When slow tourism first appeared, it was aimed at reducing environmental pollution, especially through the use of low-carbon emission vehicles, but also at tourism mobility to destinations that can save on depleted energy sources. While the focus of slow tourism was initially only environmental sustainability, over time it has evolved into a sense of travel that provides personal satisfaction and happiness, spreads, and is enjoyed by living in the moment (Parkins and Craig, 2009). For this reason, slow tourism has been a highly connected concept with sustainable tourism. At the heart of slow tourism is the philosophy that areas with rich cultural structures and attractive places should not undergo any physical changes and protect the existing structure intact (Ünal, 2016: 21).

The slow tourism is based on the idea of having a travel experience within the framework of sustainability and the need to keep the locality at the forefront in the process of these trips. Tourists who participate in slow tourism activities establish meaningful and intimate relationships with local people and nature, enjoy their lives, and get small happiness and experiences. In this tourism activity, it is not important how long you stay in a destination or how many destinations you visit. The main thing is how productive time is spent at the destination and the emotions which occur after that (Yılmaz and Çokal, 2019: 211).

As mentioned before the slow tourism has six sub-dimensions: interaction, authenticity, sustainability, slowness, time, and emotion, which are also associated with these experiences. For a tourism activity to be considered as slow tourism, it must cover these six dimensions. These dimensions of slow tourism represent the quality of the time spent in the destination rather than the quantity, which is the essence of the slow tourism movement. Among these dimensions, sustainability is the part that also forms the essence of philosophy. Sustainability means ensuring local continuity and eliminating the negative effects of tourism activity on the local environment. The concept most emphasized by slow tourism is sustainability. It demonstrates the need for a sustainable approach, that is, the need to be ecologically light, economically good, ethically, and socially fair to local people in the long term (Dall'Aglio, 2011).

The goal of slow tourism is for destinations to develop within the framework of sustainability. The main goals of destinations that have received the title of the slow city, in particular, are to ensure sustainable development with regional values, ensure income justice by contributing to local economic life, and protect the natural environment and historical values (Yurtseven et al, 2010:40). Given all this, slow tourism is a type of tourism based on sustainability, which has emerged with the philosophy of slow tourism, paying attention to the sustainability of the local rather than mass tourism movements.

SLOW TOURISM IN LOCAL SUSTAINABILITY

Natural and cultural attractions stand out when tourism attractions are considered. For tourism to be sustainable in destinations, these resources must also be sustainable. In this context, sustainability is not just a problem related to nature. The sustainability of the culture and its transfer to future generations is also very important. It is especially important to make local cultural values and practices sustainable.

Before addressing the relationship between slow tourism and sustainability, it is worth mentioning the concept of glocalization. Glocalization means, global and local integration is the creation of dissimilar settlements in different geographical areas. For this reason, globalization can reveal abstract and concrete consequences, such as the creation of heterogeneous and hybrid cultures, identities, and communities. Considering local, social and cultural features and traditions together in destinations can enable destinations to reach the globe by representing the local. This refers to the marketing of destinations with an emphasis on the points that distinguish them from the other (Ritzer, 2004). Concerns about the long-term effects of capitalism-based development driving growth and the homogenizing effects of globalization have led to important movements such as the slow city movement encouraging alternative strategies. Emerging movements in small cities and towns focus on locality, the mediocrity of the place, sustainable and local economies (Servon and Pink, 2015).

Respectively, these can be understood as culture "in", "for" and "as" sustainable development, a result of cost-action research on cultural sustainability. Three roles devise different perspectives for culture's integration in the sustainability paradigm (Dessein et al., 2015).:

1. Culture "in" sustainable development, with a supportive and self-promoting role adding culture as a self-standing fourth pillar;
2. Culture "for" sustainable development, with a more influential force that can operate beyond itself, which balances the existing pillars and guides sustainable development between their pressures and needs;
3. Culture "as" sustainable development, with an even more fundamental role as the essential foundation and structure for achieving the aims of sustainable development since it integrates, coordinates, and guides all aspects of sustainable action.

Furthermore, according to Soini and Birkeland's (2014) analysis of the scientific discourse on cultural sustainability, seven principle story-lines can be identified: heritage, vitality, economic viability, diversity, locality, eco-cultural resilience, and eco-cultural civilization (Nunes, Söderström and Hipke, 2017).

As traditional knowledge is integral to the cultures and identities of indigenous peoples, language use, ceremonial practices and opportunities for intergenerational knowledge transfer are important indicators. In addition, the identification and presence of species, flora and fauna at particular seasons as well as access to sites for hunting, harvesting, production of food (among other things) are indicators of whether economic development is sustainable from an indigenous perspective (Watene and Yap, 2015; 51-55).

Slow tourism is also a tourism activity that embraces the sustainability of local culture. This type of tourism is based on the consumption of local food, environmental and economic sustainability and adopts calm at its core. As traditional knowledge is integral to the cultures and identities of indigenous peoples, language use, ceremonial practices, and opportunities for intergenerational knowledge transfer are important indicators. Also, the identification and presence of species, flora, and fauna at particular seasons as well as access to sites for hunting, harvesting, production of food (among other things) are indicators of whether economic development is sustainable from an indigenous perspective (Watene and Yap, 2015; 51-55). In this context, slow tourism takes care of such local activities and includes them in tourism activities. Slow tourism focuses on and supports local sustainability, especially folklore, local gastronomy, the sustainability of the environment, and sustainability of infrastructure.

Local Identity

As the continuous interaction between destinations, and local people develops, culture allows not only the development of local identity in a cultural sense but also the creation of the identity of other resources. Many elements, such as centennial trees, houses, urban areas, eating and drinking culture, form a local identity and survive with sustainability. These elements provide society with important information and experiences about previously lived cultures and future cultures. Local identity represents a small area and represents a strong and independent image that distinguishes that place from other places. Small-scale elements of local identity come together to form the national identity of a country and create an image of a destination (Shao, 2014). Slow cities need to be supported by local governments, non-governmental organizations, local commercial support, alternative development plans, and, most importantly, a strong community awareness to create a local identity. The slow tourism movement develops depending on the true local identity. In this way, integrity is formed in which all stakeholders have a say, and a quiet city is introduced in a way that takes place in the mind of the tourist (Semmens and Freeman, 2012).

Folklore/Authenticity

It is aimed to make local uniqueness sustainable through slow tourism activities. In this context, locality and authenticity are very important. With slow tourism activities, tourists get the chance to experience the local culture in the destination. Tourists especially get the chance to identify themselves with them by participating in activities done by local people. This supports the adoption and continuity of local people and local life. What makes the destination visited within the scope of slow tourism different from other destinations can be characterized as natural features specific to that destination. Slow tourism with its authenticity dimension, destination offers visitors an experience reflecting the character of the region, encompassing local culture and traditions (Dall'Aglio, 2011). The preference for local products mobilizes local producers, encourages organic production, and ensures that these products gain value, thus causing a revival in the local economy. In addition to local products and production, slow tourism destinations are a center of attraction for tourism with their different and original styles (Keskin, 2012). Through slow tourism, there is a tourism movement that prevents uniformity (mediocrity, uniformity) by creating local - regional and cultural awareness, while making local uniqueness sustainable (Deniz, 2016). This highlights the sustainable authenticity and locality aspect of slow tourism.

Local Gastronomy/ Food Identity

The slow food movement is the starting point of slow tourism. Slow food movement is a movement that underlines the negative effects of globalization, advocates for the preservation and ownership of the local culture of every region of the world, care about quality and calm life, eating style, regardless of the time and without constant haste (Caffyn, 2007:3-10). Today, the slow food movement, which has over 100,000 members in more than 160 countries use local seeds to support local food production, authenticity protection, the economic income of the people to obtain the activity on the provision of sustainable local development by showing it is a rebellion against being the same (Kinley, 2012: 2).

Regional dishes play an important role in people's choice of destinations. It helps tourists to gain a beautiful holiday experience, but is also an effective factor in tourists revisiting the destination. It turned out that local food, which is a cultural heritage, has an importance that increases the value of the culture that the region has. Local food increases the cultural value of the region, while local food and beverages continue to be planted and harvested, and the local economy also revives and contributes to the development of employment opportunities. In this way, it is ensured that tourism is maintained along with the preservation and use of local food, which is one of the sources of cultural heritage (Çapar ve Yenipınar, 2016).

In the destinations local markets, local cheese producers, vineyards, olive oil producers, food festivals, beekeeping, cropping regions such as several that are unique to truffle varieties, as well as its own unique food culture of almost every town tourists ‘ destinations with those around you and share this experience allows you to experience the food culture (Ardıç Yetiş, 2015:13). Slow tourism places emphasis on the sustainability of local foods found in destinations. Slow tourism aims to ensure that local markets are established and restaurants serving local food are located, thus supporting local urban economies and ensuring their sustainability.

Local Festivals

Slow tourism activities also support local activities in the sustainability of local culture. Local festivals play a very important role in transmitting local culture. Especially in the context of slow food, many slow food festivals are held around the world. The rapid growth of slow food as internationally, whether as a theme for a stand-alone event or as a moment of elements of various events, such as ecofests, ethnic, community, sustainability or food festivals, makes it a useful frame for delivering persuasive messages via a discussion of the role and desirability of the event (Frost and Laing, 2013: 68). Since slow food promotes an idealized future world, slow-oriented festaivals can be popularized. These festivals showcase and highlight distinctive aspects of local heritage and local culture (Caffyn, 2012). In addition, slow concept festivals are organized within the scope of slow movement. Festivals are held where local products are exhibited, sometimes with room for local dishes. Tourists participating in slow tourism activity have different motivations. One of these elements of motivation is escape motivation. With this escape motivation, slow tourists also want to escape the intensity of city life and participate in local festivals (Özdemir and Çelebi, 2018). Local festivals allow slow tourists to promote and experiment with both local products and local culture. In this context, local festivals, slow tourism and sustainable local culture meet on a common ground.

Local Handcrafts

The philosophy of slow tourism also adopts principles for the preservation of these arts. The main features and principles that should be found in a slow city, which is a slow tourism destination, have been identified. At the beginning of these are the creation of calmer and less polluted physical environments; the preservation of local aesthetic traditions, local crafts and local cuisine (Unal, 2016). Among the criteria for being a quiet city, the requirements for the preservation of local crafts are important. It is expected that calm cities will develop programs for the protection of crafts that are in danger of extinction, and projects for the protection of traditional professions that are in danger of disappearing. However, in cities that are members of the slow city network, the quality of products, goods and crafts produced by artisans and artisans must be certified (Cittaslow, 2021). In this way, these crafts are protected and standard prescriptions are created. This, in turn, ensures that crafts that are in danger of extinction are passed on to future generations. For this reason, slow tourism in the sustainability of local culture and slow cities, especially slow tourism destinations, contribute to the preservation of handicrafts.

Sustainable Environment

In today's societies, the rapid consumption of the environment, that is, natural resources, and the inability of nature to renew itself at the same speed have become an important problem. But its slow philosophy is an understanding that has taken action to ensure the sustainability of food. Although the exit point is the sustainability of local foods, the environment has also been one of the points of importance of this philosophy. In particular, environmental criteria are at the first beginning of the criteria based on determining slow cities, which is an extension of this philosophy. Because at the point where the sustainability of the environment is not achieved, sustainability activities in other areas of life are also interrupted. However, economic growth achieved by local and organic products and production methods also coincides with the principle of "balance between environment and development" of sustainability, as it does not harm the environment (Keskin, 2012: 96).

Slow cities constitute the main destination of slow tourism activities. For this reason, destinations that are slow cities are required to meet certain environmentally-oriented criteria. Under environmental policies (Cittaslow, 2021).;

- Certification of compliance with the laws of the values indicating the quality of air, water, and soil belonging to the destination that wants to become a member of the union,
- Development of projects for urban and domestic waste collection,
- Making plans for sorting and collecting garbage and recycling it on various roads,
- Development of projects for urban and domestic waste collection,
- Making plans for sorting and collecting garbage and recycling it on various roads,
- Establishment of a wastewater treatment plant for the treatment of the city's sewage system,
- Ensuring energy production and saving energy by using renewable energy sources with plans to be prepared under the leadership of local government organizations,
- Prevention of the use of genetically modified products during agricultural activities,
- Regulation of advertising signs, banners, and traffic signs in the city under the rules established by local governments to ensure the aesthetics of the city,
- Taking measures to prevent noise and light pollution in the city,
- EMAS, ECOLABEL, ISO 9001, ISO 14000, SA 8000 international environmental management and quality standards are subject to compliance.

In this context, slow tourism aims to mobilize local dynamics by emphasizing the original values (nature, crafts, architecture) of the city and its surroundings (Keskin, 2012: 96).

Sustainable Infrastructure

As with many issues, slow philosophy, which is the starting point of slow tourism, is based on sustainability in infrastructure. Among the main criteria that slow cities, which are slow tourism destinations, must meet are infrastructure criteria. The infrastructure criteria that slow cities that adopt this philosophy must meet are divided into four basic titles. Located in the city urban to provide protection, improvement, or rehabilitation efforts in areas of historical and cultural value. These works of the city's inflexible, unhealthy and unattractive to be addressing the viability of the reuse of areas to ensure that. It gets to the city plan in the plan of reorganization of the creation of green fields and must be qualified for this area to be able to host a variety of activities (Cittaslow, 2021). As for infrastructure, one of the points that should be based on slow cities and slow tourist destinations is the subject of transport. In this context, making plans for the regulation of urban traffic and transportation issues, public buildings, schools population density is high, such as daily access to the bike path, such as the provision of alternative transportation systems to buildings, public transport instead of private vehicles to use the system needs to be extended. For this purpose, alternative systems such as buses, subways, trains, trams, bicycle paths, pedestrian paths should be established and connected (Cittaslow, 2021) and made suitable for slow tourism activities. Third, the introduction to the topic at hand; to prepare a program for the provision of the city's image and branding of the city as urban renewal, which fulfill the function of the introduction of the program offices and the office Cittaslow prepared for information on integrating with the city produced in natural or centers need to be established for the provision of the promotion and marketing of local products (Cittaslow, 2021). Finally, are substances that are prepared for disadvantaged groups living in the city; the elimination of factors that make difficult the transportation of the disabled in public areas, pavement, stairs, and paths of the application to be an obstacle with the introduction of various architectural warning signs and prevention programs that will allow the development of both individual and social life (regulation of sports and entertainment activities, the provision of home care services for people in need of care, benefits to be done) should be provided with opportunities to produce (Cittaslow, 2021). All these criteria are intended to be provided, especially transportation, to make all infrastructure activities suitable for slow tourism and to ensure that the local area does not deteriorate. In particular, issues such as ensuring the sustainability of urban architecture without deterioration are very important both in terms of maintaining tourism attractiveness as a destination and in terms of transferring it to future generations. When these criteria are met, the sustainability of slow tourism destinations and the continuity of tourism activities will be ensured.

Sustainable Architecture

Sustainable construction and architecture in tourism play a vital role in the conservation and development of the region. On the other hand, it is known that for the growth of tourism, it must have a capacity in terms of infrastructure. Buildings that allow the tourism industry to flourish must reflect the natural resources and identity of local communities. Sustainable architectural structures should include elements of satisfaction while observing the comfort of both the visitor and the region. Reflecting on the architectural elements of real culture and transferring them to the future by preserving them will shed light on the sustainability of architectural culture in the future. For this, it is necessary to adopt slow city strategies and rules (Feio and Guedes, 2015).

The buildings that are the subject of tourism and are used have a significant impact on the natural environment and social fabric of the society. Sustainable architecture helps calm urban practices and even promotes a sustainable lifestyle. How buildings are designed and built in the context of making a positive contribution to the sustainability agenda to reach economically strong, socially inclusive, stable communities while minimizing the impact on the environment (Sassi, 2006).

The postmodern architecture includes organic approaches to the construction of new buildings. This means structures built taking into account local, environmental and historical features. In the context of global tourism, local capture is also possible with buildings, allowing visitors from many countries to contact the community and therefore intercultural interaction. An element of local character, which is one of the elements that reflect local identity, is the key to sustainability, protecting buildings and emphasizing their importance, preferring globalization rather than grobalization (Ritzer, 2004).

THE IMPLEMENTATIONS FOR THE SUSTAINABILITY OF LOCAL CULTURE IN SLOW TOURISM

It is very important for the continuity of these cultures that the local culture is sustainable and carried to future generations. In this context, activities are carried out in many areas and local culture is tried to be preserved. Some activities blend local culture with modern culture, bringing it into harmony with today's life. Some activities, on the other hand, focus only on local culture and ensuring its continuity without disrupting this culture. Slow tourism activities are also based on activities aimed at preventing the deterioration of this local culture. In particular, it strives to ensure the sustainability of local products, local crafts, and local food by protecting destinations where local cultures are not disrupted. In slow cities, which are slow tourism destinations, many activities are carried out in this context and local culture is tried to be preserved. There are examples of applications related to these activities in Turkey.

Slow Tourism and Sustainable Local Cultural Implementations in Turkey

By 2021 there are 18 slow tourism destinations in Turkey. A different local identity comes to the fore in each of these destinations, and different examples are seen in the context of preserving the local culture. Of these examples, the most notable are; producer markets and Seed Exchange Festival established in Seferihisar, the slow city capital of Turkey, to ensure the sustainability of local products, organic farming activities in Çanakkale Gökçeada, Taraklı houses reflecting local architectural examples in Sakarya Taraklı, Trak village project reflecting local life example in Kırklareli Vize, continuing the touch of "örtmeler"[1] which are traditional handicrafts in Bolu Göynük, it is a project of teaching traditional wood carving to young people in Sinop Gerze and reusing tile and ornamentation art in architecture in Bitlis Ahlat (Cittaslow Türkiye, 2021).

Seferihisar became the first slow city in Turkey by taking the title of Cittaslow in 2009. Because of this, cittaslow is considered the capital of Turkey. Because of this, cittaslow is considered the capital of Turkey. Many projects have been carried out to be a slow city and to ensure its continuity. But two projects, especially to support local production and be sustainable, have attracted a lot of attention. Primarily within this scope, producer markets were established to develop and support local production. To open stalls in these markets, only the products produced in Seferihisar must be sold. Thus, producers of Seferihisar can reach consumers without intermediaries. Another project is the "Seed Exchange Festival". As a result of the law that came into force, the ban on the sale of domestic seeds has endangered the existence of thousands of varieties of vegetables and fruits that have existed on Anatolian soil for centuries. For this reason, Seferihisar Municipality organizes "Seed Exchange Festivals" for the protection of seeds, which are one of the most sacred legacies. In this way, after the sale of domestic seeds was banned by the "Seed Law", producers share the seeds they have stored in the method they have practiced for centuries (Cittaslow Türkiye, 2021, Seferihisar Municipality, 2021).

Gökçeada is a district located in the westernmost part of Turkey in Çanakkale province. Since 2002, it has become an important slow tourism destination with its importance to organic agriculture and its projects in this context. The district, which received the title of slow city in 2011, attaches great importance to local production and organic agriculture within the scope of sustainability. In this context, Gökçeada is the place where the first project model consolidated with the "Organic Agriculture Project" is implemented. Organic agriculture first started in olive groves in Gökçeada and over time its scope was expanded and honey, wine, and dairy products were produced by organic agriculture. Honey is one of the important organic products in Gökçeada, which offers unique opportunities in terms of its natural geography and the production of healthy products. During archaeological excavations on the island, vine saplings are distributed in Gökçeada, where grape-kernels dating back three thousand years from today, to develop viticulture and wine production. In addition to the production of household wines that have been going on for centuries in Gökçeada, barba yorgo, etis and timbre produce wine from local and foreign grape types. Gökçeada is one of the places where dairy products, whose production is limited among organic agricultural products, are produced for the first time in Turkey. Milk obtained from animals fed with organic feed is used in the production of feta cheese, cheddar cheese and goat cheese, as well as yogurt (Cittaslow Türkiye, 2021; Çanakkale General Directorate of Agriculture and Forestry, 2021).

Taraklı District of Sakarya province is a place that has been registered as a slow tourism destination by taking the title of cittaslow in 2011. Taraklı houses, which reflect the local architectural examples found in Taraklı, are one of the most important parts of the local culture of the destination. The most important feature of scalloped houses is that there are planning and design by the traditional family type in which three generations live together. In addition to being a traditional Turkish house, Taraklı every has three main features as "conservation of resources", "building life cycle" and "livable design" in terms of sustainability. Preserving traditional residential architecture, which is our cultural heritage, is important in terms of carrying our architectural values to future generations. The preservation of historical urban texture features of Taraklı city is also possible with sustainable conservation. Taraklı destination is trying to successfully transfer the property of being a city with traditional architecture that has preserved its existence for many years to the future (Turgay and Erkuş Buyruk, 2017: 38, Cittaslow Türkiye, 2021). Sustainability, which is also the essence of slow tourism, aims to make the local architecture of the destinations sustainable and attaches importance to this issue.

The only slow city of Thrace region is the Vize district of Kırıkkale province. Like other slow tourism destinations, many sustainability-based projects have been put forward in Vize county. But one of the most important projects that focus on the sustainability of local culture in the region and attract attention is the ancient Trak Village project, which reflects local life. As part of the project, it is aimed to revive the ancient Trak period and give the region an identity. The overall objectives of the project are to design under the life of the ancient Trak village, to contribute to the development of historical and cultural tourism, to bring the rich and multicultural historical heritage of the Vize to the attention of our country and Universal stakeholders, to protect the existing property and cadastral structure, to protect the plant tissue in the current situation and to use natural and local materials suitable for the historical tissue in practice (Ahmet, 2012; Cittaslow Türkiye, 2021).

Göynük district of Bolu, a city famous for its cooks in the Central Black Sea Region of Turkey, was awarded the title of slow city in 2014. Göynük's area of sustainability of the local culture has been crafts. Continuing the touch of "örtmeler", which are traditional crafts in Göynük, is very important in terms of sustainable local culture. "Örtmeler", which have characteristics specific to the Göynük region and are called buckled covers, are also widely used by local women today. The edge embroideries and buckles are completely handcrafted. It is used as a hijab in its neighborhood and as accessories and clothing throughout the country. On "örtmeler" used colored silk and cotton yarns until 30 years ago, while today orlon is used in patterns and wicks (Soysaldı, 2010: 162). For this rare craft to be sustainable, the "Local Craft Education Center" serving within the municipality of Göynük ensures that the crafts that face oblivion are passed on to future generations and that cultural values are maintained. In this center, women weave geographically marked buckle covers that are unique to Göynük and are registered by the Turkish Patent Institute. In this context, various such projects are implemented in slow cities, which are slow tourism destinations (Cittaslow Türkiye, 2021).

The Gerze District of Sinop, which received the nomination of Cittaslow in 2016, basically adopts the philosophy of sustainability in all areas, while especially crafts stand out more in this regard. Activities are carried out in Gerze to teach traditional wood carving to young people and to ensure sustainability. In cooperation with Gerze Municipality and ISKUR Sinop Provincial Directorate, various vocational training courses related to traditional crafts that have been forgotten are given and economic contribution is made along with the development of the dexterity of the people who have graduated from these courses. In the established workshops, vocational training courses in wooden toys, wooden ship model construction, wood carving are regularly given, and products produced are exhibited and marketed. Especially developing the hand skills of local women and young people, ensuring that they are brought into the economy is one of the goals of the project (Cittaslow Türkiye, 2021). However, the masters and trainees at the wood carving workshop established made specially designed wooden signs for the urban aesthetics and placed them in all the parks in the district. This project aims to reduce image pollution within the scope of urban sustainability (Gerzeninsesi, 2017).

One of the youngest slow cities in Turkey is the Ahlat District of Bitlis, which registered as a slow tourism destination in 2017. As in other slow cities, many different projects are being designed in Ahlat, where sustainability-based projects are planned to be implemented. Among these projects, the project of reusing tile and ornamentation art in architecture draws attention. The project aims to rework the Seljuk tiles and ceramics that form the basis of Turkish tile art, whose form and type can be determined among the finds in the excavations carried out in Ahlat, following the old form by the hands of 25 women between the ages of 18 and 29 who have been trained in the production of tile ceramics (Cittaslow Türkiye, 2021; Ahlat Kaymakamlığı, 2017). Slow cities work on the axis of sustainability in direct proportion to slow tourism philosophies. The core of these projects is to increase the sustainability and awareness of local culture. Similar and more supportive qualities of these projects made in Turkey are also applied in other slow tourism destinations around the world.

CONCLUSION

The sustainability of local culture has become even more important after globalization. Similar building trends in most of the cities on earth lead to the fact that cities are increasingly similar to each other. The most obvious effects of globalization are felt on cities and local cultures. In a way, prototype structures produced with similar designs and the same building material resemble the cities of the world, while destroying the urban identities determined by local originality (Keskin, 2012: 85). Along with globalization and grobalization, maintaining the culture of similar destinations has brought about the concept of glocalization, and it has been noticed that a healthy process of glocalization emphasizes the point where local communities and foreign cultures that enrich them to merge. However, it has been learned that any process that negatively affects their traditions or identities must be rejected.

Although globalization offers great opportunities for change and spread, it tends to eliminate the distinctive differences of local communities, and the characteristics of a society. In other words, he proposes an ordinary model that does not belong to any community, does not carry an identity (Uslu,

2009:53). These models bring with them destinations that no longer have any charm and remain out of the area of interest of tourists looking for differences. Introducing traditions in tourism, protecting local destinations, and providing traces of their identity to the world has now taken its place among tourism goals. Owning the past and cultural heritage and passing it on to future generations has become one of the sustainability goals of a nation and the destinations within that nation.

Sustainability-oriented approaches such as the philosophy of slow tourism have begun to be adopted to eliminate the negatives. Its philosophy of slow began with slow food, and it pioneered the slow urban current. Cittaslow movement the characteristics of cities that make up their traditional identities (local cuisine, local music, and folklore, traditional civic architectural texture, traditional crafts, local lifestyle, etc.) it is a long-term approach based on conservation and survival with an ecological and sustainability approach (Keskin, 2012: 86; Karakurt Tosun, 2013:231; Şahinkaya, 2010: 1; Sırım, 2012: 123-124; Ergüven, 2011: 204). Slow cities do sustainability-oriented work. Slow cities use local products, social and cultural differences, and sustainability as local economic mediators, focusing on resources and only part of the concern for environmental quality (Mayer and Knox, 2006:328).

As with the slow philosophy, the main point in the slow tourism movement is sustainability. Not only visiting quiet cities, tasting local delicacies, but also contributing to ensuring their continuity is the essence of the tourism genre. In order for the tourist to be able to come into contact with a society, destinations can take place in their minds, the existing culture must be developed intact. The element of cultural identity that is often mentioned in policy studies should be preserved and arrangements should be made in the functioning of communication processes. Both tourism stakeholders and participants in tourism activities should adopt sustainability and tourism strategies should be established based on this point. For this purpose, it is necessary to first plan sustainable tourism activities and to gain awareness of slow movement as a sub-element of this. In addition, destinations that are one of the most important elements of tourism also need to adopt the slow city philosophy. The main aim of this should not be to become a tourist attraction by taking the title of slow city, but to ensure the sustainability of natural and cultural resources in the local area. Slow tourism should not only be seen as a type of tourism and an element of supply, but the essence of slow philosophy should be adopted. It is possible to open up to the world by protecting the local. To achieve this, the creation of the right policies must consist of steps that can prevent the bad direction of similar destinations.

In addition to playing an important role in the sustainability of local cultures, slow tourism today lead to the competition for destinations. Especially with today's covid-19 pandemic, the importance of slow tourism and slow cities has been understood. Constant closures to homes and restrictions on movements have led people to nature and calmer tourism movements. However, acquiring deeper knowledge where people go takes a more natural, more modest, more meaningful, more authentic, more local, healthier, and more experience-oriented, i.e. utilitarian approach. Holidays intertwined with nature, where local experiences can be experienced, especially away from crowds, have become very interesting. People have sought an experience that is safe, but can also convey psychophysical well-being, contact with nature. In this context, slow tourism is seen as one of the most consistent solutions to these new needs. Tourism will be reflected in the slowing trends and new lifestyles that people who get many experiences, become conscious, read and research in the epidemic process will show in their normal lives after the pandemic. Slow tourism and slow cities will come to the first after pandemic. It is very important for sustainability that destinations evaluate these trends well and focus on slow tourism while maintaining local residents.

REFERENCES

Ahmet, S. (2012). *Vize sakin şehir, Kırklareli için turizmde yeni bir marka ve kalkınma modeli cittaslow sakin şehir Vize*. Trakya Kalkınma Ajansı Kırklareli Yatırım Destek Ofisi.

Beeton, S. (2006). *Community Development through Tourism*. Landlinks Press. doi:10.1071/9780643093881

BelediyesiS. (2021). http://seferihisar.bel.tr/seferihisarda-9-tohum-takas-senligi-basliyor

Bruner, R. F. (2004). *Think Globally Act Locally*. Understanding Strategy.

Caffyn, A. (2012). Advocating and implementing slow tourism. *Tourism Recreation Research*, *37*(1), 77–80. doi:10.1080/02508281.2012.11081690

Çağlı, I. B. (2012). *Türkiye'de yerel kültürün turizm odaklı kalkınmadaki rolü: Gastronomi turizmi örneği* (Doctoral dissertation). Institute of Science And Technology.

Çanakkale Tarım ve Orman Genel müdürlüğü. (2021). www.canakkale.tarimorman.gov.tr

Çapar, G., & Yenipınar, U. (2016). Somut olmayan kültürel miras kaynağı olarak yöresel yiyeceklerin turizm. *Journal of Tourism and Gastronomy Studies*, *4*, 100–115.

Cittaslow. (2021). *Uluslararası Yavaş Şehir Ağı.* https://www.cittaslow.org/content/organization

Cittaslowturkiye. (2021). *Yavaş Şehir Türkiye Ağı.* https://cittaslowturkiye.org/

Dall'Aglio, S. (2011). *ECONSTAT strategies for the travel industry*. Slow Tourism Seminar, Bled, Slovenia.

Deniz, T. (2017). Hızlı dünyada sürdürülebilir mekânlar: Sakin kentler. *İnsan ve Toplum Bilimleri Araştırmaları Dergisi, 6*(3), 1399-1412.

Dessein, J., Soini, K., Fairclough, G., & Horlings, L. (2015). *Culture in, for and as sustainable development: Conclusions from the COST action IS1007 investigating cultural sustainability.* Jyväskylä, Finland: University of Jyväskylä. Retrieved from https://www.cost.eu/media/publications/Culture-in-for-and-as-Sustainable-Development-Conclusions-from-the-COST-Action-IS1007-Investigating-Cultural-Sustainability

Dickinson, J., & Lumsdon, L. (2010). *Slow Travel and Tourism*. Earthscan Publishing. doi:10.4324/9781849776493

Ergüven, M. H. (2011). Cittaslow – yaşamaya değer şehirlerin uluslararası birliği: Vize örneği. *Organizasyon ve Yönetim Bilimleri Dergisi*, *3*(2), 201–210.

Feio, A., & Guedes, M. C. (2013). Architecture, tourism and sustainable development for the Douro region. *Renewable Energy*, *49*, 72–76. doi:10.1016/j.renene.2012.01.063

Frost, W., & Laing, J. (2013). Communicating persuasive messages through slow food festivals. *Journal of Vacation Marketing*, *19*(1), 67–74. doi:10.1177/1356766712461403

Fullagar, S., Wilson, E., & Markwell, K. (2012). Starting slow: Thinking through slow mobilities and experiences. In S. Fullagar, K. Markwell & E. Wilson (Eds.), Slow Tourism Experiences and Mobilities (pp. 1-10). Channel View Publications.

Gerzeninsesi. (2017). (https://www.gerzeninsesi.com/guncel/gerzenin-parklarina-el-yapimi-ahsap-tabelalar-asildi-h6945.html

Giampiccoli, A., & Kalis, J. H. (2012). Community-based tourism and local culture: the case of the amaMpondo. *Revista de Turismo y Patrimonio Cultural*, 173-188.

Heitmann, S. (2011). *Tourist Behaviour and Tourism Motivation. P. Robinson, S. Heitmann, & P. Dieke in, Research Themes for Tourism*. CAB International.

Jang, H. J., & Jung, T. (2015). *Go slow and curvy: Understanding the philosophy of the Cittaslow slowcity phenomenon*. Springer.

Jenkins, R. (1994). Rethinking ethnicity: Identity, categorization and power. *Ethnic and Racial Studies*, *17*(2), 197–223. doi:10.1080/01419870.1994.9993821

Karakurt Tosun, E. (2013). Yaşam kalitesi ekseninde şekillenen alternatif bir kentsel yaşam modeli: Yavaş kentleşme hareketi. *Uludağ Üniversitesi İktisadi ve İdari Bilimler Fakültesi Dergisi*, *32*(1), 215–237.

Keskin, E. B. (2012). Sürdürülebilir kent kavramına farklı bir bakış: Yavaş şehirler (Cittaslow). *Paradoks Ekonomi. Sosyoloji ve Politika Dergisi*, *8*(1), 81–99.

Kinley, A. (2012). Local food on a global scale: Anexploration of the ınternational slow food movement. *Journal of International Students*, *1*(3), 1–14.

Markwell, K., & Fullagar, S. & Wilson, E. (2012), Reflecting Upon Slow Travel and Tourism Experiences. In S. Fullagar, K. Markwell & E. Wilson (Eds.), Slow Tourism Experiences and Mobilities (pp. 227-233). Channel View Publications.

Mayer, H., & Knox, P. L. (2005). Slow cities: Sustainable places in a fast world. *Journal of Urban Affairs*, *28*(4), 321–334. doi:10.1111/j.1467-9906.2006.00298.x

Meydan Uygur, S., & Baykan, E. (2007). Kültür Turizmi ve Turizmin Kültürel Varlıklar Üzerindeki Etkileri. *Ticaret ve Turizm Eğitim Fakültesi Dergisi Yıl*, 30-49.

MunicipalityA. (2021). http://www.ahlat.gov.tr/kaymakamlik-birimleri

Nunes, N., Söderström, H., & Hipke, S. (2017). Culture In Sustainability Towards A Transdisciplinary Approach, In S. Asikainen, C.Brites, K. Plebańczyk, L.Rogač Mijatović, & K.Soini (Eds.), Understanding Cultural Sustainability (pp. 28-44). Sophi Publishing.

Özdemir, G., & Çelebi, D. (2018). Exploring dimensions of slow tourism motivation. *Anatolia*, *29*(4), 540–552. doi:10.1080/13032917.2018.1460854

Parkins, W., & Craig, G. (2009). Culture and the politics of alternative food networks. *Food, Culture, & Society*, *12*(1), 77–103. doi:10.2752/155280109X368679

Ritzer, G. (2003). Rethinking globalization: Glocalization/grobalization and something/nothing. *Sociological Theory*, *21*(3), 193–209. doi:10.1111/1467-9558.00185

Şahinkaya, S. (2010). *Bir Yerel Kalkınma Modeli: Cittaslow ve Seferihisar Üzerine Değerlendirmeler*. http://www.bagimsizsosyalbilimciler.org/ Yazilar_Uye/Sahin Tem10.pdf

Sassi, P. (2006). *Strategies For Sustainable Architecture*. Taylor & Francis. doi:10.4324/9780203480106

Semmens, J., & Freeman, C. (2012). The value of Cittaslow as an approach to local sustainable development: A New Zealand perspective. *International Planning Studies*, *17*(4), 353–375. doi:10.1080/13563475.2012.726851

Servon, L. J., & Pink, S. (2015). Cittaslow: Going Glocal In Spain. *Journal of Urban Affairs*, *37*(3), 327–340. doi:10.1111/juaf.12169

Shao, Y. (2014). *Local Identity Identification & Assessment. University of Sheffield, Department of Landscape*. University of Sheffield.

Sırım, V. (2012). Çevreyle Bütünleşmiş Bir Yerel Yönetim Örneği Olarak "Sakin Şehir" Hareketi ve Türkiye'nin Potansiyeli. *Tarih Kültür ve Sanat Araştırmaları*, *1*(4), 119–131.

Slowmotionporto. (2018). *Yavaş turlar*. https://slowmotionporto.wordpress.com/about/

Soini, K., & Birkeland, I. (2014). Exploring the scientific discourse on cultural sustainability. *Geoforum*, *51*, 213–223. doi:10.1016/j.geoforum.2013.12.001

Soysaldı, A. (2010). Bolu-Göynük Tokalı Örtmeleri. *Motif Academy Journal of Folklore*, *3*(5), 150–163.

Tasci, A. D., Semrad, K. J., & Yilmaz, S. S. (2013). *Community basted tourism: Finding the equilibrium in the COMCEC context*. Academic Press.

Turgay, T., & Erkuş Buyruk, P. (2017). Geleneksel Tarakli Evleri Ve Yapilarda Bozulma Nedenleri. *Journal of Advanced Technology Sciences*, *6*(2), 35–46.

Ünal, Ç. (2016). Turizm Coğrafyasinda Yeni Bir Kavram "Yavaş Şehirler". *Doğu Coğrafya Dergisi*, *21*(36), 13–28. doi:10.17295/dcd.09928

Uslu, A. (2009). *"Sürdürülebilir Yeşil Kent Fikirleri, Örnekleri ve Türkiye İçin Dersler", XXI. Uluslararası Yapı ve Yaşam Kongresi Bildiriler Kitabı*. TMMOB Mimarlar Odası Yayını.

Watene, K., & Yap, M. (2015). Culture and sustainable development: Indigenous contributions. *Journal of Global Ethics*, *11*(1), 51–55. doi:10.1080/17449626.2015.1010099

Watson, J. L. (2020, November 4). *Cultural globalization: Additional Information*. Britannica: https://www.britannica.com/science/cultural-globalization/The-persistence-of-local-culture

Wearing, S., Wearing, M., & McDonald, M. (2012). *Slow'n Down the Town to Let Nature Grow: Ecotourism, Social Justice and Sustainability. In Slow Tourism:Experiences and Mobilities*. Channel View Publications.

Yetiş, Ardıç, Ş. (2015). Kapadokya yemek kültürü ve Mustafapaşa beldesi (Sinasos) örneği. *Journal of Tourism and Gastronomy Studies*, *3*(12), 12–19.

Yılmaz, İ., & Çokal, Z. (2019). *Alternatif Turizm*. Detay Yayıncılık.

Yurtseven, H. R., Kaya, O., & Harman, S. (2010). Yavaş hareketi. Ankara: Detay Yayıncılık.

KEY TERMS AND DEFINITIONS

Authenticity: It refers to objects and people who do not lose their local characteristics.

Cittaslow: It is the philosophy that has evolved to prevent globalization from standardizing the fabric and lifestyle of cities, eliminating their local characteristics.

Culture: Culture is all of the tools used to use them with all kinds of values created in the process of historical and social development, to pass them on to subsequent generations, showing the measure of human sovereignty to its natural and social environment.

Local Culture: It is the life of a nation in the form of attitudes, values, traditions and customs that arise over time.

Local Identity: All of the cultural and natural values that the city has.

Slow Tourism: To protect nature, landscape, culture, gastronomy, history, and traditions to enjoy each individual's rights, responsible and innovative efforts to promote tourism.

Sustainability: It is that the life of humanity can be made permanent while ensuring the continuity of production and diversity. It is the ability of people to meet their own needs without compromising the needs of future generations.

ENDNOTE

1 A hijab is a kind of covering and clothing that covers most or all of the top of the head especially used to protect the head from external factors that wear out the hair, ensure covering, not being recognized.

Chapter 9
Fluvial Resources and Recreational and Sports Activities:
Contributions of Sport Fishing to the Tourist Attraction of Serra da Estrela

Gonçalo Poeta Fernandes
https://orcid.org/0000-0003-1278-0369
CITUR, Polytechnic Institute of Guarda, Portugal

Elsa Ventura Ramos
Polytechnic Institute of Guarda, Portugal

ABSTRACT

This research aims to reflect on the importance of inland surface waters for the development of local communities and their potential to foster tourism and leisure activities, promoting differentiated tourism products that project this inland tourist destination. It seeks to inquire into new business opportunities, while simultaneously reflecting on ways of controlling uses and issues of preservation and environmental sustainability. The study is developed in the Serra da Estrela territory, framing the destination and establishing an approach to the evolution of recreational fishing and its contributions to the strengthening of tourism and leisure, given the exceptional conditions of water mass supply and its environmental value for an activity that has an intense relationship with nature. The purpose is also to establish lines of research to be followed and methodologies to be implemented for a better knowledge of this activity and identify action strategies for the future.

DOI: 10.4018/978-1-7998-7339-6.ch009

INTRODUCTION

Water is a valuable resource and provides the basis for the implementation and development of tourist, recreational and sports activities, promoting its use in the promotion of the destination and generating new empowerment for the expansion of demand and its differentiation throughout the year. The expansion of tourism and recreational activities based on river resources implies a greater responsibility in their management and use, requiring the promotion of conditions, materials and regulations, which allow their enjoyment in a safe, sustainable and satisfactory way for their users.

In Serra da Estrela – the largest mountain in the Portuguese territory – the main national rivers originate, along with a number of lakes and reservoirs of relevant importance for water supply to the population and economic activities. In the last two decades, these have come to constitute important bathing, leisure and sport spaces for the inland regions, promoting different activities and with different demands throughout the year.

Recreational fishing has traditionally been seen and managed as a secondary and even residual activity in tourism, without there being an effective strategy for its valorization as a tourism development asset in the form of concerted strategies or it being promoted in an integrated way with other products.

The lakes and rivers of Serra da Estrela constitute a relevant resource for the promotion of recreational and sports activities related to fishing. The attraction of the surface waters of this mountain territory for sport fishing, in particular salmonids, represents a tourist and recreational resource for the region with concrete effects on local commerce, hotels and restaurants, given the practitioners involved and the natural and cultural contexts that the territory offers. The environmental value associated with its mountain geography and the existing ecosystems project Serra da Estrela as a fishing destination that is increasingly valued, either as a main motivation or as a complement to the recreational and tourist activities developed.

In this context it can be seen that the structuring of recreational and sport fishing in Serra da Estrela lacks a greater promotional development of the product, qualifying and regulating the bodies of water for the reception of this activity, which constitutes a market niche of interest in combating the seasonality of the destination, in the territorial promotion of the region, in the valuation of water resources and in the promotion of associated services.

The connection of this activity with the existing trout farms should be noted, stimulating their activity both through a greater commercialization of fish for gastronomic purposes, as well as for the repopulation of the region's lakes and rivers.

This research aims to reflect on the importance of inland surface waters for the development of local communities and their potential to foster tourism and leisure activities, promoting differentiated tourism products that project this inland tourist destination. It seeks to inquire into new business opportunities, while simultaneously reflecting on ways of controlling uses and issues of preservation and environmental sustainability. The study is developed in the Serra da Estrela territory, framing the destination and establishing an approach to the evolution of recreational fishing and its contributions to the strengthening of tourism and leisure, given the exceptional conditions of water mass supply and its environmental value for an activity that has an intense relationship with nature. The purpose is also to establish lines of research to be followed and methodologies to be implemented for a better knowledge of this activity and its practitioners. The aim is to identify action strategies for the future, regarding the organization of the tourism product, the promotion of the destination and the definition of procedures to regulate and operationalise this activity in a sustainable way.

FLUVIAL WATERS AND THEIR RECREATIONAL AND SPORTING FUNCTIONS – FISHING IN INLAND WATERS

Water is a valuable resource and can be the basis for the implementation and development of tourism, recreation and sports activities, constituting an asset in the promotion of the destination and generating new capacities to increase demand throughout the year. Water-based tourism activities are developed in connection with adjacent territorial areas, promoting a strong relationship with these in terms of existing infrastructures, population settlements, heritage elements and forms of occupation. The recreational and sporting practices based on water resources use the hydro-tourism potential to diversify the offer and foster activities with different intensities of water use, from contemplation and bathing to sporting activities of greater physical intensity (Bătinaş & Sorocovschi, 2012).

It should be noted that water is not consumed in recreational activities or sport tourism, but these activities can have a major impact on the quality of water resources, affecting their development and attracting tourists. (Gunn & Var, 2002). Aquatic environments such as rivers, lakes, estuaries, margins, wetlands, among others, are systems with high ecological-natural value, offering specific opportunities for education and the understanding of nature, recreation and leisure, while at the same time being vulnerable to tourism activities (Grecu & Iosif, 2014).

Protected areas are, in their globality, in close connection with water, constituting fragile environments that are valuable assets for local communities and should be used to promote forms of sustainable development, articulating tourism, education, conservation and protection in the valorization of the resource and in the relationship with local communities (Eagles et al., 2002).

In this context, sport and recreational fishing is an opportunity to promote leisure and tourism activities in inland areas, particularly mountain areas, adding to these activities the value of the heritage of these territories, fostering the connection to the ecocultural values of mountain ecosystems and promoting economic benefits for the region and its inhabitants. In fishing, in addition to relaxation and recreation, the educational focus should be promoted, as protected areas should be conserved and valued for the heritage resources they contain and the fragility of their ecosystems. Thus, the articulation between the resources and the different forms of enjoyment should promote new types of tourism in the region. At the same time there is an increased awareness of the environmental values and uses associated with local people, particularly considering their natural assets (Dax, 2019; Higginbottom, 2008).

Recreational fishing is one of the dominant or even exclusive uses of many aquaculture resources in inland waters (Arlinghaus, 2008). In mountainous regions. where water resources are abundant, income growth and employment opportunities can be fostered from fishing tourism, supported by fish farming and associated services (Hoitsy et al., 2012).

Inland waters include river ecosystems and the artificial or heavily modified bodies of water associated with them (reservoirs and canals), as well as other ecosystems where the aquatic component is decisive, whether permanent or temporary, such as marshland, mudflats or ponds (Ferreira & Guerreiro de Brito, 2009). These resources and their ecosystems are extremely fragile due to the growing demand for water and associated uses. They require careful and regulated management that enables the maintenance of their quality and the development of different uses and functions.

Fishing tourism may be developed according to the following variables that should be observed:

- Existence of surface water bodies – rivers, lakes, dams, canals, ponds, weirs, among others;
- Specific fish species, generating attraction and interest for recreational and sport fishing;

- Basic infrastructure and support for the development of fishing tourism – commerce and services;
- Access conditions to fishing places – roads, paths and trails;
- Capacity of the water body and its banks for a safe and sustainable activity;
- Environmental and landscape quality for an activity that values the contact with and contemplation of nature;
- Adequate legislation and regulation for a sustainable fishing practice and control of the activities to be promoted;
- Complementary services that guarantee the tourist fishing activity and its accomplishment with comfort and safety – accommodation, catering and support services.

For the development of fishing, it is also crucial to know the characteristics and trends of the segment that seeks it and about the region that hosts and develops it, allowing the correct decision-making and definition of strategies to improve tourism performance and the protection of water resources that support the different practices (FAO, 2008; Taylor et al., 2016). In addition to the income generated by fishing activity and associated services, benefits can also be expected from the conservation of species and their habitats in terms of biodiversity and geoconservation (Brainerd, 2010). Fishermen will search for an area according to its species and the abundance of catches. The existence of a recognized and valued fishing destination will be achieved if natural resources are properly managed, ensuring the maintenance of fish stocks and promoting the attractiveness to anglers.

Consider that fishing has been an activity linked to man since its origins and, around it, other activities have been developed, which implies understanding its practice as something that characterizes a society or gives it an identity.

Recreational fishing is divided into two categories (Amaral & Ferreira, 2010): recreational fishing and sport fishing. Recreational fishing is considered to be 'fishing carried out as a leisure or recreational activity in which the fish caught cannot be sold. Sport fishing is recreational fishing carried out in organized competition with a view to obtaining sporting marks, including championships, learning techniques and developing equipment. Recreational fishing has traditionally been seen and managed as a secondary or even residual activity within the scope of tourism, without an effective strategy for its enhancement as a tourism development asset, through strategies that articulate and enhance the economic and social factors involved. According to IBAMA, (2010, p.46), fishing tourism by itself is capable of attracting tourists from all over the world, and should add other attractions and other product options, without losing the essence, the practice of fishing and interaction with nature.

Thus, it is urgent to recognize the importance of aquatic resources, the valorization of ecosystems and effective management of the conservation of species, their natural environments and fishing activities. In this context, the regulation of recreational fishing should provide a balance between the conservation of the environment and natural resources and the economic safeguard that sustains the activity. It becomes relevant to develop studies and investigations that allow knowing the facts and the perceptions of the activity, either by anglers, local communities and administrative entities, or in biological, socio-economic and regulatory terms. In this context, regulatory measures must be taken that are dynamic and sensitive to changes in ecosystems, to the interests of fishermen/women and to the economic sectors for which the activity is important. The importance of the community and its active participation in the process of developing tourism activity must be recognized. The local community is the great ally of the fishing tourism activity, as its knowledge, respect for the environment and capacity to control activities adds value to the activity and contributes to the solution of social, economic, cultural and political problems

Figure 1. Natural resources and potential of the fishing tourism territory
(Source: Authors elaboration).

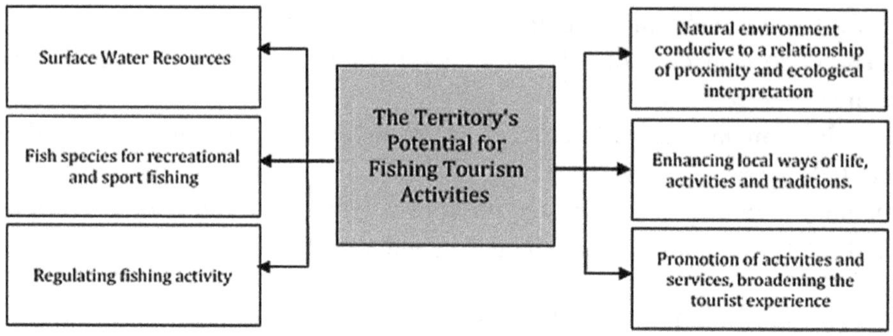

that an activity in contact with sensitive environmental resources generates (ANA, 2005). Their involvement in tourism practices is relevant for the strengthening of relations with tourists and the maintenance of local identity values and environmental conservation. This relationship encourages efforts and pressure to upgrade the water lines, their banks and the heritage elements present, particularly those that represent the connection between the communities and their economic activities with water (watermills, bridges, weirs, water reservoirs, river basins, among others). Adding attractiveness to ecocultural assets is a strategy to differentiate tourism products by incorporating activities, services, value and other attributes to natural resources. These actions become more important when considered in inland areas, with socio-economic weaknesses to be reversed, and in order to enhance the natural elements and traditional activities. In a context of growing demand for areas of greater natural value, outdoor experiences, appreciation of authenticity and recognition of environmental preservation, activities for contact with nature and generating recreation are gaining significance and promote value to spaces of great natural integrity, such as mountain territories.

In this context, there is an attempt to associate benefits to the tourist destination, such as differentiated products (like fishing), offering new possibilities of use, different demands and new consumption (Figure 1). Simultaneously, to generate attractiveness to the fishing tourism product, the integration of environmental interpretation activities, preservation of ecosystems and valorization of local identity is considered (Ministry of Tourism, 2010). Furthermore, the aggregation of attractiveness brings benefits to the destination, as it helps to minimize seasonality and creates new opportunities for the territory, such as sports competitions, nautical activities, adventure sports, ecotourism, cultural and gastronomic activities, among others. This potential implies careful management of water resources and their biology, so as to strengthen the links between conservation and new tourism opportunities for natural areas (Figure 1).

Fishing Management and Its Regulation in Portugal

In Portugal, the regulation and supervision, interventions in small bodies of water or rivers of a small order, or interventions to repopulate certain species are routinely undertaken by the Directorate General of Forests (DGF) and the Regional Directorates of Agriculture (DRA) (Ferreira, 2009). Decree-Law No. 222/2015, of 8 October, establishes the legal regime for the planning and sustainable management of aquaculture resources in inland waters, regulating fishing and aquaculture in those waters. Article 5 provides that salmonid fishing waters shall be classified by deliberation of the management board of the

Instituto da Conservação da Natureza e das Florestas, I. P. (ICNF, I. P.), to be published on its website, on the basis of the presence of species of the salmonid family or the potential of such waters to support specimens of that family. In accordance with the regulations in force, professional fishing is forbidden in salmonid waters. It also considers that in salmonid waters, during the period in which fishing for river or brown trout is forbidden, fishing for all other species in those waters is also banned.

Recreational fishing licences authorize their holders to engage in recreational and sport fishing in the bodies of water where these are permitted under the terms of the law and in accordance with the fishing Schedule, annually approved. In practice, the vast majority of aquaculture resources are a public good managed by the state, and their use is subject to fishing licences and compliance with the applicable legislation. However, there are areas managed by anglers' associations and municipalities – sport fishing concessions.

Whereas the legislation on fishing in inland waters is an old one, dating back more than four decades, and remains an important tool for the protection and conservation of fish communities in these waters. It incorporates concepts of sustainable management and maintenance of biodiversity and establishes regulatory standards for the capture of species and for the planning and management of aquatic resources.

Portugal's surface water bodies have a high potential for supporting a range of aquaculture species that constitute a valuable renewable natural resource from a social, cultural and economic point of view. For this reason, its management, protection and exploitation must be guided by principles of sustainability and biodiversity maintenance (Bochechas, 2000; Gonçalves et al., 2008), generating uses that, in a regulated form, allow its recreational enjoyment, promoting value for local communities and, simultaneously, developing an environmental awareness that generates the protection of ecosystems.

For fishing tourism, the interpretation of bio- and geodiversity can stimulate a new look at the rational use and conservation of the resources involved, constituting an awareness-raising tool for both fishermen and local communities. Anglers may become elements of interest in the control of the quality of resources and contribute to the operationalization of studies that contribute to the monitoring of fishing activity and the preservation of the environment. Environmental interpretation, promoting knowledge and an awareness of value about ecosystems and regulations makes the visitor aware of socio-environmental issues, contributing to the preservation of flora and fauna (Morar & Pop, 2016). The interpretative process may be aided by informative material, in addition to adequate tourist signing, as well as by interaction with the community, which implies its awareness and recognition of the activity as a promoter of development. Fish marking and other scientific initiatives are also commonly promoted through sport fishing clubs, enhancing the role of these tournaments in research and the conservation of fish resources and collaboration with management and conservation entities (Gonçalves et al., 2008). Travelling for the practice of fishing implies contemplating the appreciation of cultural diversity in trips with an emphasis on nature and cultural identities associated with the tourist experience. Each water body and hydrographic basin has specific characteristics and its own ecological individuality, and it is now recognized that there is a need for targeted management, on a local or regional basis, in addition to the general guidelines and strategy that may be adopted for the sector (Templeton, 1995, Ellender et al., 2009).

Fishing tourism, recreational or sportive, requires individuals to travel from their usual residence to destinations with water bodies, where they can successfully develop their activity in terms of catches and positive environmental experience. This mobility requires products and services such as transport, accommodation, food, entertainment and events, fishing equipment and other complementary activities that exist as a function of fishing tourism (workshops, workshops and ateliers, visits, among others). In accordance with good sustainable recreational fishing practices, catch and release should be considered in order to promote the conservation of good populations of fish species in water bodies, ensuring the attractiveness and competitiveness of the destination. Trophies and fishing days should be recorded in the memory and in photos, not sacrificing the specimens and reducing the existing populations. Fishing and release is, in practice, besides a form of leisure, a way to ensure the reproduction of species and the sustainability of fishing tourism (Hoitsy et al., 2012).

For tourism purposes, competitions and meetings (tournaments, championships and fishing days) organized by representative bodies (associations, federations, confederations) are considered to be of a competitive nature, as 'friendly' competitions, whether practised by professionals or amateurs, promoted by clubs, local development associations and municipalities. Thus, amateur competitions and gatherings are also aligned with sports tourism. As Ferreira (2009) points out, a large part of recreational fishing activity is carried out within the scope of recreational fishing competitions, mobilized mainly by fishing groups, associations and clubs, followed by groups and sports sections of some institutions, recreational centres, cultural groups and INATEL, municipal councils and parish councils and also other institutions such as trade unions, football clubs, residents' and friends' groups, etc. Although the practice of fishing and release is not yet a mandatory legal requirement, we have observed an increase in its regulation and recognition of interest by its practitioners, as well as an increase in the strictness of the legislation that deals with fishing in general. Fishing and releasing is not only a form of leisure, but also a way of guaranteeing the reproduction of the species and the sustainability of fishing tourism.

SERRA DA ESTRELA - TERRITORY AND FLUVIAL RESOURCES

The Serra da Estrela, at 1993 m, is the highest mountain range in mainland Portugal and is part of the Iberian Central Cordillera (Fernandes et al., 2016; Gomes et al., 2017). Serra da Estrela is a territory with geological and geomorphological resources, consisting of glacial, periglacial, and granite landforms whose singularity and heritage value bring to this territory an ecocultural dimension of great relevance (Figure 2).

The existence of unique natural resources of international relevance, with Natural Park status and recognized by UNESCO giving rise to the classification of Geopark, makes this destination a privileged area with high tourism potential in Portugal. The Serra da Estrela area encompasses markedly natural spaces with others of great historical and cultural value, generating an ecocultural mosaic of great environmental relevance (Fernandes et al., 2016; Gomes et al., 2017). The geopark constitutes a 'living laboratory' in which the natural resources, the ways of life of the population and the biological and geological diversity blend into cultural landscapes. Bounded by fault scarps, a granite massif occupies the central area forming a summit plateau between 1400 and 2000 m, while in its surrounding core area, there is an interplay between schists and greywackes (Castro et al., 2017). The uniqueness of these spaces contributes to the strong symbolism regarding their morphological structure, the richness of fauna and flora, the built heritage and the cultural and ethnographic dimension. These elements, associated with the different settlement forms of the Serra da Estrela region and its production systems, accentuate the genuineness and authenticity of mountain spaces, promoting tourist attractiveness (Fernandes, 2015; Silva et al., 2018).

In this context, Serra da Estrela has its own meaning and dynamics, which have been conferred over time by its mountainous dimension and differentiated resources related to tourism. Serra da Estrela, as a geographical and environmental space, has geological, geomorphological, fluvial, forestry, biological,

Figure 2. Localization of Serra da Estrela (UNESCO Global Geopark) in Portugal and and representation of its altimetry
(Source: AGE – Associação Geopark Estrela).

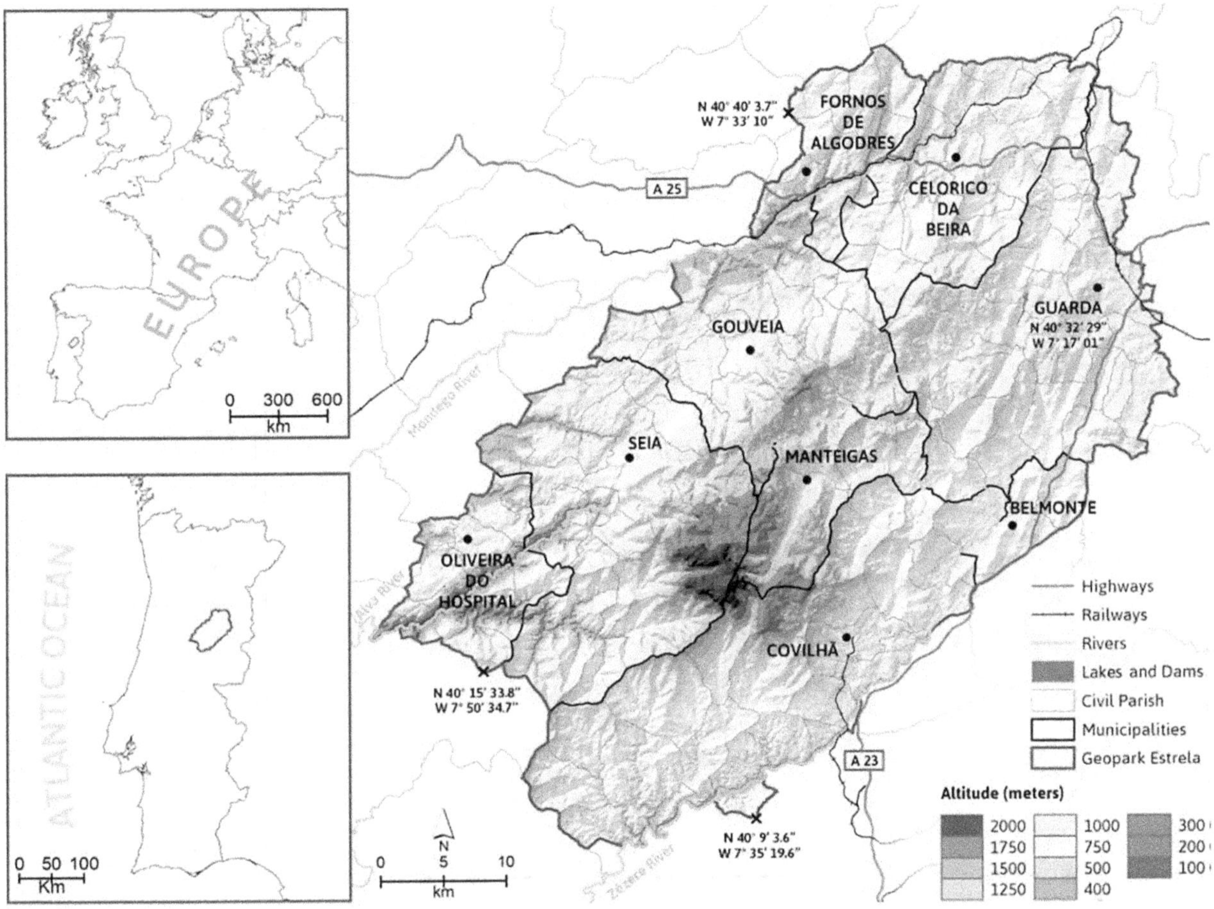

ecological and landscape organization and climatic resources, which make it the largest mountain in the national territory. The differentiating features that characterize the tourism potential of the region can be structured into:

- Natural resources (geological, geomorphological, climatic and fluvial);
- Landscape and environmental resources (mountain ecosystems, biodiversity and ecological thresholds);
- Resources of historical, monumental heritage and traditional architecture;
- Cultural resources linked to traditions and festivities;
- Gastronomy and wine resources;
- Rural resources associated with livelihoods, customs and agro-pastoral activities.

The resources, products and tourism activities that this space holds and promotes, are currently revalued and are linked to the perceptions and new motivations of tourism demand, associated with the

Figure 3. Map of water resources in the Estrela Mountain and indication of protected wetland areas (RAMSAR)
(Source: AGE: Associação Geopark Estrela –AGE).

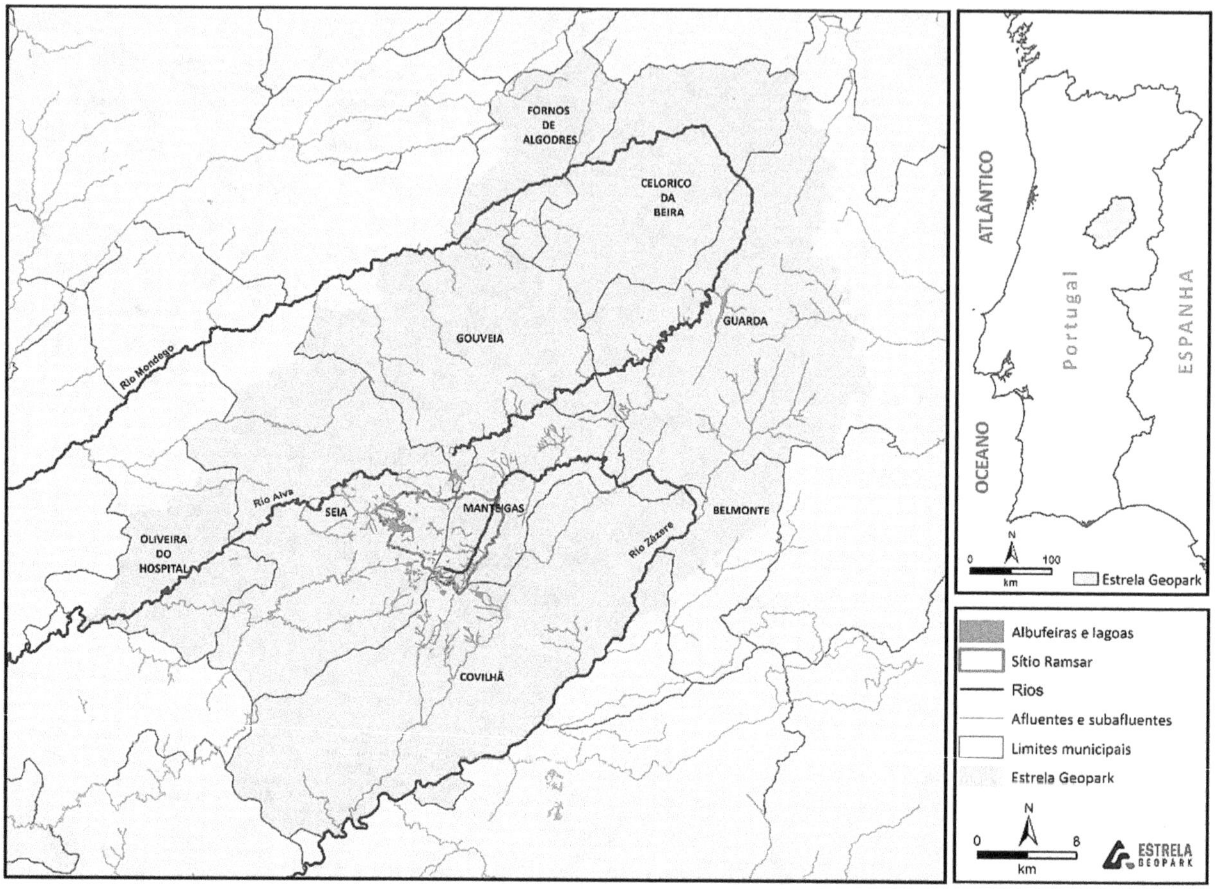

characteristics of mountain regions, either associated with their natural conditions or with its historical heritage and culture. According to Silva et al. (2018), tourist destinations have a meaning of their own for the tourists who visit them and, in the case of Serra da Estrela, these meanings are associated with affective images, nature and ecology, sport, leisure and adventure, culture and tradition, life and health, mystique and sacredness, which is consistent with the image associated with mountain tourist destinations and gives them their special attractiveness.

The surface waters of the Estrela Geopark territory are composed of river, lakes and reservoirs, corresponding in total to a surface divided by 53.7 km^2 of retained waters (lakes, dams, rivers and RAMSAR site - Wetlands of International Importance) and 826.8 km of running water, corresponding to rivers, their tributaries and sub-tributaries (Figure 3 and Table 1). The diversity of water lines, and the ecosystems they integrate, together with lakes, provide the territory with strong potential for the development of water-based tourism and leisure activities. The main activities carried out on these water surfaces, given their characteristics in terms of geomorphology, climate, ecosystems present and recreational / sporting interests of local communities and tourists, can be systematized in Table 1.

Table 1. Surface water resources and associated recreational and sport uses

Water Resources	Entertainment Uses and Sports	Restrictions
Rivers: *Zezere and Afluents* *Mondego and afluents* *Alva and afluents*	*Recreational (contemplation and picknicks)* *Bathing,* *Canoeing* *Rafting* *Recreatinal and sport fishing*	*- Only river beaches are properly trained for bathing practices, with means of support, access and control.* *- Sport fishing requires licenses for its practice due to its performance in rivers or in the protected fishing zone of Serra da Estrela.* *1.*
Dams: Lagoa Comprida; Lagoa Escura; Lagoa do Vale Rossim; Lagoa do Viriato; Lagoa do Covão de Ferro; Albufeira da Barragem do Lagoacho; Lagoa do Covão do Forno; Poços de Loriga; Lagoa Seca Lagoa dos Conchos; Lagoa Redonda; Lagoa da Erva da Fome Lagoa do Covão do Curral; Lagoa do Vale do Conde; Lagoa dos Cântaros	*Recreational (contemplation and picknicks)* *Bathing* *Canoeing* *Windsurf and Padel* *Recreational and sport fishing* *Scientific Activities*	*- Bathing practices limited to lagoons* *- Use of propelled vessels conditioned;* *- Fishing subject to a specific calendar and obtaining special licenses for daily fishing in the protected fishing zone in the lagoons of Serra da Estrela.*
RAMSAR zone – Protection zone on the Torre plateau relative to wet areas and small natural lakes	*Walking and observation* *Bathing* *Scientific activities*	*Area of great environmental sensitivity, with human activities severely limited in this area of maximum restrictions under the PNSE regulations.* *2.*

(Source: Authors elaboration)

FISHING ACTIVITIES AND WATER RESOURCES IN THE SERRA DA ESTRELA

Recreational fishing is one of the few forms of nature-based recreation enjoyed from childhood to adulthood (McManus et al., 2011). The Food and Agriculture Organization (FAO) of the United Nations considers recreational or sport fishing as the form of fishing that catches aquatic animals, particularly fish, that are not the primary resource for individuals to meet their own needs (basic nutrition). Fishing contemplates several known forms of its practice, both in terms of techniques and the context in which it is carried out, thus recognizing the existence of different types of recreational and sport anglers (EMBRATUR, 1996; Miranda, 2012). Within this framework, we highlight, for example, 'the family angler', who fishes to complement his family's leisure activities; 'the competition angler', whose interest lies in the fish biomass obtained in competition; and 'the targeted angler', who is particularly devoted to a given species (in Portugal, we consider the attraction of trout *Samo trutta* in the north of the country and of the largemouth bass *Micropterus salmoides* in the centre and south). Thus, recreational fishing presents itself as a leisure option, during weekends and holidays, for those who live in areas that do not have water bodies, and who like to maintain a certain contact with nature. It also corresponds to a healthy occupation of available time walking the footpaths that are interconnected with the circuit of the lagoons, promoting photographs, observation of the existing flora and fauna and interacting with local communities (Lemus et al., 2018).

Sport fishing has the potential to significantly stimulate economic development, highlighting the contribution of this activity in the development of various sectors of the economy, directly or indirectly (Cowx et al., 2010). Of note are: the sale of fishing licences; trade in equipment and materials for the practice of fishing; trade in literature of the speciality; television channels; transport; hotel and restaurant industry; local and regional trade; tourism in rural areas among others.

Table 2. Water bodies in the territory of Serra da Estrela

Mass of Water	Name	Regime	Main Fish Species
Lake	Lagoa Comprida;	Fishing with death [1]	Brown Trout; Rainbow trout; Bordalo; Sunfish
Lake	Lagoa Escura;	Fishing with death	Brown Trout; Rainbow trout; Bordalo; Sunfish
Lake	Lagoa do Vale Rossim;	Fishing with death	Brown Trout; Rainbow trout; Bordalo; Sunfish
Lake	Lagoa do Viriato;	Pesca sem morte [2]	Brown Trout; Rainbow trout; Bordalo; Sunfish
Lake	Lagoa do Covão de Ferro ou Barragem do Padre Alfredo	Fishing with death	Brown Trout; Rainbow trout; Bordalo; Sunfish
Dam	Albufeira da Barragem do Lagoacho	Fishing with death	Brown Trout; Rainbow trout; Bordalo; Sunfish
Lake	Lagoa do Covão do Forno	Fishing with death	Brown trout, Rainbow trout, Bordalos, Sunfish, Redfin tuna, Stone bass.
Wells	Poços de Loriga: Lagoa do Covão do Meio, Lagoa da Francelha, Lagoa Serrano e Lagoa Covão das Quelhas	Fishing with death	Brown Trout; Rainbow trout; Bordalo; Sunfish
Lake	Lagoa Seca	Fishing with death	Brown Trout; Rainbow trout; Bordalo; Sunfish
Lake	Lagoa dos Conchos	Fishing with death	Brown Trout; Rainbow trout; Bordalo; Sunfish
Lake	Lagoa Redonda	Fishing with death	Brown Trout; Rainbow trout; Bordalo; Sunfish
Lake	Lagoa da Erva da Fome	Fishing with death	Brown Trout; Rainbow trout; Bordalo; Sunfish
Lake	Lagoa do Covão do Curral	Fishing with death	Brown Trout; Rainbow trout; Bordalo; Sunfish
Lake	Lagoa do Vale do Conde	Pesca com morte	Brown Trout; Rainbow trout; Bordalo; Sunfish
Lake	Lagoa dos Cântaros	Pesca com morte	Brown Trout; Rainbow trout; Bordalo; Sunfish

(source: Authors elaboration)

The lakes and rivers of the Serra da Estrela constitute a relevant resource for the promotion of recreational and sporting activities, making various practices possible throughout the year, of which fishing assumes significance both for its effect in attracting practitioners and for the need for its control and regulation in benefit of its sustainability. The environmental value of this territory, framed in a singular mountain geography and in its ecosystems, projects Estrela into a space of great attractiveness for the practitioners of fishing, especially directed to the trout species (brown and rainbow). Fishing in this territory can be subdivided into components according to the bodies of water in which it is practised. In this context we differentiate between fishing in rivers and streams, which in general can be practised by holders of general fishing licences (national, regional or local), according to the hunting seasons of the different species throughout the year. And fishing in lakes and dams, which is subject to special fishing permits restricted to each particular body of water and practised in accordance with the days of the week

Table 3. Principal fish species in the water bodies of the Serra da Estrela

Species	Geographic Expression/Water Body	Recreational Fishing
Brown Trout (Salmo trutta)	Ponds, Lagoons, Dams and Wells	Fishing with/without death
Raimbow Trout (Oncorhyncus mykiss),	Ponds, Lagoons, Dams and Wells	Fishing with/without death
bordalo (Squalius alburnoides)	Ponds, Lagoons, Dams and Wells	Catch and release
Sunfish (Lepomis gibbosus)	Ponds, Lagoons, Dams and Wells	Fishing with death
Ruivacos (Achondrostoma oligolepis	Covão do Forno Lagoon	Fishing with death
Escalos (Leuciscus idus)	Covão do Forno Lagoon	Fishing with death

and fishing period established especially for the purpose in each fishing season. Besides this practice, we must consider the fishing competitions promoted by the federation or by fishing associations/clubs and that require permission by the ICNF, in order to manage the bodies of water (ICNF, 2020).

The demands on these lakes are quite high because they are controlled and restocked by ICNF and are inserted in the protected area of the Serra da Estrela Natural Park. Each year the ICNF establishes a notice – *Zona de Pesca Reservada das Lagoas da Serra da Estrela* (Reserved Fishing Area of the Serra da Estrela Lagoons) – setting out the rules and restrictions for the fishing season (Tables 2 and 3). The contributions of fishing activities to the territory are diverse and allow highlighting the importance of this market for rural areas, given its environmental, economic and social impacts. Thus, fishing may constitute a differentiated tourist product for Serra da Estrela with positive impacts for the economy and the environment.

In this context we can observe the evolution of fishing licences in the Serra da Estrela lakes in this decade, which saw an increasing trend of practitioners, which, according to the 2017 register, accounted for 918 licences (Figure 4).

A growing trend over the last five years has been noted, revealing an increase in demand for this destination and its water resources for fishing, particularly for salmonids. This enables us to understand that

Figure 4. Evolution of the number of special licences in the Serra da Estrela lakes
(Source: ICNF – Instituto de Conservação da Natureza e Florestas – Forest and Nature Conservation Institute)

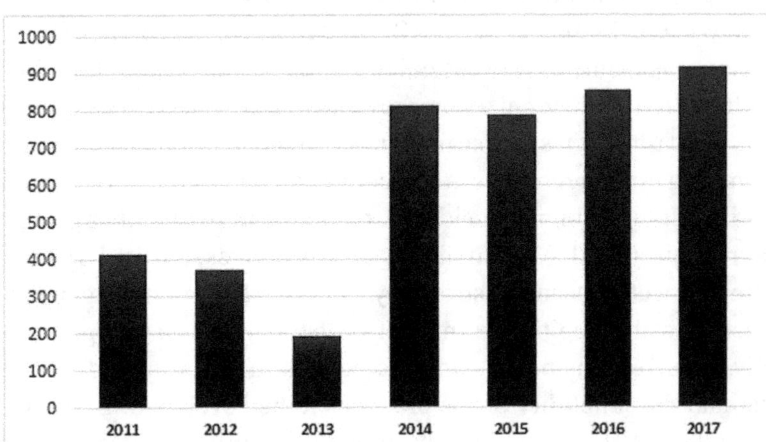

Figure 5. Factors for the promotion of fishing as a tourism product.
(Source: Authors elabortaion)

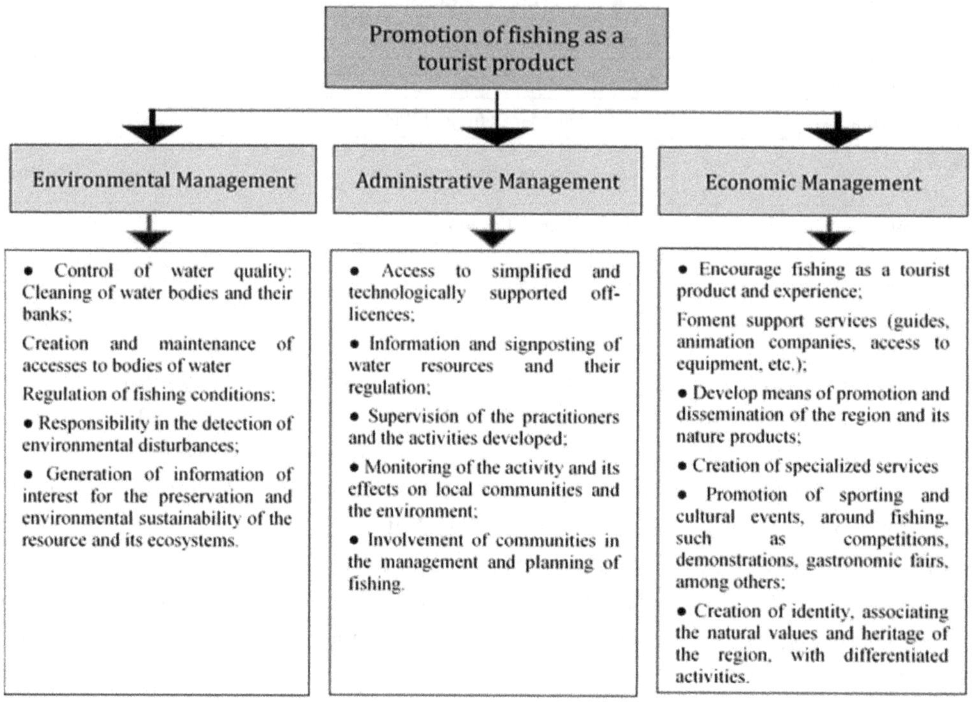

the enlargement of the water masses for fishing activity may increase the market, which currently benefits from the status of the Estrela UNESCO Global Geopark, constituting an increase of responsibility for the managing entities, but also an attraction for a tendentially growing and international demand, with direct effects on the local economy. The connection of this activity with the existing trout farms should be noted, stimulating their activity both through a greater commercialization of fish for gastronomic purposes, as well as for the repopulation of the region's lakes and rivers.

The management policy to be developed implies an increasing degree of knowledge and responsibility of all the actors involved, from the anglers themselves, the researchers and officials of the supervisory authority, the tourism operators involved and other interest groups, such as non-governmental organizations and civil society itself (Lemus et al., 2018; Madariaga & Garcia del Hoyo, 2014). Therefore, for an effective fisheries management, it is essential that the sector – which includes all actors with a legitimate interest in the activity – recognize their responsibilities, share good practices and sustainably value the water bodies and fish species. In particular, the practice of quality recreational fishing should be promoted, whose management recognizes the ecological, economic and social impacts (Figure 5). Long-term guiding principles should be considered, explicit and public, that promote this product as a complement to the attractions that this mountainous area generates, particularly associated with nature tourism, sports, health and wellness and gastronomy.Effective fishing management requires the sector (which includes all actors with a legitimate interest in the activity) to recognize not only their rights but, more importantly, their responsibilities and the actions to be promoted in defence of the quality of the bodies of water and the suitability of the practices to be developed. In particular, the practice of recreational catch-and-release fishing should be promoted, underpinned by long-term guiding principles

Figure 6. Structuring the factors promoting recreational and sport fishing as a tourism product.
(Source; Authors elaboration)

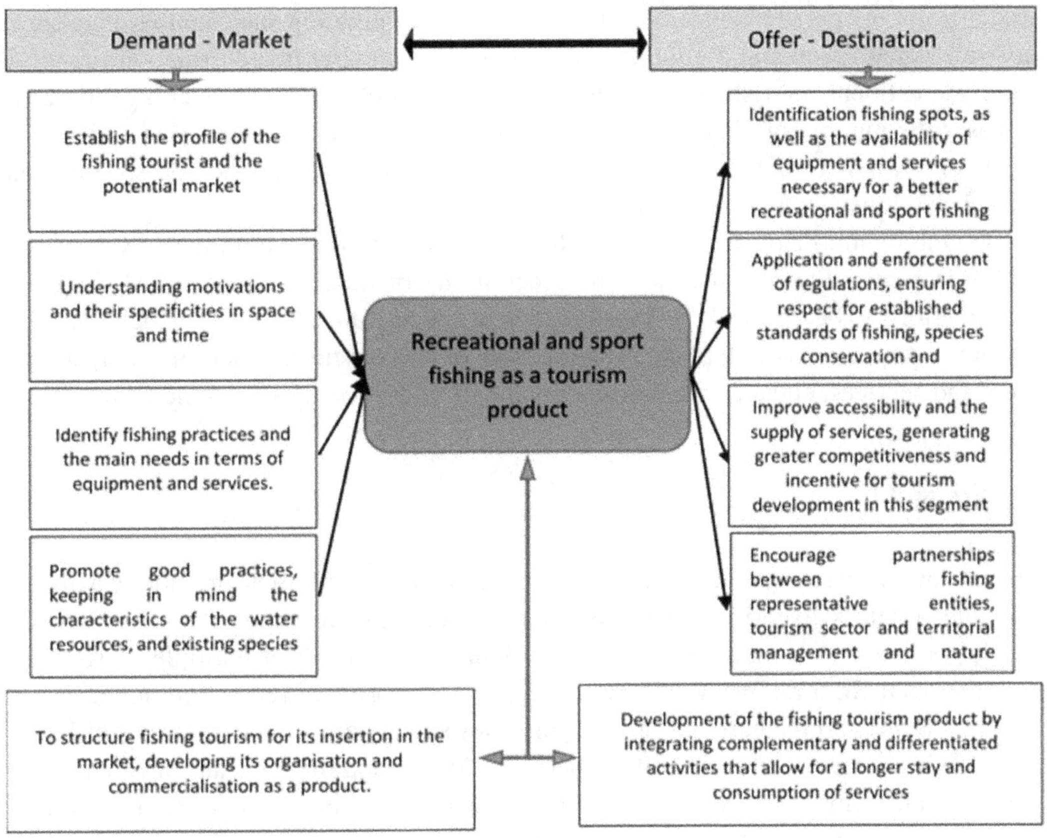

in which the ecological, economic and social objectives pursued are clearly established, resulting from an integrated management strategy.

The recreational and sport fishing structuring as a touristic product in Serra da Estrela needs a bigger promotional development of the product, qualifying and regulating the water bodies to host this activity. The fishing constitutes a niche of market of interest in combating the seasonality of the destination, in the territorial promotion of the region, in the valuation of the hydric resources and promotion of the associated services. In this context, cooperation between the bodies of water, fishing associations and clubs, hoteliers and tourist service companies is crucial. The known developments and the different activities that are promoted throughout the year on these waters require an efficient management. Namely their control, the stimulation of their practice and learning, their connection with the hotel and catering industry and their consideration in the tourism strategies of the municipalities and the regional tourism authority, for which the clubs and associations play an important role.

It is essential to know the characteristics of the demand, their motivations and perception of the destination Serra da Estrela in order to promote a structured product recognized by its organization and the value of the experience created. The development of information through various media, the existence of illustrative maps of fishing locations, information on regulations and access to licences, the promotion of specific offers for fish, will strongly contribute to the affirmation of this product (Figure 6).

The organization of guided routes along rivers or through lakes, gastronomic festivals, fishing tastings and competitions, as well as the production of films, documentaries and thematic publications are all valuable aspects of sport fishing activities. It is worth highlighting the role that magazines on the subject, as well as television and digital channels, play in promoting this destination, giving visibility to the scenic, recreational and fishing value of the Serra da Estrela. Organizing competitions or tournaments increases the number of visitors to a region, boosting the income generated and promoting the extension of services. The success of competitions and tournaments depends on species populations and their catches (Hoitsy et al., 2012). There is a need to adapt infrastructures for people with reduced mobility, allowing them to have full access to the activity and to make it an inclusive tourist product. To this end, it is essential to map and plan the accessibility of water surfaces, which helps to identify the means and efforts to be undertaken in the practice of sports. The survey of financing sources for the adequacy of the equipment, as well as the establishment of partnerships, generating a greater number of practitioners and services, enhance the destination and promote greater economic flows.

CONCLUSION

Inland waters show an increase in demand for outdoor recreational activities, promoting a diversification of segments and practices, fostering new tourism opportunities and the expansion of associated services. On the whole, we are observing the valorization of inland territories, for multiple tourism and leisure activities in the open air, for contact with nature, for diverse forms of recreation, for which water is a crucial element. Sport and recreational fishing generates social and economic effects of interest to the destinations, having the capacity to constitute, besides a leisure activity, a means of enhancing the value of these resources in terms of control of their environmental conditions and interaction with local communities. The rivers, dams and lakes have the ability to attract sport anglers, enhancing a complementary demand market to traditional products and seasons of greater attractiveness. In some cases, the lack of collaboration between stakeholders – local communities, administrative bodies, tourists – puts sustainable development mechanisms and processes at risk. Given the characteristics of the tourism product, all elements must be interconnected, spatially and functionally, in order to work together in an integrated manner for the benefit and regulation of water-based activities and existing visitor flows. Recreational fishing has traditionally been seen and managed as a secondary or even residual activity within the scope of tourism, without an effective strategy for its valorization as a tourism development asset. The action of fishing and contact with nature – the main attraction of the fishing tourism product – motivates the tourist to go to the destination and requires satisfactory conditions for the tourist experience, as well as its regulation and sustainable management. The proximity and contact with nature and the sensation of peace and rest that it provides is what motivates the choice of this type of travel by its practitioners, enhancing tourist services, whose adequate organization will allow extending the stay.

Sport fishing, as mentioned, corresponds to an activity with recreational and leisure purposes, generating regional economic value resulting not only from direct benefits, such as the sale of fishing licences, but also indirect ones related to restaurants, accommodation and the acquisition of different fishing equipment, appropriate for each modality. Its structuring as a tourist product in the Serra da Estrela lacks a greater promotional development, qualifying and regulating the bodies of water to host this activity and creating specific offers for travel, accommodation and assistance. The quality and diversity of water resources, the patrimonial value of this territory, and the existing bio- and geodiversity, allow the construction of a differentiated product in collaboration with the several entities linked to activities and those that promote the governance of this territory. The connection of this activity with the existing trout farms should be noted, stimulating their activity, both through a greater commercialization of fish for gastronomic purposes and for the repopulation of the region's lakes and rivers. It is worth noting the increasing number of sport fishing competitions organized by the portuguese sport fishing federation, the European championships and various meetings and gatherings that boost the image of the destination and the development of a differentiated product.

REFERENCES

Agência Nacional de Águas (ANA). (2005). *O Turismo e o Lazer e sua Interface com o Setor de Recursos Hídricos*. ANA.

Amaral, S. & Ferreira, M.T. (2010) *Caracterização da Pesca Recreativa nas águas interiores em Portugal Continental - Pesca Lúdica e Desportiva Pesca Desportiva em Albufeiras do Centro e Sul de Portuga.l Contribuição para a redução da eutrofização*. Instituto Superior de Agronomia Autoridade Florestal Nacional.

Arlinghaus, R. (2008). *Code of Practice for Recreational Fisheries*. EIFAC 251h session, Antalya, Turquia.

Bătinaú, R., & Sorocovschi, V. (2012). *The Hydrographic Network – Mobilisation Vector of Pollutants in Roúia Montana Area, Roúia Montană in Universal History*. Cluj University Press.

Bochechas, J. (2000). Pesca nas águas interiores. Floresas de Portugal, 217-221.

Brainerd, S. M. (2010). *European Charter on Angling and Biodiversity. Presentation of NINA2 and ESUSG3 specialist on 9 April 2010*.

Castro, E., Vieira, G., Loureiro, F., Patrocínio, F., Firmino, G., Gomes, H. & Fernandes, M. (2017). Managing the Estrela Aspiring Geopark: A framework for promoting geoheritage conservation. *Conference Ciência 2017*.

Cowx, I. G., Arlinghaus, R., & Cooke, S. J. (2010). Harmonis-ing recreational fisheries and conservation objectives foraquatic biodiversity in inland waters. *Journal of Fish Biology, 76*(9), 2194–2215. doi:10.1111/j.1095-8649.2010.02686.x PMID:20557659

Dax, T. (2019). Development of mountainous regions: smart specialization approaches as a means to overcoming peripheralization. In I. Kristensen, A. Alexandre Dubois, & J. Teräs (Eds.), *Strategic Approaches to Regional Development. Smart Experimentationin Less-Favoured* (pp. 52–67). Routledge.

Ditton, R. B., Holland, S. M., & Anderson, D. K. (2002). Recreational fishing as tourism. Fisheries. *Afsjournals, 27*(3), 17-24. http://www.rbff.org/uploads/Research_section/fishing_as_tourism.pdf

Eagles, F. J. P., McCool, F. S. & Haynes, C. D. (2002). *Sustainable Tourism in Protected Areas: Guidelines for Planning and Management*. The United Nations Environment Programme, World Tourism Organization, International Union for Conservation of Nature and Natural Resources, World Commission on Protected Areas (WCPA) Best Practice Protected Area Guidelines Series No. 8.

Ellender, B., Weyl, O., & Winke, R. H. (2009). Who uses the fishery resources in South Africa's largest impoundment? Characterising subsistence and recreational sectors on Lake Gariep. *Water S.A., 35*(5), 677–684. doi:10.4314/wsa.v35i5.49194

Embratur (Brasil). (1996). *Guia Brasileiro de Pesca*. Almeida Prado & Conceição.

FAO. (2008). Codigo de conducta de la pesca recreativa. *Version en espanhol aceptada y presentada por Confederacion Espanola de Pesca Maritima de Recreo Responsable*. http://www.fao.org/3/a-i0363s.pdf

Fernandes, G. (2015). Geografia, Cultura e Desenvolvimento Turístico em Territórios de Montanha. Das Perspetivas às Logicas de Atuação e Governação Territorial, Coleção Instituto Politécnico da Guarda, Guarda.

Fernandes, G., Castro, E., Firmino, G., & Ferreira, A. (2016). Geoparks and regional and local development: aspiring Estrela Geopark. *Abstracts European Geoparks Conference*. 10.13140/RG.2.2.28770.94405

Ferreira. M. T. (2009). *Estudo estratégico para a gestão das pescas continentais*. PAMAF medida 4 – ied, acção 4.4: estudos estratégicos. ICNF.

Ferreira, M. T., & Guerreiro De Brito, A. (2009). Águas interiores supericiais. In H. Pereira, T. Domingos, L. Vicente, & V. Proença (Eds.), *Ecossistemas e Bem-Estar Humano em Portugal*. Fundação da Faculdade de Ciências da U. L. e Escolar Editora.

Gomes, H., Fernandes, M., Castro, E., & Vieira, G. (2017). The (Serra da) Estrela Aspiring Geopark (Portugal): preserving geoheritage, while promoting science and its links to local communities. In *Geophysical Research Abstracts* Vol. 19. EGU General Assembly.

Gonçalves, A., Barbosa, A., Maia, A., Martinho, A., Domingues, D., Alexandrino, P., d'Eça, P., Vieira, S., & Varandas, S. (2008). *A pesca nas águas interiores do Entre Douro e Minho*. Associação Regional do Norte da Pesca Desportiva. Jódique- Artes Gráficas, Lda.

Grecu, F., & Iosif, D. (2014). *La notion du Géosite et son pertinence dans une étude sur la région des Gorges du Danube en Roumanie*. Analele Universităţii Bucureşti: Seria Geografie LXIII.

Gunn, C. A., & Var, T. (2002). *Tourism Planning: Basics, Concepts and Cases* (4th ed.). Routledge.

Higginbottom, K. (2004). *Wildlife Tourism: Impacts, management and planning*. Cooperative Research Centre for Sustainable Tourism.

Hoitsy, G., Woynarovich, A., & Moth-Poulsen, T. (2012). *Guide to trout farming based angling tourism*. FAO.

IBAMA (Brasil). (2010). *Sensibilização da Comunidade para o Turismo de Pesca*. Programa Nacional do Desenvolvimento da Pesca Amadora.

ICNF. (2020). *Edital – Zona de Pesca Reservada das Lagoas da Serra da Estrela 2020*. ICNF.

Lemus, J., Dagostino, R., Olivas, M., & Huízar, A. (2018). *La pesca Deportiva em Bahía de Banderas*. Universidad de Guadalajara.

Madariaga, C., & García del Hoyo, J. J. (2014). Cultura de la pesca, patrimonio y turismo. In *Cultura, mercados y gestión de la pesca artesanal en el Golfo de Cádiz* (pp. 109–131). Universidad de Huelva. doi:10.13140/2.1.3189.5047

McManus, P. A., Hunt, W., Storey, J., & White, J. (2011). *Identifying the health and well-being benefits of recreational fishing*. Retrieved from http://www. tarfish.org/documents/Health and Wellbeing of Rec Fishing 2012.pdf

Ministério do Turismo. (2010). *Turismo de Pesca: orientações básicas. Ministério do Turismo, Secretaria Nacional de Políticas de Turismo, Departamento de Estruturação, Articulação e Ordenamento Turístico, Coordenação Geral de Segmentação. – 2.* Ministério do Turismo.

Miranda, F. (2012). *A Pesca Lúdica e Desportiva no Nordeste Transmontano (Bacia do Douro, Portugal).* Dissertação de Mestrado em Gestão de Recursos Florestais Escola Superior Agrária de Bragança, IPB.

Morar, C., & Pop, A.-C. (2016, November). Water, Tourism and Sport. A Conceptual Approach. *GeoJournal of Tourism and Geosites Year IX, 18*(2), 249–258.

Silva, C., Abrantes, J., & Kastenholz, E. (2018). A Imagem da Serra da Estrela na per-spetiva dos turistas. In *Turismo no Centro de Portugal – Potencialidades e Tendências* (pp. 279–298). Actual.

Taylor, W., Bartley, D., Goddard, C., Leonard, N., & Welcomme, R. (Eds.). (2016). Freshwater, fish and the future: Proceedings of the global cross-sectoral conference. Food and Agriculture Organization of the United.

Templeton, R. (1995). *Freshwater Fisheries Management*. Fishing News Book.

KEY TERMS AND DEFINTIONS

Fishing Tourism: Tourism developed by fishermen who go to destinations where superficial fluvial resources - sea, lakes, dams, rivers, among others - are home to fish populations that provide an experience in contact with nature and consume varied local goods and services.

Fluvial Resources: A set of surface or underground water reservoirs, such as rivers, lakes, reservoirs, canals, marshes, and other aquifers, which have different functions and uses and which require careful management in view of their importance and interest for the population.

Serra da Estrela: The main mountain in Portugal, with a unique geography due to its altitude, geological and geomorphological evolution, constituting an important nature tourism destination, classified as a UNESCO World Geopark.

ENDNOTES

1 Except Bordalo, a Mandatory Discard Species (DO), which must be immediately returned to the water, except during sport fishing competitions when it may be returned at the end of the competition.

2 Except the Sun Perch, a prohibited discard species (DP) whose retention is mandatory, and the specimens cannot be kept or transported alive.

Section 3

Culture and Heritage in the Construction of Differentiated Tourism Products

Chapter 10
Evolution of the Literature on Social Responsibility in the Tourism Sector:
A Systematic Literature Review

María Paula Lechuga Sancho
INDESS, University of Cádiz, Spain

Alicia Martín Navarro
https://orcid.org/0000-0002-9443-6491
INDESS, University of Cádiz, Spain

ABSTRACT

This chapter describes the evolution of research on social responsibility in the tourism sector until November 2017. The authors have compiled a database of 66 articles focused on the field published in academic journals. Bibliometric methods and techniques were used to describe the evolution of scientific activity, countries and active institutions, most productive authors, applied methodologies, and the main topics researched to determine the state and describe the evolution of literature in this field. Based on these analyses, possible ways are proposed for future research to advance this consolidation as a scientific discipline.

INTRODUCTION

At present, society does not perceive companies only from a financial perspective. Changes brought about by the environment, increased globalization of markets, greater social repercussions, increased contribution of companies to community development through wealth creation, as well as other factors, such as the necessity to incorporate good governance practices or carry out socially responsible investments, have motivated organizations to adopt changes in their management employing the implementation of socially responsible practices (Moneva and Hernández, 2009).

DOI: 10.4018/978-1-7998-7339-6.ch010

One of the most important economic sectors with significant growth rates today is tourism. Specifically, according to the Synthetic Index of PIB Spanish Turístico (ISTE)1, the tourist GDP (Gross Domestic Product) in Spain has grown to a rate of 4.2% during the third trimester of 2017. This data shows that the behaviour of the companies that form this sector significantly influences the development of a better, more equitable and more sustainable society, thus contributing to its overall well-being (Golja and Krstinić, 2010; Peña et al., 2016). By its magnitude, tourism is a sector capable of generating social change since it can alleviate problems, such as unemployment and unequal opportunity, and impel the creation of social value (Sigala, 2016). In this context, according to Medrado and Jackson (2015), generally, hospitality companies tend to invest more in environmental initiatives concerning other economic sectors and worry less about environmental issues, such as the carbon footprint.

There is a growing number of studies in the academic literature that address issues related to Corporate Social Responsibility (CSR) in the tourism sector. However, no studies found in the literature make a systematic review of the literature regarding the number of scientific articles produced in this area. In this sense, in an ample attempt to expand on the literature and show a more realistic state of affairs, the objective of the present work is to conduct a thorough review of the scientific production of articles on CSR and tourism sector, using a bibliometric analysis, a methodology that has proliferated in the last decades2 (López and Tortosa, 2002; Agudelo et al., 2003). Essentially, we tried to provide answers to the following questions:

- How has the literature evolved?
- In which journals are these articles published?
- What countries show a more significant concern for this type of research?
- Who are the most productive authors?
- What methodologies were used?
- What has been published in the field of CSR?

Based on the above, this work is developed in the following form: on the one hand, we describe the methodology used to conduct the bibliometric study and, on the other hand, we analyze the main results obtained from the review of the literature.

METHOD

To carry out a qualitative analysis of the scientific production of a particular research topic, an analysis of the number of articles published in different scientific journals has to be conducted (Férnandez, 1998). The bibliometric studies aim to examine works published in scientific journals, capturing information about science through the bibliography analysis (Moravsick, 1989). In this way, a realistic measurement of the scientific production can be obtained through the simple publication count, independent of the length and the nature of the article (Ferreiro, 1993).

In effect, identifying the scientist productivity of any academic discipline requires the study of the different works published on the matter, using the bibliometric methodology. According to Carrizo-Sainero (2000), bibliometrics is now known as bibliometrics from the term Statistical Bibliography applied by Hulme in 1923. It was based on the necessity to conduct a count of the existing publications that began to be prohibitive to researchers due to increased volume. However, it is true that quantity is not the main problem in the subject that concerns us, its dispersion in means of publication.

Based on the contributions of Pritchard (1969), bibliometrics is the science that studies the nature and course of a discipline using the calculation and analysis of the different facets of written communication. In contrast, the bibliometric methodology would handle the quantitative analysis of certain variables present in a published text to approach the situation scientifically at any given time: main topics that dominate, most relevant authors and other standard indicators. To sum up, bibliometrics refers to studying the quantitative aspects of the production, dispersion and use of the bibliography used in science (Carrizo-Sainero, 2000).

This type of work has boomed in recent years, and, in the field of CSR, we found a series of sharp bibliometric studies that have promoted discussion on the level of development achieved in this regard. In Table 1 we show the principal bibliometric analysis found in this area.

The exact and correct application of the bibliometric methodology to the set of works that conform to the research conducted on a particular subject of study requires examining the main variables relating to the bibliometric research. Following the proposals exhibited in literature (Agudelo et al., 2003; Flores et al., 2004; Zubeidat et al., 2004; Martín-Navarro et al., 2018, among others), the selection of the main indices to analyze in the bibliometric study of any area of knowledge could revolve around the following items:

- Titles of the journals in which the articles have been published
- Most productive authors
- Relevant content of the works
- Index of collaboration in publications
- Main languages in which the works have been published
- Year of publication of the works
- Type of research conducted
- Percentage of contribution by countries

If we stay focused on the objective of our study, it would be interesting to determine the sample of scientific journals that will conform to the bibliometric study. Indeed, to find out the state of social responsibility in the tourism sector, in terms of research, we have decided to focus our bibliometric analysis on indexed scientific journals to the year 20173 in the database ABI INFORM of social sciences within the virtual library of the University of Cadiz. The search was achieved through abstracts, using keywords. Specifically, we used the following (combined by pairs):

- Responsabilidad Social (Social Responsibility)
- Responsabilidad Social Corporativa (Corporate Social Responsibility)
- RSE (CSR)
- Turismo (Tourism)
- Turístico/a (Tourist)
- Destino turístico (Tourist destination)

Therefore, this analysis has allowed us to select journals with sharp empirical works focused on social responsibility and/or ethics, which have been framed in the tourism sector. The application of the previous filters has led us to identify 66 works of empirical research in social responsibility. Mainly to carry out this study correctly and have a representative database of the scientific production of CSR articles

Table 1. Antecedents of bibliometric studies in CSR

Authors	Scope Study	Years of Study	Objective	Databases Used	Keywords	Findings
Sabrin (2002)	Business ethics	The period from 1995 to 1999	To know the scientific productivity of the different universities and/or research centres. To know who are the most prolific authors in the field of business ethics	Accounting & Tax Index, Index to Legal Periodicals, Business Periodicals Index, Journal of Economic Literature and the Philosopher's Index and ABI/ INFORM	Not specified	- The professors associated with the area of economy and the company are the researchers with the highest index of scientific productivity (they publish three or more articles in this matter to the year). - The number of specialized professors in business ethics who frequently publishes in this field is scarce. - The most prestigious Business Schools are not the institutions with the highest index of scientific productivity in this area as they are more focused on offering their students a good education in the field of economics and business, with business ethics being considered as a topic secondary from both a teaching and a scientific point of view.
De Bakker *et al.* (2005)	CSR	Period from 1970 to 2002	Try to establish consensus on the degree of scientific development in CSR	ISI Web of Social Science Citation Index and in ABI/ Inform Archive Complete, Global, and Trade&Industry	*CSR, corporate social responsib*, corporate social responsive*, corporate social performance,* and *CSP*	- They found 155 articles centred on CSP, 505 on CSR and 549 on a combination of CSP and CSR. - Since the 1990s, there has been a considerable increase in the literature on CSR. - The Journal of Business Ethics and Business & Society journals have the highest scientific productivity indexes. - Most of the works found are theoretical or descriptive.
Özmen (2010)	Business ethics with an accounting orientation	The period from 1988 to 2007	To examine empirically the scientific production related to accounting ethics	EBSCO Business Source Complete	*Accounting, ethics*	- A total of 248 were found, with the Journal of Business Ethics being the most productive and with the highest number of citations. - Most of the studies focused on the incorporation of ethics and morals in decision-making processes, training in business ethics in the field of accounting or the implementation of codes of conduct in business management

continued on following page

in the tourism sector. Each one of the articles selected has been included in the MS Excel software. Subsequently, different variables or main indexes have been delimited to analyze each one of the studies; these are: the number of articles published by a journal; the year of publication; the topic of study; the institutional affiliation of the authors to departments or research centres; and the applied research methodology or geographic affiliation of authors (Agudelo et al., 2003; Flores et al., 2004; Perez-Serrano et al., 2004). In addition, a count also has been made of the most productive authors, index of collaboration in publications, and studies in the field of CSR by continents.

Table 1. Continued

Authors	Scope Study	Years of Study	Objective	Databases Used	Keywords	Findings
Fraj-Andrés *et al.* (2012)	CSR in SMEs	Period from 2006 to october 2011	Analyze, through a systematic analysis of the literature, the state of the issue of CSR in the context of SMEs. Specifically, they study aspects such as terminology and language, background, barriers and accelerators of CSR in the context of SMEs.	SSCI/JCR Index	*CSR* y *SMEs*	- Most of the articles are based on qualitative studies, being less common in articles based on quantitative methodologies. - Most SMEs are becoming more aware of their RS but do not use CSR when defining them. - SME owners/managers generally find it challenging to identify their stakeholders. - Among the SME accelerators, CSR stands out: improving image/reputation, improving market positioning, increasing personal motivation, qualified personal attraction and increased efficiency. - Barriers include limited perception of commercial utility, implementation costs and budgetary constraints and a general lack of knowledge of aspects such as tools related to CSR and standards.
Herrera et al., (2015)	CSR in SMEs	Period from 1976 to December 2013	To evaluate the general state of CSR in the field of SMEs (Small and Medium Enterprises) from a bibliometric study	Social Science Citation Index	Combination by pair: First group of words (question of search): corporate responsibility, social responsibility, CSR, ethics. - The second group of words (organizational size): small, medium.- Third group of words (the type of organization): enterprise, business, SME.	- Most of the articles are based on qualitative studies, being less common in articles based on quantitative methodologies. - Most SMEs are becoming more aware of their RS but do not use CSR when defining them. - SME owners/managers generally find it difficult to identify their stakeholders. - Among the murillo accelerators, CSR stands out: improving image/reputation, improving market positioning, increasing personal motivation, qualified personal attraction and increased efficiency. - Barriers include limited perception of commercial utility, implementation costs and budgetary constraints and a general lack of knowledge of aspects such as tools related to CSR and standards.

Source: "Authors elaboration"

The codification of the information of each of the selected articles has been done through the technique of content analysis, which is a method of codification of a text (or content) in several groups (or categories), following a selection criterion (Hackston and Milne, 1996). Krippendorf (1980) defines it as a research technique aimed to formulate, from specific data, reproducible and valid inferences that can be applied to their context. Nevertheless, for the objective of our research, the application of content analysis shows a problem regarding possible subjectivity and the appearance of difficulties at the time of assigning some content to the different informative items defined previously. To avoid or to diminish this possible subjectivity, this analysis has been conducted by two researchers, and a third researcher has taken part to coordinate the resolution of discrepancies or different interpretations when it comes to quantifying the information.

RESULTS AND DISCUSSION

Evolution of Publications per Years

As has been shown above, the role of companies in the tourism sector play in the economy is increasingly important. In Table 2, where the distribution per year of publications related to our base study is shown, it is denoted that most studies focusing on the tourism sector and CSR appear in the last decade, with 2014 being the most productive year. This may be due, among other things, to the recent concern about climate change, and the increasing demand, on the part of the tourist, for the involvement of tourism companies in matters of social responsibility. The current tourist, without a doubt, is more sensitized to the environment that surrounds him, which leads him to consider social, environmental or ethical aspects in the selection processes of products and brands (known phenomenon as "responsible consumption") (Galán Ladero et al., 2006). On the other hand, the increase in publications in this area coincides with the end of the global crisis that began in 2008. In order to be faithful to the expectations of stakeholders, companies now feel the need to be transparent and more socially responsible.

Table 2. Publications per year

Year	Articles	No.	%
2004	[62]	1	0,02
2005		0	0,00
2006	[7][52]	1	0,02
2007	[1][15][63]	3	0,05
2008	[18][56]	2	0,03
2009	[2][3}	2	0,03
2010	[8][16][17][41]	4	0,06
2011	[10][29][45][46][48][50]	6	0,09
2012	[20][30][58]	3	0,05
2013	[5][7][22][28][36][37][55][60][65]	9	0,14
2014	[6][11][12][25]27][31][32][39][40][49][61][64][66]	13	0,20
2015	[23][24][35][38][44][51][53]	7	0,11
2016	[4][13][14][21][33][34][42][43][57]	9	0,14
2017	[9][19][26][47][54][59]	6	0,09

Source: "Authors elaboration"

Main Topics

Despite the low yields in the last year and the boom in this field of study in previous years, we can appreciate that the research covers multiple directions due to the lack of an established paradigm. Table 3 presents those conclusions and shows that, among the topics covered in the various papers, the most discussed were those related to "Sustainable tourism", just over 25%. These works focused on determining the different respectful tourist practices, not only concerning the environment but culturally and socially, as well as with the values of the community in which the company operates, allowing residents and tourists to enjoy a positive exchange of experiences.

The second most analyzed topic was "Environmental impact" (15,15%). Some of the papers related to this second topic are focus on prevention systems, identification and evaluation of environmental impact activities of the companies in this sector.

At the same rate, the other subject most analyzed in CSR and tourism literature is "Tourism development". These studies discuss the major active contributions tourism companies make voluntarily to improve society, the economic and environmental scope of the locality in which they operate. Although this is positive in all the economic sectors, it acquires an exceptional value in the case of tourism.

Table 3. Main topics

Main Topics	Articles	No.	%
Sustainable tourism	[3][7][13][14][19][20][21][30][33][34][42] [43][45][52] [56][58][62]	17	25,76
Environmental impact	[5][9][17][22][28][36][37][47][59][60]	10	15,15
Tourism development	[23][24][35][38][39][40][44][51][53][66]	10	15,15
Entrepreneurship	[6][11][12][31][32][49][61][64]	8	12,12
Quality management	[2][8][15][18][27][48][57]	7	10,61
Innovation	[1][25][26][29]	4	6,06
CSR and tourist destinations	[16][46][50][55]	4	6,06
Hotel industry	[4][41][54][63]	4	6,06
CSR and financial performance	[65]	1	1,52
Airline sector	[10]	1	1,52

Source: "Authors elaboration"

Table 4. Article distribution according to the number of authors

No. of Authors per Article	No. Articles	%
1	12	18,18
2	21	31,81
3	17	27,75
4	11	16,66
5	5	7,57

Source: "Authors elaboration"

As Leira (2009) well explains, tourism is, without a doubt, an activity very dependent on the socio-environmental development of the surroundings. Also, it is interesting to note the work related to "Entrepreneurship" (12,12%) and "Quality management" (10,61%), where the CSR is a subject integrated into the concept of quality management and authors examine to what extent a quality model can help companies establish a socially responsible perspective in the way of managing their businesses.

Index of Collaboration and Most Productive Authors

As far as the rate of collaboration in publications, following the same methodology used by Pérez Andrés et al. (2002) or Zubeidat et al., (2004), Table 4 shows that almost all of the articles (75.74%) are written by 1, 2 or 3 authors.

Table 5 includes the most productive authors. Those who have published at least two articles in CSR in the tourism sector. The most productive authors are Ovidiu Ioan Moisescu (3 papers), Lujun His (2 documents), Yan Zhu (2 articles) and Jasveen K. Rattan (2 articles). In the studies of review of accumulated research on a particular topic of study, it is typically found that most of the authors have only published one work (Quiles et al., 2000; Cantón and Domingo, 2008). Indeed, we can observe how, of the 134 located authors, 130 met this characteristic, forming 97.01% of the sample's total. In contrast, the remaining authors, 2.99%, have published two or more articles, with three being the maximum number of articles by the authors found. Considering the classic classification by productivity of authors proposed by Crane (1969), large producers are those authors with more than ten articles, moderate producers are those who present between 5 and 9 works, aspirings are those with between 2 and 4, and passers-by are those with only a single work mentioned. Of the four proposed categories, only two are represented in this study, as shown in Table 5. Concretely, in our sample, there do not appear any large or moderate producers. Of the 134 authors, four authors are in the aspiring category, and 130 (97.01%) are in the passers-by category. Table 5 shows the list of aspiring authors.

Institutional Author Affiliation to Departments and/or Research Centres

To group these authors, following the methodology of Sabrin (2002), the graph shown in Figure 1 shown the division according to the same affiliation.

Focusing on the different academic departments we find a multidisciplinary character of social responsibility that is of interest to a diverse academic community. We found a particular dominion by the authors in the field of economic and social sciences, particularly in business organization (36%), tourism

Table 5. Most productive authors

Author	No. Articles	Category
Moisescu, O.I.	3	Aspiring author
Su, L.	2	Aspiring author
Zhu, Y.	2	Aspiring author
Rattan, J.K.	2	Aspiring author

Source: "Authors elaboration"

Figure 1. Affiliation of the authors whom they research in CSR
Source: "Authors elaboration"

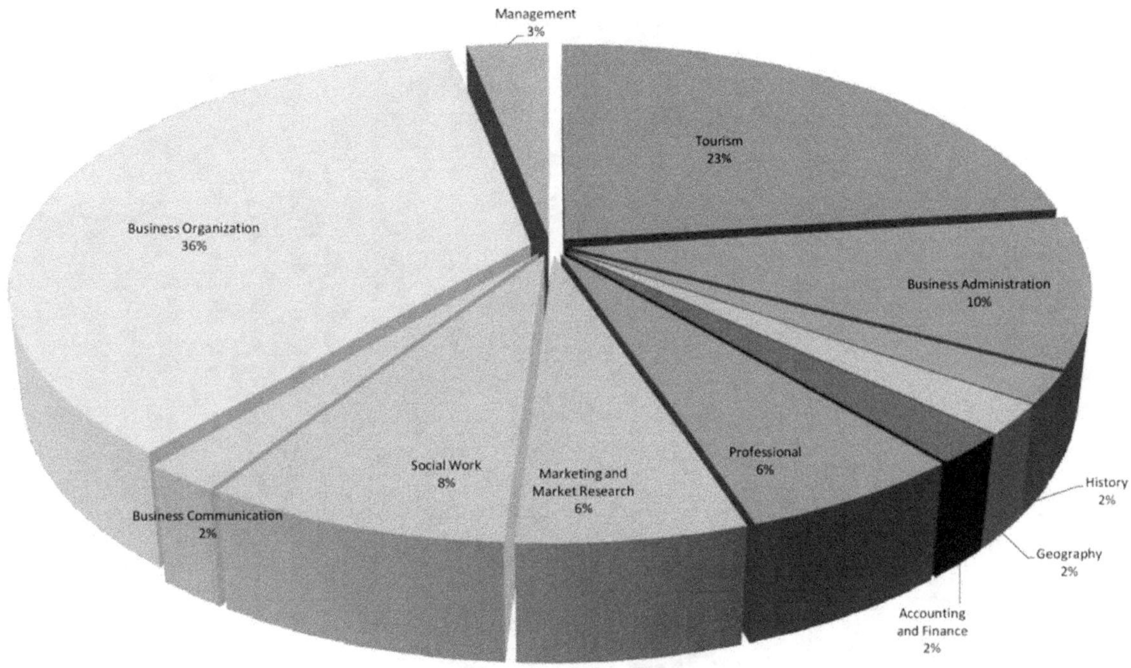

(23%), management (10%), social work (8%), and marketing and market research (6%). This was already revealed in the work of Sabrin (2002), who found that the researchers with the highest rate of scientific productivity in the publication of articles related to business ethics are the authors associated with the area of economics and business. However, in our case, this interest is not confined only to mentioned sciences, as it is possible to observe the participation of very heterogeneous areas of knowledge, such as history, geography, accounting and finances, and business management.

Publications by Continents

In Figure 2, we present the publication location by continents. Interesting enough, most of the studies were developed from samples of European companies (68%), followed by the American continent (28%). It is important to note, at this point, that, despite everything, the first empirical papers found in the field of CSR in the tourism sector were carried out in the United States at the end of the 70s (Gomolka, 1978) and especially the beginning of the 80s (Wilson, 1980; Brown et al., 1982; Chrisman et., 1982; Chrisman et al., 1984).

Let's analyzed the previous results in greater detail (Figure 5). The data obtained show essential differences in the countries that have shown more significant concern for this type of study.

Indeed, most of the works have been concentrated in the United Kingdom (33%), followed by the USA (20%), Croatia and Romania (11% respectively), Spain (6%), Brazil (5%), Venezuela (4%) and finally, Singapore, France, Germany, China, Slovenia, Italy, South Africa and Switzerland, all of them with 2% of the total.

Figure 2. Articles by continents
Source: "Authors elaboration"

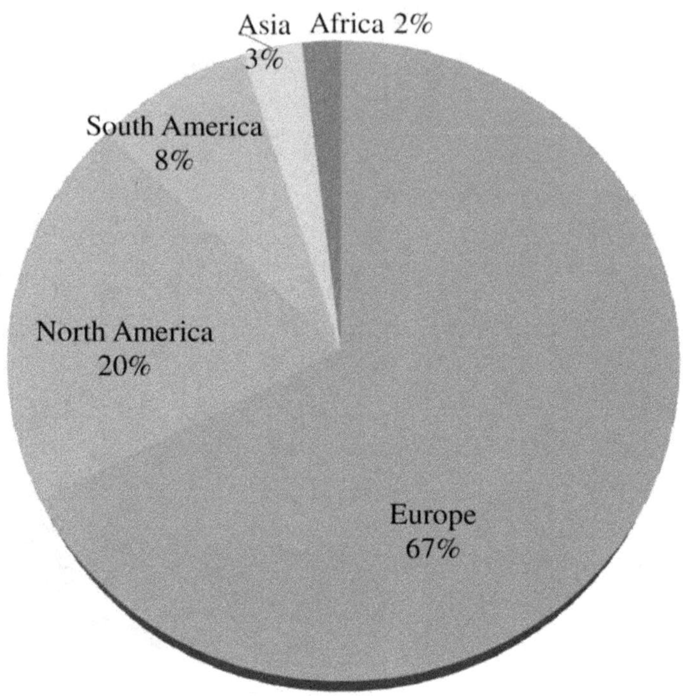

In this sense, in Prado-Lorenzo et al., (2008), whose purpose was to analyze the CSR practices implemented by the large Spanish companies and their relation to business financial results, it is noted that Spain is behind the USA and the United Kingdom in such practices. As these authors show, the reason for this support can be based on culture, which, to date, has not given sufficient or as much importance to responsible business behaviour. They also explain the need for government intervention, as in the United Kingdom, where there is a Ministry of CSR that coordinates the efforts in this area. In the Spanish state context, efforts also have been made (as seen in figure 3), although minor and occasional. In particular, note the report on CSR prepared by the Subcommission of the Congress of the Deputies, the constitution of the Commission of Experts on CSR or the start-up of the roundtable dialogue on CSR within the framework of the processes of social dialogue sponsored by the Ministry of Work (Uriolagoitia et al., 2009).

Main Journals

Another aspect that we have considered in carrying out a study of this nature has been analyzing the journals in which these works are published. We agree with Cox (2009) when affirming that one of the main problems in reuniting literature is a specialization of journals. This is why we have denoted a broad spectrum of journals, which are not limited to specific publications in this field. In this sense, Vázquez-Carrasco and Lopez-Perez (2012) suggest that one of the main limitations or difficulties in access to the works related to CSR is the wide variety of journals in which they are published (Figure 4). On this basis, we found that the journal *International Journal of Contemporary Hospitality Management* is the

Figure 3. Articles by countries
Source: "Authors elaboration"

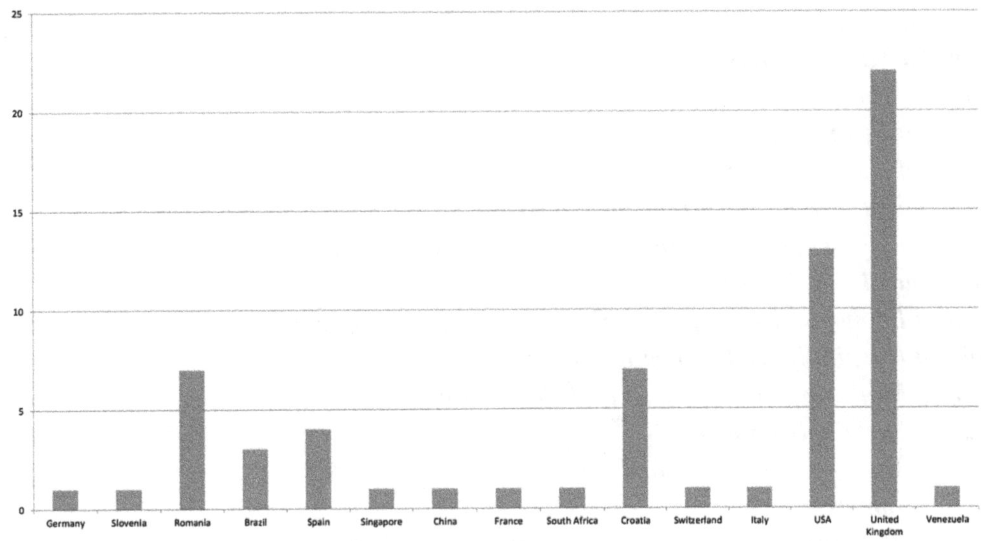

Figure 4. Main journals in which the articles are published
Source: "Authors elaboration"

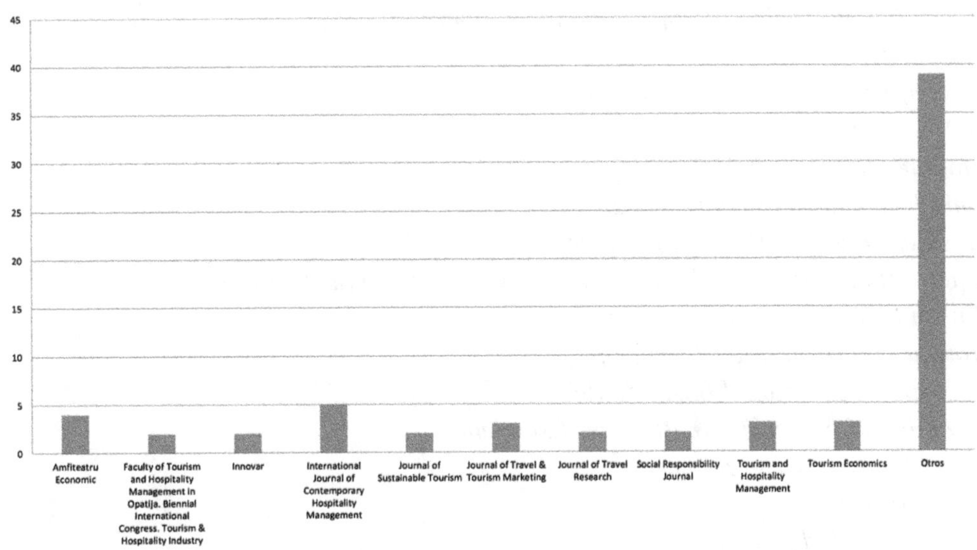

one that presents most of the contributions in this area, followed by *Amfiteatru Economic*; *Tourism and Hospitality Management*; *Tourism Economics*; and *Journal of Travel & Tourism Marketing*.

In the section, others highlight the following journals:

- *Acta Turística*
- *Asia Pacific Journal of Management*
- *Clean Technologies and Environmental Policy*

- *Decisions Marketing*
- *Economics, Management and Financial Markets*
- *Ekonomska Istrazivanja*
- *Ekonomska Misao i Praksa*
- *European Planning Studies*
- *Frontiers of Business Research in China*
- *Frontiers of Health Services Management*
- *International Journal for Responsible Tourism*
- *International Journal of Culture, Tourism and Hospitality Research*
- *International Trade Forum*
- *Journal of Economic Development, Environment and People*
- *Journal of Hospitality & Tourism Research*
- *Journal of Marketing Development and Competitiveness*
- *Journal of Sustainable Tourism*
- *Journal of Tourism Challenges and Trends*
- *Local Economy: LE*
- *Management: Journal of Contemporary Management Issues*
- *Management of Environmental Quality*
- *Pasos*
- *Revista de Administração Pública*
- *Revista de Globalización, Competitividad y Gobernabilidad*
- *Revista Turismo em Análise*
- *RUO. Revija za Univerzalno Odlicnost*
- *South African Journal of Business Management*
- *Sustainable Development*
- *Systemic Practice and Action Research*
- *Systems Research and Behavioral Science*
- *The International Business & Economics Research Journal (Online)*
- *The International Journal of Tourism Research*
- *The Journal of Corporate Citizenship*
- *Thunderbird International Business Review*
- *Total Quality Management & Business Excellence*
- *Tourism Planning & Development*
- *Trziste = Market*
- *Turismo y Sociedad*
- *Worldwide Hospitality and Tourism Themes*

Methodologies

The main methodology used in articles of CSR in the tourism field is the study of cases (Figure 5), which affirms that the main methods of collection of selected data, whether used, individually or together, have been content analysis4 (27%), questionnaires (24%), theoretical studies (18%), and interviews and case studies (15% respectively).

Figure 5. Main used methodologies
Source: "Authors elaboration"

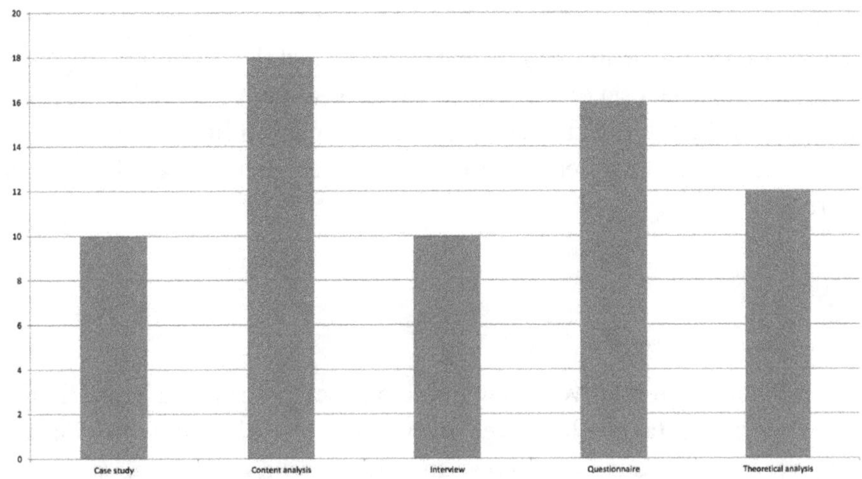

SOLUTIONS AND RECOMMENDATIONS

The contribution of this work to other researchers is two-fold. On the one hand, researchers working in corporate responsibility in the tourism sector have an updated state of the art. They can find out which are the current and emerging schools of knowledge, which topics are the most widely treated and have gaps that could be explored in future lines of research. On the other hand, this work will also be very useful for those researchers and organizations related to documentation services or social science library services, to know what topics and research areas within this area should be promoted or what type of journals should be acquired, which will be useful in future lines of research.

FUTURE RESEARCH DIRECTIONS

As a future line of research, and since Vázquez-Carrasco and the López-Pérez (2012) and Herrera et al. (2015) already did it in their studies, we propose analyzing the main activities of the tourism companies regarding CSR about different stakeholders, according to other analyzed publications. In the same way, we propose determining the main drivers that make tourism companies interested in developing CSR practices and identifying the main factors affecting the implementation of socially responsible practices. We also suggest that the main barriers and accelerating CSR practices tourism companies have implemented be analyzed. And finally, we suggest a study of the intellectual and social structure of the field of using science mapping techniques.

However, the present study shows some limitations that must be mentioned. First of all, the group of indicators and bibliometric techniques used for the content analysis of published articles is an issue. It would be interesting to use other data analysis techniques with different objectives that complement this work, as, for example, the method of co-citations, (Benavides et al., 2011). In the same way, one could expand data searches to new search engines and other data bases, as well as use other terms to search for the resulting articles. Also, content analysis can be subjective, to some extent, since it is not always easy to attribute some of the contents to the new previously defined elements. To avoid or to minimize the possible subjectivity, this analysis has been codified by two different researchers, using a third researcher to resolve discrepancies.

Finally, another important limitation may be linked to the fact that the study used abstracts and titles instead of in full text for the selection of works. However, as already pointed out by Vázquez-Carrasco and Lopez-Perez (2013), although it is necessary to mention certain limitations of the research, it must also be taken into account that the defects of these studies are inherent to the systematic reviews of literature, as a genre.

CONCLUSION

Whereas the term CSR can seem relatively new in the business world, academic literature reveals that the evolution of the concept has taken place over several decades. Although most empirical research on CSR focuses on large companies, our study shows an upward trend in the publication of articles centred on the tourism sector in prestigious journals within the field of business management and business ethics (Taneja et al., 2011).

Notable is the role and position of the United Kingdom in research on an international level, occupying the first position as far as many published works. Nevertheless, the studies carried out in America by pioneers in this field should also be noted. On the other hand, it is necessary to affect the current dispersion, in terms of diversity of scopes of origin of the different authors and the multiple publications that, to an increasingly greater extent, incorporate the various articles related to the question that concerns us. However, the tourism departments and business administration can be identified as those to which a percentage of authors is assigned, and the International Journal of Contemporary Hospitality Management as the publication that offers the most significant amount of articles on CSR and tourism.

This article also suggests many potential research areas related to the subjects frequently discussed in CSR literature in tourism, paying attention to methodological issues. In this sense, the most researched topic in this field has been tourist development and environmental impact. The papers found in these topics focused on identifying different respectful tourism practices. In addition to generating environmental, cultural and social effects, tourism activities allow visitors to integrate with the community. Among the positive aspects developed, it stimulates job creation and the attraction of new investments at the economic level. On a social level, there are exchanges between residents and visitors and promoting the local community's interest in its culture, traditions, and historical heritage. And on the natural environment, its conservation and improvement are stimulated, creating protected natural areas or preserving places of interest.

Of the 66 selected works, 75.74% of articles were written by 1, 2 or 3 authors. However, authorship analysis removed most of them (97.01%.), leaving only those authors who have published in this field and are generally affiliates to academic departments within the scope of economic and business sciences, social sciences, and communication, in general. The fact that we found so many empirical works on CSR-related matters, and the widespread use of qualitative research methods (40.45%), as well as quantitative methods (32.43%), for the analysis of CSR in the tourism sector, could help strengthen the knowledge dimensionally smaller companies have on the subject (Taneja et al., 2011).

REFERENCES

Agudelo, D., Bretón-López, J., & Buela-Casal, G. (2003). Análisis bibliométrico de las revistas de Psicología Clínica editadas en castellano. *Psicothema, 15*, 507–516.

Aláez, F. L. (1993). Bibliometría: (análisis bivariante). Madrid: EYPASA.

Benavides, C., Guzmán, V., & Quintana, C. (2011). The evolution of family firm literature as a research discipline. *Cuadernos de Economía y Dirección de la Empresa, 14*, 78–90.

Carrizo-Sainero, G. (2000). Toward a concept of bibliometrics. *Journal of Spanish Research on Information Science, 1*(2), 59–86.

Cox de Moura-Leite, R. M. (2009). ¿Cómo es la responsabilidad social empresarial en las PYMES? Administrando en entornos inciertos = managingin uncertain environment. Escuela Superior de Gestión Comercial y Marketing, ESIC.

Crane, D. (1969). Social structure in a group of scientists: A test of the "invisible college" hypothesis. *American Sociological Review, 34*(3), 335–352. doi:10.2307/2092499

De Bakker, F. G., Groenewegen, P., & Den Hond, F. (2005). A bibliometric analysis of 30 years of research and theory on corporate social responsibility and corporate social performance. *Business & Society, 44*(3), 283–317. doi:10.1177/0007650305278086

Fernández, R. (1998). Análisis bibliométrico de la Producción Científica. Revista de Divulgación Científica y Tecnológica de la Asociación Ciencia Hoy, 8(44).

Flores, M.J., Muñoz, V., & Cabañero, M. (2004). *Análisis bibliométrico comparativo entre tres disciplinas del ámbito de la salud.* Comunicación presentada en la XV Jornadas de Interrelación de Enfermería, La Cañada, 5 y 6 Mayo.

Fraj-Andrés, E., López-Pérez, M. E., Melero-Polo, I., & Vázquez-Carrasco, R. (2012). Company image and corporate social responsibility: Reflecting with SMEs' managers. *Marketing Intelligence & Planning, 30*(2), 266–280. doi:10.1108/02634501211212019

Galán Ladero, M., Galera Casquet, C., Valero Amaro, V., & Díaz Méndez, M. (2006). Cause-related marketing in Spain. A consumer behaviour perspective. *35th EMAC Conference Proceedings.*

Golja, T., & Krstinić, M. N. (2010). Corporate social responsibility in tourism - the most popular tourism destinations in Croatia: Comparative analysis. *Management, 15*(2), 107-121.

Hackston, D., & Milne, M. J. (1996). Some determinants of social and environmental disclosures in New Zealand companies. *Accounting, Auditing & Accountability Journal, 9*(1), 77–108. doi:10.1108/09513579610109987

Herrera Madueño, J., Larran Jorge, M., & Lechuga Sancho, M. P. (2014). Evolución de la literatura sobre la responsabilidad social en pymes como disciplina científica. *Revista Europea de Dirección y Economía de la Empresa, 24*(2015), 117–128.

Krippendorf, J. (1980). *Marketing im Fremdenverkehr*. Academic Press.

Leira, G. (2009). Responsabilidad Social y Desarrollo Turístico. Oportunidades en Ferrol-Ortegal. *Revista de Ocio y Turismo*, *2*(1), 195–228. doi:10.17979/rotur.2009.2.1.1241

López López, P., & Tortosa Gil, F. M. (2002). Los métodos bibliométricos en Psicología. In *Nuevas Tecnologías de la Información y Documentación en Psicología* (pp. 199–225). Ariel Psicología.

Martín-Navarro, A., Sancho, M. P. L., & Medina-Garrido, J. A. (2018). BPMS para la gestión: Una revisión sistemática de la literatura. *Revista Espanola la de Documentacion Cientifica*, *41*(3), 213. doi:10.3989/redc.2018.3.1532

Medrado, L., & Jackson, L. A. (2015). (in press). Corporate nonfinancial disclosures: An illuminating look at the corporate social responsibility and sustainability reporting practices of hospitality and tourism firms. *Tourism and Hospitality Research*. Advance online publication. doi:10.1177/1467358415600210

Moneva, J. M., & Hernández, J. C. (2009). Responsabilidad social corporativa e información de sostenibilidad en la Pyme. *Revista Internacional de la Pequeña y Mediana Empresa*, *1*(2), 23–41.

Moravsick, M. J. (1989). ¿Cómo evaluar a la Ciencia y a los Científicos? *Revista Espanola la de Documentacion Cientifica*, *12*, 313–325.

Organización Mundial del Turismo. (2007). *Tourism Development and Climate Change: Understanding, Anticipating, Adapting, Participating in the Common Effort*. https://www.un.org/apps/sg/sgstats.asp?nid=2603

Özmen Uysal, Ö. (2010). Business Ethics Research with an Accounting Focus: A Bibliometric Analysis from 1988 to 2007. *Journal of Business Ethics*, *93*(1), 137–160. doi:10.100710551-009-0187-9

Peña, D., Guevara, A., & Fraiz, J. (2016). La investigación de la responsabilidad social empresarial en el sector hotelero. Análisis y revisión de la literatura científica. *Turismo e Sociedade*, *XVIII*, 137–158. doi:10.18601/01207555.n18.08

Pérez Andrés, C., Estrada Lorenzo, J. M., Villar Álvarez, F., & Rebollo Rodríguez, M. (2002). Estudio bibliométrico de los artículos originales de la Revista Española de Salud Pública (1991-2000). Parte Primera: Indicadores generales. *Revista Espanola de Salud Publica*, *76*(6), 659–672. doi:10.1590/S1135-57272002000600003 PMID:12602135

Pérez-Serrano, G., Anta Cabreros, C., Miguel Badesa, S., García Llamas, J. L., Pérez de Guzmán, M. V., & Pérez Serrano, M. (2004). Análisis bibliométrico en educación. Incidencia en la calidad universitaria. Proyecto subvencionado por la Subdirección General de Estudios y Análisis; España. *Referência (Coimbra)*, EA2003–EA0007.

Prado-Lorenzo, J. M., Gallego-Álvarez, I., García-Sánchez, I. M., & Rodríguez-Domínguez, L. (2008). Social responsibility in Spain: Practices and motivations in firms. *Management Decision*, *46*(8), 1247–1271. doi:10.1108/00251740810901417

Pritchard, A. (1969). Statistical bibliography or bibliometrics. *The Journal of Documentation*, *25*(4), 348–349.

Sabrin, M. (2002). A ranking of the most productive business ethics scholars: A five years study. *Journal of Business Ethics*, *36*(4), 355–379. doi:10.1023/A:1014432004643

Sigala, M. (2016). Learning with the market A market approach and framework for developing social entrepreneurship in tourism and hospitality. *International Journal of Contemporary Hospitality Management*, *28*(6), 1245–1286. doi:10.1108/IJCHM-06-2014-0285

Spinak, E. (1996). *Diccionario enciclopédico de bibliometría, cienciometría e informetría.* UNESCO.

Taneja, S., Taneja, P., & Gupta, R. (2011). Researches in corporate social responsibility: A review of shifting focus, paradigms and methodologies. *Journal of Business Ethics, 101*(3), 343–364. doi:10.100710551-010-0732-6

UNWTO. (2004). *ST-EP Programme Sustainable Tourism Eliminating Poverty.* UNWTO, OMT and IOHBTO.

Urriolagoitia, L., Murillo, D., & Lozano, J.M. (2009). *CSR and SMEs: speech to the implementation, a European perspective, contains major European perspective, contains major policies to bring CSR to SMEs developed in Germany, Austria, Denmark, France, Greece, Italy, Netherlands and United Kingdom, and analyze the underlying central lines.* Ministry of Industry, Tourism and Trade of Spain.

Vázquez-Carrasco y López-Pérez, M. (2013). Small & medium-sized enterprises and corporate social responsibility: A systematic review of the literature. *Quality & Quantity, 47*(6), 3205–3218. doi:10.100711135-012-9713-4

Zubeidat, I., Desvarieux, A. R., Salamanca, Y., & Sierra, J. C. (2004). Análisis bibliométrico de la revista journal of sex research (1980-2003). *Universitas Psychologica, 3*(1), 47–54.

ADDITIONAL READING

Apospori, E. (2018). Regional CSR policies and SMEs' CSR actions: Mind the gap—The case of the tourism SMEs in Crete. *Sustainability, 10*(7), 2197. doi:10.3390u10072197

Camilleri, M. A. (2020). Strategic corporate social responsibility in tourism and hospitality. Camilleri, MA, Strategic Corporate Social Responsibility in Tourism and Hospitality. *Sustainable Development*, 1–3.

Chi, C. G. Q., Zhang, C., & Liu, Y. (2019). Determinants of corporate social responsibility (CSR) attitudes: Perspective of travel and tourism managers at world heritage sites. *International Journal of Contemporary Hospitality Management, 31*(6), 2253–2269. doi:10.1108/IJCHM-03-2018-0217

Gligor-Cimpoieru, D. C., Munteanu, V. P., Nițu-Antonie, R. D., Schneider, A., & Preda, G. (2017). Perceptions of future employees toward CSR environmental practices in tourism. *Sustainability, 9*(9), 1631. doi:10.3390u9091631

Kamanga, G., & Bello, F. G. (2018). Exploring corporate social responsibility (CSR) adoption in the tourism industry: The case of Malawi. *Tourism Review International, 22*(2), 117–130. doi:10.3727/154427218X15319286372270

Koseoglu, M. A., Uyar, A., Kilic, M., Kuzey, C., & Karaman, A. S. (2021). Exploring the connections among CSR performance, reporting, and external assurance: Evidence from the hospitality and tourism industry. *International Journal of Hospitality Management, 94*, 102819. doi:10.1016/j.ijhm.2020.102819

Lechuga Sancho, M. P., Martín-Navarro, A., & Ramos-Rodríguez, A. R. (2020). Information Systems Management Tools: An Application of Bibliometrics to CSR in the Tourism Sector. *Sustainability, 12*(20), 8697. doi:10.3390u12208697

Mao, Y., He, J., Morrison, A. M., & Andres Coca-Stefaniak, J. (2020). Effects of tourism CSR on employee psychological capital in the COVID-19 crisis: From the perspective of conservation of resources theory. *Current Issues in Tourism*, •••, 1–19. doi:10.1080/13683500.2020.1770706

Uduji, J. I., Okolo-Obasi, E. N., & Asongu, S. A. (2019). Does CSR contribute to the development of rural young people in cultural tourism of sub-Saharan Africa? Evidence from the Niger Delta in Nigeria. *Journal of Tourism and Cultural Change*, *17*(6), 725–757. doi:10.1080/14766825.2018.1502777

Volgger, M., & Huang, S. S. (2019). Scoping irresponsible behaviour in hospitality and tourism: Widening the perspective of CSR. *International Journal of Contemporary Hospitality Management*, *31*(6), 2526–2543. doi:10.1108/IJCHM-05-2018-0367

KEY TERMS AND DEFINITIONS

Bibliometrics: Part of scientometrics that applies mathematical and statistical methods to all scientific literature and the authors who produce it to study and analyze scientific activity.

CSR: Corporate Social Responsibility. The responsibility that each organization has towards the environment in which it operates and towards its society is a part.

Hospitality: How the guest is treated with kindness and generosity from the welcome at the reception desk, to the way, the staff treats the guest, to the food and beverage service, etc.

SLR: Systematic Literature Review. A synthesis of available evidence article, in which a review of quantitative and qualitative aspects of primary studies is carried out, to summarize the current information on a particular topic.

Social Responsibility: Commitment or obligation of the members of a society either as individuals or as members of a group, both to each other and to society as a whole.

Tourism: Activities that people do when travelling or staying away from their environment. Usually for leisure, business or other purposes and a short period.

Tourism Industry: Business fabric made up of industries dedicated to tourism demand, providing service to tourists.

ENDNOTES

1. Available at: http://www.exceltur.org/indice-sintetico-del-pib-turistico-espanol-iste/ (consulted November 20, 2017).
2. In this regard, interest in bibliometric research has increased as a result of the improvement in electronic repositories of academic publications and access through a greater amount of databases (De Bakker et al., 2005).
3. It is necessary to clarify that all items of the year 2017 that appear in this metasearch engine to current date, i.e. November 15, 2017, are included.
4. Content analysis consists of the study and comparison of sustainability posted information.

Chapter 11
Literary Tourism as a Good Practice to Promote Inland Tourism:
The Case of the Eça de Queiroz Foundation in Portugal

Anabela Sardo
School of Tourism and Hospitality, Polytechnic of Guarda, Portugal

Rita Arala Chaves
Polytechnic of Guarda, Portugal

ABSTRACT

This study is carried out within the scope of the topic "innovative projects and good practices in inland tourism," and it will present the case of the Portuguese project Eça de Queiroz Foundation. The Eça de Queiroz Foundation was founded in 1990 on some of the properties left by the 19th-century Portuguese writer, Eça de Queiroz: the Vila Nova farm and house in Santa Cruz do Douro (Tormes/Baião/Portugal), a small inland village in the Douro, a Portuguese region whose landscape is classified as World Heritage. Currently, the organization has several areas of activity, such as cultural, agricultural, and commercial intervention. The foundation also offers numerous products and services that enhance the promotion of Tourism in the Northern Portugal Inland Region.

DOI: 10.4018/978-1-7998-7339-6.ch011

INTRODUCTION

The World Tourism Organization (UNWTO, 2018) defends the comprehensive and multidimensional character of Tourism, drawing attention to its importance at the most diverse levels, such as cultural preservation, protection of the environment, economic growth and development.

The UNWTO also presents operational definitions on tourism types and says that

Cultural tourism is a type of tourism activity in which the visitor's essential motivation is to learn, discover, experience and consume the tangible and intangible cultural attractions/products in a tourism destination.

These attractions/products relate to a set of distinctive material, intellectual, spiritual and emotional features of a society that encompasses arts and architecture, historical and cultural heritage, culinary heritage, literature, music, creative industries and the living cultures with their lifestyles, value systems, beliefs and traditions (2019, p. 30).

Bearing in mind that literary tourism can be placed in the realm of cultural and heritage tourism and that Culture is also a factor of economic development that produces wealth, generates investment, creates employment, sustainable development and internationalization, this chapter revisits the phenomenon of literary tourism and explores the means by which an inland region of Portugal benefits from it in the form of destination branding, based on the life and work of a national writer.

Thus, this study will present the case of the Portuguese project Fundação Eça de Queiroz / 'Eça de Queiroz Foundation', showing how this organization, starting from the personal patrimony and the work of a renowned Portuguese writer from the last half of the 19th century, managed to preserve part of his patrimonial assets, as well as his recognized literary work, for the pursuit of lasting and socially relevant ends. Those constitute, in the present case, projects of a cultural, commercial and tourism nature that contribute to the development of the Portuguese statistical sub-region (NUTS III[1]), Tâmega e Sousa, included in the statistical North, a territory close to the internationally recognized Alto Douro Wine Region, whose landscape has been a World Heritage Site by UNESCO since 2001.

The methodology was founded firstly on a bibliographic review, in order to provide contextualization and theoretical foundation of the concept of Literary Tourism, based on recent studies. Afterwards, the contents of the Eça de Queiroz Foundation Web Portal (https://feq.pt) were analyzed, namely its curriculum and management reports, the activities and accounts for 2018 and 2019 (2020 not yet available). Therefore, in this chapter, after presenting the necessary theoretical framework regarding the concept of Literary tourism and its typologies, the issue of marketing literary destinations, as well as tourism products and experiences, is addressed, albeit briefly, followed by the concept of literary tourism in Portugal.

Before proceeding to the presentation and characterization of the Eça de Queiroz Foundation, and because it started as a house museum, the relationship between this typology of museums and literature is analyzed, based on those that are the house museums of writers and poets in Portugal.

In the last part of the chapter, the Eça de Queiroz Foundation is introduced, starting with its location and characterization, and then attempting to illustrate how its different intervention areas (Culture, Education, Agriculture and Trade) and the various services and products it offers are relevant for regional development and for the promotion of Tourism in areas of inland Portugal, in this case, in the Northern region of Portugal.

LITERARY TOURISM

Definition

Authors such as Butler (2000), Herbert (2001), Robinson e Andersen (2002, 2004), Sardo (2008, 2009), Hendrix (2014), Baleiro e Quinteiro (2014, 2016), among others, have shown that the places where a writer lives are a source of tourist attraction.

As Quinteiro & Baleiro refer, in their work *Key concepts in literature and tourism studies* (2018), research in literature and tourism requires a fundamental distinction from that of the research field of studies in travel literature, "although they both focus on the act of travelling, there are fundamental differences between these two research areas" (p. 1). On the other hand, and as noted by the authors cited, "furthermore, unlike the travel literature research field, which has existed for a long time, literature and tourism research is recent and, as such, not many scientific studies have been conducted" (Quinteiro & Baleiro, 2018, p.1), especially in Portugal, a country where researchers have only recently become interested in this topic.

Elsewhere, however, this type of tourism has long been studied, as is the example of the United Kingdom, as shown by Richard Butler's investigations, dating from 1986, in which he refers that Literature has an effect in shaping the image of tourism destinations.

As far as the conceptual definition of Literary Tourism is concerned, our study is theoretically based on the recent studies by the Portuguese researchers Quinteiro and Baleiro, who define literary tourism, based on the articulation of the basic concepts of Literature and Tourism, both of which are hard, but not impossible, to outline: "both are difficult concepts to define: literature due to its fluid boundaries and because it is a contextually determined concept; tourism due to its multidisciplinary dimension and because it is still a recent research area that does not yet have an autonomous body of scientific knowledge" (Quinteiro & Baleiro, 2018, p. 26).

Placing the concept of literary tourism in the broader context of cultural tourism, Quinteiro and Baleiro (2018) base themselves on Anne Hoppen (2011, pp. 12-13), the author who integrates it "in the realm of cultural and heritage tourism, highlighting the strong link between literary tourism and cultural tourism, and presenting the former as a subsidiary form of the latter (…)" (p. 29).

Hence, these authors conclude that "literary tourism can be placed both in the context of cultural tourism and heritage tourism, and potentially in the context of creative tourism" (Quinteiro & Baleiro, 2018, p. 31), in the wake of several authors, such as:

1. Greg Richards, who extends the notion of cultural tourism to art tourism, in his book Cultural tourism in Europe, including literature in the sphere of cultural tourism (literary museums, literary landscapes, weeks of literature);
2. Mike Robinson and Hans-Christian Andersen, who understand literary tourism as a niche in cultural tourism or as a type of tourism that appears associated with cultural tourism and creative tourism (because it is based on creative art, literature);

And, also,

3. Shelagh J. Squire and David Herbert, who place literary tourism in the field of heritage tourism.

In an attempt to simplify this "complex network of associations", Quinteiro & Baleiro (2018), as Sardo had already mentioned in 2009, define literary tourism "as a niche of cultural tourism that involves travelling to places that somehow have a connection with literature" (p. 31), adding, in a similar form to Squire (1996) and Nicola Watson (2009), "that literary tourism refers to a type of tourism associated with the places that were celebrated in literary representations, or places that bear a connection to the literary realm" (Quinteiro & Baleiro, 2018, p.33)

In conclusion, literary tourism "occurs when authors or their literature become so popular that people are drawn to either those locations associated with the author (e.g., birthplace, home, graveside) or those featured within their writings" (Busby & Klug, 2001, cited by Hoppen *et al.* 2014, p. 37).

Quinteiro and Baleiro (2018) also draw attention to the question of the evolution of literary tourism, which has been determined, as confirmed by the history of this type of tourism, "by the motivations of those who go on these types of journeys: in search of the author, in search of the literary text and in search of an intellectual and cultural status" (p. 37).

Supported by Mansfield (2015, pp. 78-79) and Quinteiro & Baleiro (2018, p. 37), we may infer that there are circumstances in which the authors themselves intervene by opening their homes to visitors, and others in which local authorities play a crucial role in the creation of literary places.

When creating literary tourism products, there is however another dimension to consider, namely "to create supply that in itself is capable of generating a previously non-existent demand" (Quinteiro & Baleiro, 2018, p.37).

Types of Literary Tourism

Buttler (2006, cited by Hoppen *et al.*, 2014, p. 40) identifies four types of literary tourism. Busby and Klug (2001, also cited by Hoppen *et al.*, 2014, p. 40) mentioned two further types (Hoppen *et al.*, 2014, p.40):

1. Homage to an actual location. This form of literary tourism refers to "a mixture of both fiction-related literary tourism and author-related literary tourism, as, even though it refers to an actual location, it is also about the author's connection to that particular location. Watson (cited by Hoppen *et al.*, 2014, p.40) notes that "the literary tourist's experience is likely to be most powerfully compounded when both the place of composition and that of the fictional setting coincide".
2. Fiction related to literary tourism. It is related merely with places of significance within the work of fiction (Busby, 2004) and thus less focused on the author.
3. Concerned with the appeal of areas because they were appealing to literary (and other) figures.
4. When an area becomes a tourism destination in its own right based on the popularity of an author. This fourth form of literary tourism also applies when an area becomes a "country".
5. Travel writing. It is considered a channel through which both places as well as people have been re-interpreted and communicated to wider audiences.
6. Film-induced literary tourism. It is based on film-induced tourism, whereby a tourist has read an author's work after having viewed the film based on the work.

LITERARY DESTINATIONS AND THEIR MARKETING

For the promotion of literary tourism destinations, it will be necessary to distinguish the qualities of literary sites, which can have two types, according to Herbert (2001, cited by Hoppen *et al.,* 2014): general qualities and exceptional qualities.

The destination has exceptional qualities when the places have connections with the writer's lives, form the settings for novels, when they arouse broader and deeper emotion than the specific writer or the story (e.g., nostalgia) or when they are related to some dramatic event in the writer's life. These exceptional qualities can be associated with general qualities that can also be used to promote their attractiveness. This happens, as Herbert refers (2001, cited by Hoppen *et al.*, 2014), when those literary places are located in attractive settings, such as scenic environments; offer a range of facilities (afternoon teas, souvenirs); and become stopping points along a more general tourism itinerary because of their geographically convenient location (e.g., locations that fit into a route along cathedrals, churches, country houses and gardens).

As a place marketing tool, such sites of interest can also be linked through the development of literary heritage trails, which provide a purposeful and interpreted route for visitors to follow on foot, by car, by bike or even on horseback.

The more exceptional qualities a literary place possesses, the easier it is to develop and exploit for tourism purposes. As Hoppen *et al.* (2014, p.43) refer, a "literary place's existing exceptional and general qualities can arguably form the basis of the particular place's image, which can be exploited for marketing purposes".

Products and Literary Tourism Experiences

In the context of the study carried out, it is necessary to clarify the importance of creating tourism-literary products which, as Martino Alba mentions (2017, p. 60, cited by Quinteira & Baleiro, 2018), does not lie in matching supply to demand but does instead

lie in creating supply that in itself is capable of generating a previously non-existent demand: [...] the need to create a specific demand leads us to conclude that it is extremely useful to encourage the creation of new products based on the great literary authors. For example, by creatively implementing products that may be enhanced by staging and representing parts of texts, and through the playful use of new technologies such as holograms to imaginary conversations with the authors, whose clues we must follow to create new tourism products (p. 37).

As mentioned by Connell (2005, cited by Hoppen *et al*, 2014), there are examples of countless countries in which the "literary tourism has grown into a commercially significant phenomenon" (p.45) and, as a result of this, places connected to literature are used to promote destinations.

Literary Destination: Marketing and Branding Strategies

Literary tourism has been understood as niche tourism (Sardo, 2009). However, more recently, it has migrated "from niche to mass tourism opportunities with literary festivals now expanding their scale and reaching many countries around the world" (Hoppen *et al*, 2014, p.43). Some festivals have reached record numbers of hundreds of thousands of entries, millions of tweets and tens of thousands of books sold.

In recent years, the mass appeal of literature is perhaps best personified in by the global success of the literary figure Harry Potter and Downton Abbey. There was a change from cultural/heritage tourism to international themed development (Hoppen *et al*, 2014), with the creation of theme parks, for example, based on literary characters.

Another very important aspect these days, when you want to promote a literary destination or literary products are "the collaborative approaches to destination marketing via tourist 'routes', 'trails' or 'clusters' offer significant opportunity for new product and experiential destination development" (Hoppen *et al*, 2014, p. 44).

In any literary tourism initiative, it is important to ensure (Hoppen *et al.*, 2014):

1. That all stakeholders are in agreement with the quality of the literary tourism product and its ability to meet, if not exceed, the expectations of visitors.
2. At the same time, authors and publishers will be keen to maintain ownership of the destination's touristic interpretation of their literary content and the degree to which it is commodified for the mass, rather than a niche, market.

LITERARY TOURISM IN PORTUGAL

If it is certain that literary tourism in Portugal does not yet have the same projection that it has in other countries, it is slowly becoming adopted as a cultural phenomenon that one can find in innumerous literary sites, literary routes and literary museums. As a consequence, the number of tourism and literary guides has been increasing.

Moreover, literary tourism has proven to be an important strategy in the development and international projection of Portugal, a country that is already on the list of the United Nations Educational, Scientific and Cultural Organization (UNESCO), Creative Cities of Literature, with the town of Óbidos, for example, recognized for its "vibrant literary community" (https://www.afar.com/magazine/unesco-cities-of-literature-every-book-lover-should-visit). According to Lev-Tov (2019),

This ancient walled city on Portugal's west coast, 50 miles north of Lisbon drew international attention in 2012 when it transformed its Gothic church of Santiago into a massive library and a bookshop (now called Livraria de Santiago), cementing its commitment to literature and its dedication to creative urban renewal. From there, bookstores opened in places like an old fire station, a produce market, and a former wine cellar. Casa José Saramago, the headquarters of Óbidos City of Literature, named after the famous Portuguese author, has information on all the other cities of literature, including guidebooks and a library of Saramago's works for sale in different languages.

Óbidos hosts the annual FOLIO festival, an international literary event that brings authors from across Portugal and the world. Bookworms will dig a stay at the Literary Man, a literary-themed hotel with more than 65,000 books stacked throughout the 20 rooms and common areas. The dining room and gin bar incorporate books into their designs, and guests can schedule a "bibliotherapy" massage in the ancient wine cellar, which is, of course, lined with books. (https://www.afar.com/magazine/unesco-cities-of-literature-every-book-lover-should-visit)

The UNESCO Creative Cities Network (UCCN) was launched in 2004 and aims

to strengthen cooperation with and among cities that have recognized creativity as a strategic factor of sustainable development as regards economic, social, cultural and environmental aspects. By joining the Network, cities acknowledge their commitment to sharing best practices, developing partnerships that promote creativity and the cultural industries, strengthening participation in cultural life and integrating culture in urban development plans (https://en.unesco.org/creative-cities/sites/creative-cities/files/Mission_Statement_UNESCO_Creative_Cities_Network_1.pdf).

This Network "covers seven creative fields: Crafts and Folk Arts, Design, Film, Gastronomy, Literature, Media Arts, and Music" (https://en.unesco.org/creative-cities/sites/creative-cities/files/Mission_Statement_UNESCO_Creative_Cities_Network_1.pdf).

According to the UNESCO,

the Creative Cities Network is a privileged partner of UNESCO, not only as a platform for reflection on the role of creativity as a lever for sustainable development but also as a breeding ground of action and innovation, notably for the implementation of the 2030 Agenda for Sustainable Development (https://en.unesco.org/creative-cities/content/about-us).

Óbidos is representative of the Portuguese usage of "creativity as a strategic factor of sustainable development as regards economic, social, cultural and environmental aspects" (https://en.unesco.org/creative-cities/content/about-us), based on Literature[2]. In fact, many other Portuguese towns could already be part of this international city network to improve their creative potential for sustainable urban development, exchange know-how and cooperate on an international level. To apply, municipalities must send their proposals to UNESCO, in the tab 'Join the Network', filling in the 'Applications of the UNESCO Creative Cities Network' (https://en.unesco.org/creative-cities/content/join-network).

There is a myriad of heritage places in Portugal with a creative potential, some of which have begun to be worked on. This is the case, for example, of the creation of literary museums that, as a tourist and cultural offer, are embodied in house museums, museums and foundations of writers, poets, etc., and become instruments for promoting the image of localities, regions or the country and its sustainable development.

Like any other economic initiative, the tourism offer is aimed at profit, profitability, wealth production and job creation and this is essential for the territories, in particular for the least populated (cf. Rosa, 2013).

Historic House Museums and Literary Tourism in Portugal

Vítor Rosa, in his article, From literature to the tourism itinerary: the house museums and literary tourism (2013), starts with the wide universe of houses, house museums and foundations in Portugal, to present a selection of those dedicated to Portuguese writers and poets, as can be seen in Table 1, in which the Eça de Queiroz Foundation is included.

In his study, Rosa refers the Eça de Queiroz Foundation, located in Quinta de Tormes (Baião/Douro), explaining that Tormes did not exist until it was invented by Eça in his novel A Cidade e as Serras (1901) / *The City and the Mountains*[6]. The word Tormes was adopted by the Foundation for the various

Table 1. Houses, house museums, museums and foundations of Portuguese writers

Houses, House Museums, Museums and Foundations	Location
North Region	
House Pascoaes	Amarante
House museum Camilo	São Miguel de Ceide
House museum Guerra Junqueiro	Porto
House museum José Régio	Vila do Conde
House Antero de Quental	Vila do Conde
House museum Ferreira de Castro	Oliveira de Azeméis
House museum Abel Salazar	São Mamede de Infesta
Eça de Queiroz Foundation	**Baião**
Eugénio de Andrade Foundation [4]	Porto
Center Region	
Camões Memory House [5]	Constância
House museum Afonso Lopes Vieira	São Pedro de Moel
House museum Fernando Namora	Condeixa-a-Nova
House museum Miguel Torga	Coimbra
House museum Vasco de Lima Couto	Constância
Museum Júlio Dinis	Ovar
Lisbon region	
House Bocage	Setúbal
House museum Fernando Pessoa	Lisboa
House museum João de Deus	Lisboa
José Saramago Foundation	Lisboa
Museum Ferreira de Castro	Sintra
Alentejo Region	
House museum José Régio	Portalegre
House museum José Saramago (Azinhaga)	Azinhaga
Algarve Region	
House museum João de Deus	São Bartolomeu de Messines
Azores	
House Vitorino Nemésio	Santa Cruz

(Source: Rosa, 2013, p. 17-18[3])

services and activities on offer: "A day in Tormes", Tormes Library and Archive, International Tormes Colloquia, among others.

He also mentions other literary places linked to the writer, such as Sintra, for example, emphasizing the universal character of Eça de Queiroz's work, as opposed to other Portuguese writers, which should enhance the tourism appeal of literary spaces linked to the author's work.

As Tobelem points out (2003, cited by Rosa, 2013), in Portugal, however, not all places attract tourists, namely cultural tourism fans, despite the rise in a new offer of destination tourism, specifically, that of literary places, houses and museums of writers. In this sense, those responsible should inquire in advance about the opportunity of an offer of this nature, taking into account, for example, the interest of the collections, as well as the attractiveness of the place and the territory where it is located, among other aspects.

Then again, in order to reach a wide audience, locals must have "general and exceptional qualities", according to Herbert (2001, cited by Hoppen *et al.*, 2014) and pay attention to the scenography, the entertainment program and its management (promotional, reception) in a professional way. This equipment must be registered within local tourism policies.

In addition to the house museums, literary tourism offers a wide range of other products and literary experiences, of which we can also highlight literary festivals and literary tours and itineraries. In Portugal, there are several festivals, such as: *Correntes d'Escritas* Literary Festival, in Póvoa de Varzim; *Escritaria* in Penafiel; the Literature in Travel Festival, in Matosinhos; the Douro Literary Festival, in Sabrosa; the Ovar Literary Festival; Madeira Literary Festival; Gardunha Literary Festival; *Tinto no Branco* Festival, in Viseu.

With regard to literary routes and tours, we can mention a few examples, namely

1. The *Levantado do Chão* literary route, a project promoted by the municipality of Montemor-o-Novo, in partnership with the parish councils of the municipality, the municipalities of Évora and Lisbon, and the institutions José Saramago Foundation and the Aljube Museum. It is part of the "Literary Tourism Network of Alentejo and Ribatejo", in turn promoted by the Regional Tourism Authority of Alentejo and Ribatejo;
2. Writers to the North, Lives with Work at *Casas d'Escritas* (https://escritoresanorte.pt/).

It is rather unusual to verify that, in the state document Estratégia Turismo 2027[7], which presents the main strategic orientations of the tourism sector for the coming years, at no time does it have a direct reference to literature or literary tourism. However, the same plan announces as one of the central lines of action, in the area of territory and community enhancement, the conservation and enjoyment of the historical-cultural and identity heritage (TdP, 2017, p. 51). Literary tourism is in line with this fundamental purpose and aligns itself as a potentially important tool in its execution.

This relevance that was given to Literary Tourism has also enabled Portugal to develop projects aimed at tech-savvy tourists. This is the case of the project LITESCAPE.PT - Atlas of Literary Landscapes of Continental Portugal, which developed an app, through a partnership between IELT-FCSH[8] and IHC-FCSH[9] with NOVA LINCS[10] and DI-FCT/UNL[11]. This app uses the repository of literary excerpts compiled and classified in the project's database and constitutes a tool with great pedagogical, research and decision potential, which supports leisure and tourism projects in Portugal.

In Portugal, the real impact (in economic, social and cultural terms) of the Portuguese writers and poets' houses, museum houses, museums and foundations is still unknown. There are various public, private and associative entities seeking to promote them, as well as festivals, tours and literary routes. Statistics (practically nonexistent) and small scientific and technical studies do not allow us to fully evaluate this type of niche tourism. For this reason, it is essential to encourage research within this area to fill this gap.

THE EÇA DE QUEIROZ FOUNDATION

The Eça de Queiroz Foundation (EQF Figure 1.) is a non-profit institution of public administrative utility, aimed towards national and international promotion of the work of one of the most important Portuguese writers, Eça de Queiroz.

The Foundation was created in 1990 by Maria da Graça de Almeida Salema de Castro, the widow of Eça de Queiroz's grandson, Manuel Pedro Benedito de Castro, and by the limited company "João Pires Vinhos, S.A". It started from part of the properties left by the Vila Nova farm and house in Santa Cruz do Douro (Tormes / Baião / Portugal), a small inland village in the Douro, a region from the North of Portugal whose landscape is classified as World Heritage.

Figure 1. Eça de Queiroz foundation
(Source: Susana Luzir)

In an organization like the Eça de Queiroz Foundation, human potential is a fundamental resource: Human resources, at the end of 2019 (EQF, 2019, p. 43), had a total of eight employees, distributed among the different categories / functions, as shown in Table 2.

In addition to the employees, who are part of the Foundation's staff, it is worth mentioning the important contribution of the people who, *pro bono*, collaborate with the organization, particularly the members of its governing bodies, namely the Board of Directors (7 members), the Fiscal Council (3 members) and the Cultural Council (30 members). In summary, the Foundation's bodies are as follows:

- Board of Curators;
- Administrative Council;
- CEO;
- Fiscal Council;
- Cultural Council;

Different types of entities, companies and individuals are part of the Foundation's life, which can be 'curators', 'patrons', or 'friends of Tormes'.

Table 2. EQF human resources - 2019

Function	No. of Workers
Direction	1
Senior technicians	3
Auxiliary staff	2
Agricultural staff	2

(Source: EQF, 2019, p. 43.)

Statutorily, individual curators and collective representatives constitute the Board of Curators. The Chairman of the Board of Curators, elected by his peers, is a member of the Foundation's Board of Directors. In addition to the statutory rights assigned to them, the Curators enjoy several perks, specified in the Foundation's reports and accounts. Banking institutions and city councils are curators, as follows: Banco BPI; Caixa Geral de Depósitos; Póvoa de Varzim City Council; Amarante City Council; Baião City Council; Matosinhos City Council; Sintra City Council; Vila Nova de Gaia City Council.

In 2017, a new model of participation and support for the Foundation was created, through the creation of the figure of Patron, which implies the annual donation of € 5,000 or € 1,000. Until then, only the status of Founder/Co-Founder existed, which implied a single contribution, in the amount of € 50,000, or an annual contribution of € 10,000, for 5 years. The Foundation's patrons can be individuals or collectives, who contribute € 5,000 annually and also, enjoy various benefits. The Foundation's patrons are the following: Marcelo Faria de Lima, António Manuel Queirós Vasconcelos da Mota, Mota Gestão e Participações, SGPS, S.A., Fundação Manuel António da Mota, Mota Engil, SGPS, S.A., José António Ferreira de Barros; SONAE, SGPS, S.A. e Artur Eduardo Brochado dos Santos Silva and the Aveiro Municipality.

It is also thanks to the contribution of the 'Friends of Tormes' that the Eça de Queiroz Foundation develops its activities of perpetuation, organization, promotion and dissemination of the life and work of Eça de Queiroz and, simultaneously, of the Portuguese language and culture, which have a relevant role in the socio-cultural development of its region, reflecting on the cultural, agricultural, tourism and socio-economic level.

Having been statutorily constituted for cultural, educational and artistic purposes, its goals have been subsequently broadened to the area of tourism and social development of the region. The core objectives of the Foundation are as follows:

1. Perpetuate the memory of the writer José Maria Eça de Queiroz, collaborating in the promotion of his work in Portugal and abroad;
2. Organize, maintain and, whenever possible, expand the author's library, archive and museum, installed at its headquarters;
3. Promote the holding of colloquia, conferences, study cycles, training actions or any other manifestations appropriate to the purposes in view, and may also establish prizes for literary works;
4. Encourage activities to support disadvantaged social groups and/or those at risk of exclusion in the region, through training and professional qualification actions and others that are considered adequate for the purposes in view;
5. Contribute to the cultural, agricultural, tourism and socioeconomic development of the Foundation's region, through the promotion of cultural, agricultural, tourism, recreational, leisure and other relevant activities.

In the cultural sphere, the activities are subdivided into three main areas: educational service, training activities and promotional activities. Among the activities within the scope of the educational service, the following stand out:

1. Guided visits to the House Museum (Figure 2) since 1997. Between June 1997 and December 2019, 167,682 people visited Tormes, of which 84,875 were students and teachers integrated in school visits;

Figure 2. The House Museum Eça de Queiroz
(Source: https://feq.pt/servico-educativo/)

2. Pedagogical workshops taking place since 2005.

With regard to training activities, the Foundation develops:

1. International Summer Courses since 1998;
2. International Conference Cycles (2008: Italy, Spain, France, USA, Brazil and England - 1,000 people involved);
3. *Sendas Queirosianas* /International Tormes Colloquia: between 2009 and 2014, they took place in several countries such as Argentina, Brazil (Recife and Rio de Janeiro), France (Sorbonne) and Italy (Rome and Viterbo);
4. Thematic meetings since 1998;
5. Adult Education and Training Courses between 2008 and 2014;
6. Tormes - Center for Studies and Literary Translation, since 2010, developed within the scope of the cooperation established between the Eça de Queiroz Foundation (EQF) and the University of Minho (UMinho).

The Foundation keeps a Tormes Library and Archive and has continued to pursue the work, started in 1997, of stocktaking and organizing the EQF library, with 5,023 books registered until 2019, which are available for public consultation. In addition to the physical library, Eça de Queiroz's manuscripts, in digital format, are also available, relating to the manuscripts that were in possession of the family and that were delivered to Torre do Tombo[12] in Lisbon. This heritage, in digital format, was assigned to the Foundation, by means of a protocol with the National Library of Portugal. The Foundation has a very rich documentary collection from a literary and historical point of view. Constituted by public and private documents, it is divided into three funds: Eça de Queiroz personal archive, archive of the Eça de Queiroz family (Condes de Resende) and António Eça de Queiroz archive.

Regarding promotional activities and others, the following stand out:

1. The Eça de Queiroz Foundation Literary Prize, established in 2014;
2. The Summer Concert with the Orquestra do Norte since 2007;
3. Promotion of the Eça de Queiroz Dinners in Lisbon in 2011;
4. Celebration of the Eça de Queiroz Birthday since 2000;
5. Expansion and maintenance of the Archive and Library since 1997;

Figure 3. Silvério house
(Source: https://www.rotadoromanico.com/pt/experiencias/onde-comer/restaurante-de-tormes/)

6. Participation in the "Escritores a Norte" project, promoted by the Regional Directorate for Northern Culture;
7. Participation in the "e-cultura" project, promoted by the National Centre of Culture;
8. Collaboration with various media and different entities in the dissemination of Eça's life and work since 1997;
9. The organization of cultural activities in partnership with the Municipalities of Baião, since 2006, and Amarante, since 2007;
10. Participation/collaboration in activities to promote the life and work of Eça de Queiroz, organized by other entities, since 1997;
11. The integration of associations in the sector, such as the Asociación Española de Casas-Museo y Fundaciones de Escritores ACAMFE (Spanish Association of House Museums and Foundations of Writers), which in turn belongs to the International Council of Museums (ICOM).

The tourist intervention encompasses tourism in rural areas, in the form of the renovation of an old, rustic house, the Casa do Silvério / Silvério House (Figure 3).

In order to enhance the House and the surrounding space, the Foundation offers products and services:

1. Cultural and gastronomic programs for groups (minimum 10 people) - "Um dia em Tormes". This program, subject to prior booking, includes a visit to the House of Tormes museum space, viewing of a documentary on the life and work of Eça de Queiroz and a Queirozian lunch;
2. The menus include and promote the gastronomic products of the region and are based on the dishes referenced in the works of Eça de Queiroz, particularly in *As Cidades e as Serras /The City and the Mountains* (Figure 4);
3. The spaces and services have been made available to the public. The Foundation has a mini-auditorium, equipped with audio-visual media, with a capacity for about 60 people, which can be used to hold meetings and other initiatives. The use of the auditorium and the visits may, at certain times, count not only with what the House offers but also with specific thematic exhibitions, encompassed in the general objectives and in the Foundation's Activity Plans;
4. The other spaces in Tormes can also be rented for different events, namely the restaurant, installed in the old olive press, the event hall, with capacity for about 200 people, the adaptation of the old courtyard, and the Silvério House.

Figure 4. Eça de Queiroz Menu - gastronomic products of the region based on the dishes referenced in the books of Eça de Queiroz
(Source: https://www.rotadoromanico.com/pt/experiencias/onde-comer/restaurante-de-tormes/)

Figure 5. Tormes restaurant - outdoor Queirozian lunches
(Source: https://visao.sapo.pt/visaose7e/comer-e-beber/2018-10-06-restaurante-de-tormes-em-baiao-eca-agora)

The promotion of Queirozian cuisine has been offered since 1999. Upon request, a Queirozian lunch can be served for groups. In addition, in 2014, the Tormes Restaurant opened, with a permanent Eça de Queiroz menu for visitors (Figure 5).

Since the centenary of the death of Eça de Queiroz, in 2000, the pedestrian path "Caminho de Jacinto" was created, which reflects the path the writer would have travelled during his visit to the House in 1892 and that is fictionalized, through the character Jacinto, in the work *A Cidade e as Serras / The City and the Mountains.*

The Foundation also has agricultural activity, having established, since March 2017, a partnership with the company Lima & Smith for the production, promotion and commercialization of wine products. On the website of Quinta de Covela (https://www.covela.pt/), which belongs to the company Lima & Smith, the visitors can find all the information about the available products.

The commercial activity is similarly being explored, with the Foundation having a sales shop installed in the old wine press of the house museum, where the visitor can purchase several articles: wine, books, jams, jellies, among others. It also has an online store.

Cultural and tourism entertainment is another aspect that the Foundation explores, inasmuch as the previously mentioned initiatives and projects will have a greater motivation and adhesion if they are accompanied by entertainment. Autonomous performances of a musical, theatrical nature, plastic arts events, or others are also held.

According to the *2019 Annual Report and Accounts*, the Eça de Queiroz Foundation has been investing in affirming the institution as one of the most relevant and significant cultural entities in the region where it is based, having achieved, throughout its existence, an undeniable level of acknowledgement, a widespread recognition of success and a consensus on the quality of its cultural program, since it has, over the years, developed a very wide range of activities to promote and publicize Eça's life and work, which constitutes an essential contribution to the cultural enrichment of the region where it is installed, as well as to the national and even international culture (EQF, 2019, p. 5).

Between January and December 2019, 7,907 people visited the Eça de Queiroz House Museum, coming not only from Portugal but also from countries such as Brazil, France, Spain, England, Germany, Italy, USA, Romania, Turkey, Angola, Belgium, Austria, Hungary, Argentina, Australia, Canada, Czech Republic, Netherlands, Israel, Japan, Egypt, Chile and Russia. Of the total visitors, 2,824 were students and teachers integrated in school visits and 233 enjoyed the Queirozian lunch service (FEC, 2019, p. 15).

The number of foreign tourists represented 5% of the total number of visitors to the House, showing that, compared to the previous year, there was a slight increase in visits, representing a total of over 250 visitors. With regard to the Queirozian lunch service, this continues to have little expression, as shown by the figures. Fewer and fewer groups opt for the 'A day in Tormes' program, which, in addition to the visit, includes the Queirozian lunch and the viewing of a documentary. This fact should make this offer be reconsidered, both in terms of planning and in terms of product marketing, for example.

With regard to tourism intervention, particularly with regard to Silvério House, between January and December 2019, it welcomed 174 guests, who stayed 232 nights. Of the visitors, 96% were of national origin, with the remaining 4% coming from France and Brazil. In the same period of 2018, Silvério House had a register of 173 guests, who stayed 263 nights.

The report states that the Foundation has been receiving less and less support from public bodies. However, it has managed to maintain a public service policy, evidenced by the high number of visitors who enjoy a discount on admission, such as schools, or even free admission. They have free visits to schools and groups that come from municipalities that are Curators of the Foundation.

In view of the data presented, it is considered that the Foundation should rethink its marketing and branding strategies, betting on collaborative approaches to destination marketing via tourism 'routes', 'trails' or 'clusters', that offer significant opportunity for new product and experiential destination development (Hoppen *et al.*, 2014). In the 2019 accounts report, it is stated that with regard to communication and image, the Eça de Queiroz Foundation has sought to make these two aspects more effective, taking advantage of the phenomenon of social networks, in order to obtain greater notoriety at national and international level.

Furthermore, it has tried to achieve coherence in the language used in the website and social networks, taking into account the target audience (EQF, 2019, p 44). Nevertheless, the numbers show that it will have to be implemented in an even more effective way, specifically with regard to social networks, since only the use of Facebook is mentioned. The page has about 15,000 followers, a significantly reduced number to achieve the Institution's dissemination and promotion objectives. The management of this network and the inclusion of other social networks are aspects that must be improved in order to reach a greater number of people at national and international level.

The *Annual 2019 Report* refers that another interesting form of dissemination and promotion may be an institutional support for cultural and promotional projects about Eça de Queiroz. An excellent example, to be replicated, is the support for television series and documentaries, such as the Portuguese television drama series, directed by Francisco Manso, titled *Our Consul in Havana*, which premiered on June 7, 2019, on the RTP Rádio Televisão Portuguesa[13], or the documentary *7.2 767 Miles to Bristol - Douro Museum*, directed by André C. Macedo and produced by Look Closer.

On the other hand, although the connection with the media is being worked on, according to information contained in the *Annual 2019 Report*, this is another aspect that should not be overlooked, and the Foundation should have, within the scope of its human resources, a specialist in the field of communication and marketing, with a special focus on digital marketing.

Furthermore, a greater public attraction in the different areas to which the Institution is dedicated would help to overcome the difficulties in capturing revenues, as referred to in the *Annual Report and Accounts* (EQF, 2019, p. 50), which would allow it to develop the necessary investments in maintenance and remodelling of some spaces, such as the House Museum and the Tourism Houses.

CONCLUSION

The 2021 Plan of Activities adjusts and reinforces the various existing activities in the Foundation and, on the other hand, it lays the foundation for new challenges taking into account the dynamics of the context and the conditions and emerging opportunities, always in compliance with the statutory objectives and spirit of service to the region and its development (EQF, 2020, p. 5).

The Foundation continues to pursue its mission, as well as the purposes and objectives, stated in Article 4 of its Statutes, highlighting the task of perpetuating the memory of the writer José Maria Eça de Queiroz, collaborating in the dissemination of his work and promoting the study of his life and work in Portugal and abroad. On the other hand, the organization, maintenance and, whenever possible, expansion of the Queirozian library, archive and museum, installed at its headquarters, remain fundamental objectives. Such a task is closely linked to the holding and promotion of colloquiums, conferences, study cycles, training actions, as well as any other manifestations suitable for the purposes envisaged, as well as the establishment of prizes for literary works.

It is worth mentioning the growing concern with the social responsibility of the organization, revealed in the actions for the benefit of the community, partners, environment, which is a highlight of its plan of activities for 2021, as planned:

1. Promoting activities to support disadvantaged and/or at risk of exclusion social groups, located in the region where the headquarters are located, through training and professional qualification and other actions that are considered appropriate for the purposes in view (EQF, 2020, p. 5);

2. Contributing to the cultural, agricultural, tourist and socioeconomic development of the region where the headquarters is located, through the promotion of cultural, agricultural, tourist, recreational, leisure and other activities that are considered important for the realization of the end in view (EQF, 2020, p. 5).

The educational service and cultural activities represent another important aspect of the Foundation's activities, in particular in the fulfilment of what is established in the Statutes, with regard to the promotion and dissemination of the life and work of Eça de Queiroz, as well as the contribution made to the socio-economic development of the region. Promotional and cultural activities, as well as editorial activities, continue to be developed by the Foundation.

The 2021 Plan of Activities reveals the Foundation's concern with regard to its financial situation and therefore aims to strengthen the establishment of protocols and partnerships with several national and international entities, which can represent an important contribution to the development of statutory objectives and their projection, as well as the existing partnerships, namely with Curators, Patrons, Queirozian Municipalities, Universities, Training Centres and other entities. Another objective to be achieved is the maintenance and, whenever possible, the expansion of collaborations and partnerships with local entities, fundamental cooperation for the development of the region.

With regard to the Foundation's Patrons, an initiative started in 2017, it is likely, in 2021, that a new Patron is included as a result of several contacts in 2020.

At the level of the agriculture sector, only the objectives outlined above are expected to be maintained. The same is true for the tourism sector, which would require greater investment in the planning and promotion of existing literary tourism projects and products, as well as the production and realization of new projects and products.

Although there is always the possibility of making improvements at the most diverse levels, the Eça de Queiroz Foundation has revealed the ability to intervene in different areas of activity and has been able to offer services and products that enhance the promotion of Tourism in the Northern Portugal Inland Region, showing how Culture and Literature can be factors that help promote the economic development of a country and its regions, producing wealth, generating investments, creating employment, sustainable development and internationalization.

REFERENCES

AFAR: The Experiential Travel Guide. (n.d.). Https://www.afar.com

Butler, R. (1986). Literature as an influence in shaping the image of tourist destinations: A review and case study. In J. Marsh (Ed.), Canadian studies of parks, recreation and foreign lands (pp. 111–132). Department of Geography, Trent University.

Butler, R. (2000). Literary tourism. In J. Jafari (Ed.), *Encyclopedia of tourism* (p. 360). Routledge.

Cacilhas, N. (2020). Roteiro literário Levantado do Chão. *Dos Algarves: A Multidisciplinary e-Journal, 37*, 106-122. . doi:10.18089/DAMeJ.2020.37.7

EQF - Eça De Queiroz Foundation. (2018). *Relatório e Contas*. Fundação Eça de Queiroz

EQF - Eça De Queiroz Foundation. (2020). *Plano de Actividades e Orçamento 2021*. Fundação Eça de Queiroz.

EQF - Eça De Queiroz Foundation. (2021). *Web of the Foundation*. https://feq.pt/

EQF - Eça De Queiroz Foundation. (n.d.). *Estatutos da Fundação Eça de Queiroz*. Fundação Eça de Queiroz.

Hendrix, H. (2014). Literature and tourism: Explorations, reflections, and challenges. In Lit&Tour: Ensaios sobre Literatura e Turismo. Vila Nova de Famalicão: Húmus.

Herbert, D. (2001). Literary places, tourism and the heritage experience. Annals of Tourism Research, 28(2), 312-333. doi:10.1016/S0160-7383(00)00048-7

Hoppen, A., Brown, L., & Fyall, A. (2014). Literary tourism: Opportunities and challenges for the marketing and branding of destinations? *Journal of Destination Marketing & Management*, *3*(1), 37–47. doi:10.1016/j.jdmm.2013.12.009

Lev-Tov, D. (2019). UNESCO Cities of Literature Every Bookworm Should Visit. *AFAR*. Retrieved from https://www.afar.com/magazine/unesco-cities-of-literature-every-book-lover-should-visit

LITESCAPE.PT. (n.d.). *Atlas das Paisagens Literárias de Portugal Continental*. https://litescape.ielt.fcsh.unl.pt/.

PORDATA. (2011). https://www.pordata

Queiroz, E. (2008). The City and the Mountains. New Directions Books.

Quinteiro, S. & Baleiro, R. (Orgs.). (2014a). *Lit&tour: Ensaios sobre literatura e turismo*. Vila Nova de Famalicão: Húmus.

Quinteiro, S., & Baleiro, R. (2014b). Uma personagem à procura da literatura: A ficção literária e a prática turística. *Dos Algarves. A Multidisciplinary e-Journal*, *24*, 9–27.

Quinteiro, S., & Baleiro, R. (2017a). Turismo literario. In *Tendencias actuales del turismo en el mundo* (pp. 137–144). Editorial Síntesis.

Quinteiro, S., & Baleiro, R. (2017b). *Estudos em Literatura e Turismo: conceitos Fundamentais*. Universidade de Lisboa. Faculdade de Letras. Centro de Estudos Comparatistas.

Quinteiro, S., & Baleiro, R. (2018). *Key concepts in literature and tourism studies*. University of Lisbon.

Quinteiro, S., Baleiro, R., & Santos, I. D. (Eds.). (2016). *Literatura e turismo: Turistas, viajantes e lugares literários*. Universidade do Algarve.

Richards, G. (Ed.) (1996). Cultural Tourism in Europe. Wallingford: CABI Publishing.

Robinson, M., & Andersen, H. C. (2002). Beyond and between the pages: Literature and tourism relationships. In M. Robinson & H. C. Andersen (Eds.), *Literature and Tourism: Essays in the Reading and Writing on Tourism* (pp. 39–79). Continuum.

Robinson, M., & Andersen, H. C. (2004). *Literature and tourism: Essays in the reading and writing on tourism*. Thomas Learning.

Rosa, V. (2013). Da literatura ao itinerário turístico: as casas-museu e o turismo literário. *Research Gate*. Retrieved from https://www.researchgate.net/publication/344544005_Da_literatura_ao_itinerario_turistico_as_housemuseums_e_o_turismo_literario

Saramago, J. (1989). *Viagem a Portugal*. Porto Editora.

Sardo, A. (2008). Literary Tourism: a way of enriching the local heritage and culture. Egitania Sciencia (vol. 2). Guarda: Instituto Politécnico da Guarda.

Sardo, A. (2009). Turismo literário: A importância do património e dos sítios literários para o desenvolvimento turístico regional. In J. M. Simões & C. C. Ferreira (Eds.), Turismos de nichos: Motivações, produtos, territórios (pp. 339-352). Centro de Estudos Geográficos. Lisboa: Universidade de Lisboa.

TdP - Turismo De Portugal I.P. (2017). *Estratégia Turismo 2027*. Retrieved from https://estrategia.turismodeportugal.pt/

Ugarte, I. A. (2017). Asociación Española de Casas-Museo y Fundaciones de Escritores. RdM. Revista de Museología: Publicación científica al servicio de la comunidad museológica, 70, 81-83.

UNESCO. (n.d.). *Creative Cities Network*. Retrieved from https://en.unesco.org/

Visit Portugal. (n.d.). *Roteiros literários*. Retrieved from https://www.visitportugal.com/pt-pt/content/roteiros-literarios

World Tourism Organization. (2018). *Annual Report 2017*. UNWTO.

World Tourism Organization. (2019). *UNWTO Tourism Definitions | Définitions du tourisme de l'OMT | Definiciones de turismo de la OMT*. Retrieved from www.e-unwto.org/doi/book/10.18111/9789284420858

KEY TERMS AND DEFINITIONS

Eça de Queiroz Foundation (Eqf): Is a non-profit institution of public administrative utility, aimed towards national and international promotion of the work of one of the most important Portuguese writers, Eça de Queiroz. The Eça de Queiroz Foundation has revealed the ability to intervene in different areas of activity and has been able to offer services and products that enhance the promotion of Tourism in the Northern Portugal Inland Region, showing how Culture and Literature can be factors that help promote the economic development of a country and its regions, producing wealth, generating investments, creating employment, sustainable development, and internationalization.

Literary Tourism: Is as a niche of cultural tourism that involves travelling to places that somehow have a connection with literature. It occurs when authors or their literature become so popular that people are drawn to either those locations associated with the author or those featured within their writings.

Literary Tourism in Portugal: Does not yet have the same projection that it has in other countries, it is slowly becoming adopted as a cultural phenomenon that one can find in innumerous literary sites, literary routes, and literary museums. As a consequence, the number of tourism and literary guides has been increasing. Moreover, literary tourism has proven to be an important strategy in the development and international projection of Portugal.

ENDNOTES

1 "NUTS is the Portuguese acronym for 'Classification of Territorial Units for Statistical Purposes', a hierarchical system for dividing the territory into regions. This classification was created by Eurostat, in the early 1970s, with the aim of harmonizing the statistics of the various countries in terms of the collection, compilation and dissemination of regional statistics" (Retrieved from https://www.pordata.pt/O+que+sao+NUTS, February, 2021).

2 In Portugal, other cities are part of this Creative Cities network, albeit in other fields: Music (Idanha-a-Nova and Amarante); Crafts & Folk Art (Barcelos) and Media Arts (Braga).

3 Rosa's sources: Ponte (2007), A. (2008), I*nstitute of Museums and Conservation, City Councils, Foundations and Internet.*

4 The Eugénio de Andrade Foundation was extinguished on September 23, 2011.

5 At the time of Vítor Rosa's study, the Camões Memory House was closed to the public, opening only for exhibitions.

6 QUEIROZ, E. (2008). *The City and the Mountains*. Translator: Margaret Jull Costa. New Directions Books. ISBN: 100811217019.

7 2027 Tourism Strategy.

8 Instituto de Estudos de Literatura e Tradição - NOVA University of Lisbon – School of Social Sciences and Humanities.

9 Instituto de História Contemporânea - NOVA University of Lisbon – School of Social Sciences and Humanities.

10 NOVA LINCS – Laboratory for Computer Science and Informatics, Department of Informatics – Nova School of Science and Technology, NOVA University of Lisbon.

11 Department of Informatics – Nova School of Science and Technology, NOVA University of Lisbon.

12 *Torre do Tombo* is the National Portuguese Archive.

13 Portuguese National Television.

Chapter 12
Music and Musical Heritage as Factors for Rural Development:
Searching for Good Practices

Martín Gómez-Ullate
Universidad de Extremadura, Spain

Javier Barra Sanz
Universidad de Extremadura, Spain

Manuel Rodríguez Palacios
Universidad de Extremadura, Spain

ABSTRACT

This chapter reflects on the understanding and measurement of development and rural development and on how music can influence it. Now more than ever, sustainability must direct development, and life quality indicators must be taken into consideration rather than income, growth, etc. From fieldwork, deep knowledge of different European contexts and a thorough state-of-the-art research, the chapter analyses cases and projects that have had relevant impact in their territories or may be considered for different reasons good practice cases. The aim of the chapter is to analyze those factors that may be considered to evaluate the quality and impact of a case. Researching, safeguarding, and recreating musical heritage produce impacts that go far beyond the economic aspects. Some of these aspects will be highlighted in this chapter. Results and conclusions will serve therefor to advance in research lines related to music tourism, musical heritage, and rural development, but will also be useful for managers, rural agents, local governments.

DOI: 10.4018/978-1-7998-7339-6.ch012

INTRODUCTION

Music has been scientifically proved as a very important factor in human lives since the beginning of human culture. In fact, it is intertwined with other fundamental human activities in most cultures, in a way that it is difficult to talk about music without talking, for example, about agriculture (Stobart, 1994). So does religion or a spiritual dimension of human beings, still remaining in some cultures and being revitalized in post-industrial societies, like the movement around some American cultures medicinal chants that are exchanged, taught, learned and used by more and more occidental people in and out of ceremonies celebrating human beings' connection with the Earth and the main elements and directions (fire, water, wind, ground) (Gómez-Ullate, 2009).

Indeed, music is present in the five domains of the UNESCO`s 2003 Convention for the Safeguarding of the Intangible Cultural Heritage (UNESCO, 2018):

1. Oral traditions and expressions, including language as a vehicle of the intangible cultural heritage
2. Performing arts
3. Social practices, rituals and festive events
4. Knowledge and practices concerning nature and the universe
5. Traditional craftsmanship

As a central element, we can see its importance in the actual World Intangible Heritage constellation (see figure 1).

Music is present as a central element in the categories of Vocal music, Instrumental Music, Dance and Festivals. But it is also fundamental in some cases of oral tradition as the Pansori Epic Chant, in the Republic of Korea, where "during performances lasting up to eight hours, a male or female singer, accompanied by a single barrel drum, improvises on texts that combine rural and erudite literary expres-

Figure 1. UNESCO's constellation of intangible cultural heritage
Source: UNESCO.

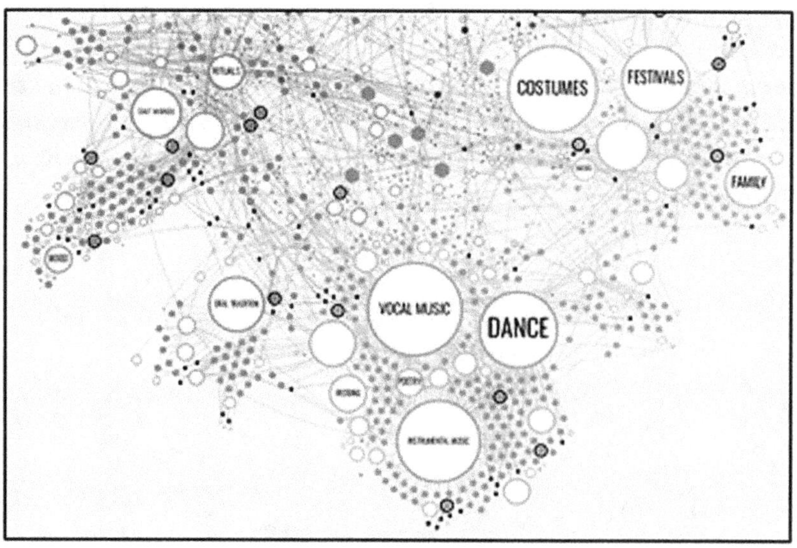

sions" or the Music of the Bakhshis of Khorasan in Iran, where the ''dotār,'' a two-stringed, long-necked lute accompanies the recounting of Islamic and Gnostic poems and epics containing mythological, historical or legendary themes (UNESCO, 2021). It is also a central element in some declared rituals as in Malawi's Vimbuza healing dance or Mexico`s Ritual ceremony of the Voladores.

In the year 2005, it took place in Paris an expert meeting on "Music, dance and theatre traditions". So, as anthropologues as Clifford Geertz (1973), have shown in the Javanese puppet theater, *Wajang*, and we can see in many of the manifestations in the UNESCO intangible cultural heritage list, theater and music are absolutely intertwined. In Portuguese Alentejo, we find the Bonecos de Santo Aleixo (Puppets from Santo Aleixo), always accompanied by a Portuguese guitar, and a repertoire composed of works from the secular tradition, of an essentially religious genre, but also from popular literature. The performances always begin with the Baile dos anjinhos (Dance of the angels) or the Contradança (Contradance). (World Encyclopedia of Puppetry Arts, 2016)

In populated developing countries, music has served as a vehicle for health and safety campaigns or to change traditions attempting against human rights as ablation or arranged child marriages. According to a Food and Agricultural Organization (FAO) study on folk and traditional media for rural development:

"If we are to properly communicate with rural communities, we must learn more about and better understand how to channel our development work through those communities' traditional communication channels" (Ng'ombe, 2000, cited in Kumar et al., 2011).

In European aged and depopulated areas, music serves as a rural development factor mainly through festivals that attract tourism. Music tourism is still an emerging sector in cultural as well as in tourism studies, but music is a constant relevant element in many tourism experiences. While in many cases it seems to be less important than other elements, in some destinations it is a unique and indispensable trait of their cultural landscape and international image. Some of the declared European Cultural Routes of the Council of Europe, as the European Mozart Ways are based on music and some cities as Viena are related to music events. So the Spanish island of Ibiza is, in a very different way. Some countries are also well known as music tourism destinations as Cuba (García de León, 2017) or the United Kingdom, where the Beatles are one of the most important tourism motivations to visit cities as Liverpool. In the latter, due to thorough reports we can find the economic impact of music tourism in cities and in rural areas (UK Music, 2013).

Music tourists are those whose motivations rely mainly in music aspects as attending to a concert, a festival or a place closely related to music (as Abbey road or Jim Morrison`s grave). A specialized report has shown that "in the UK as a whole, far more music tourists attended concerts and gigs than festivals. However, music tourists comprise a higher proportion of festival audiences (...) as some festivals are held in rural locations with very small local populations. Across the UK as a whole, the study revealed 75% of all music tourists attended a concert in 2012, while a quarter attended music festivals" (UK Music, 2013, p. 13).

But it is not only tourism what music reinforces in rural areas, it also has several impacts like enhancing cultural identity, community cohesion, interest in cultural heritage, a regional extension of music and other arts. All these are impacts that can be better related to rural development when we think in terms of a wellbeing perspective on cultural value in rural development (Scott et al., 2018) and in a shift to value quality of life and community social cohesion higher than economic growth as human development indicators. So is the European Union in its 2020 strategy doing when stating smart, sustainable and inclusive growth (Naldi et al., 2015).

BACKGROUND

The research and comparative analysis that sustains this chapter has been done and continues in the framework of two European Projects that analyze and evaluate good practices cases of cultural heritage management and governance in European areas (Spain, Portugal) facing demographic challenges and finding in cultural heritage an important factor for rural development.

For the last 2-3 years, demographical challenges and rural depopulation have been issues completely present in the public agenda and mass media in countries like Spain. The traditional music may be a powerful tractor element for dynamism in the less favored depopulated areas of deep inner Spain and Portugal, and festivals as Guitavera in Spanish Extremadura or Andanças in Portuguese Alentejo play an important role in rural development in the above-mentioned diverse dimensions.

Beyond festivals and routes, the concept of Music Cities or villages give important clues and tips to rural development agents to find in musical heritage and music a powerful lever for dinamyzing their territories. This work deepens in the key elements and strategies that can be valued in that model.

Musical heritage has been certified as world heritage in some cases as Flamenco music (and culture), Fado or, more recently, "Cante Alentejano". These declarations when properly managed may be useful also to attract musical tourism and foster rural development. Intangible Cultural Heritage is fragile, as it is composed by material and immaterial elements, facing threats as economic pressures, loss of objects or systems, emergence of new products and techniques, cultural globalization, weakened practice and transmission, decontextualization, demographic issues, negative attitudes (repressive policies, intolerance, disrespect, conflicts) UNESCO, 2021.

Rural villages worldwide find different ways to deal with their cultural heritage in which music is a central factor: organizing and hosting music and cultural festivals and historical recreations, supporting tradicional dances and singing agroupations (e.g., "Coros y Danzas" in Spain, "Ranchos folclóricos" in Portugal), hiring community music facilitators, awarding chroniclers. There are also combined efforts as the case of Idanha-a-Nova that has been certified as a Music City by UNESCO's programme on Creative Cities.

RURAL FESTIVALS AND LOCAL DEVELOPMENT

Rural festivals have often been managed by Community enterprises (CEs), non-profit community enterprises that create social value for their community attempt to stimulate social and cultural life, increase business development and strengthen community identity with the aim of building community resilience. These enterprises contribute to revitalize communities which are experiencing economic stagnation or decline or are facing the challenges arising from depopulation, (Vestrum et al., 2017)

These organizations have a significant role as turning the intangible (traditional music, folklore, landscape) into tangible (revenues from tourism), turning cultural heritage into rural development. "To do this successfully, CEs need to legitimize their role and position as perceived by different stakeholders, and often need to convince resource providers to accept non-financial gains in exchange for resources." (Vestrum et al., 2017)

Conforming to the internal environment (to the rural community as Vestrum et al. study) is seen as an essential strategy to gain legitimacy for succeeding in the creation and continuity of the rural festivals. This is especially important when the promoters are from outside the community. We can model

Figure 2. Festivals, community enterprises and legitimacy
Source: Authors.

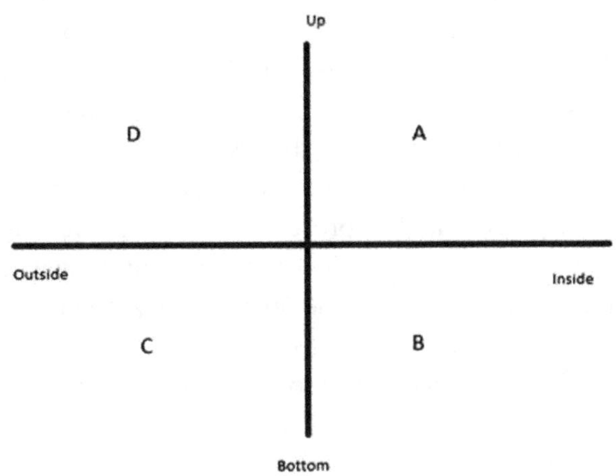

building legitimacy and ensuring the project`s sustainability in a classification of two crucial continuous axis (figure 2).

The ideal case for having legitimacy is B, a festival created and organized for people from the community, individuals and/or civil associations. While the festival can start in this position it can change over time into position A, if the challenges for managing it become too big to be handle by the initial amateur group. This happens in one of our case studies Guitarvera Festival (Gómez-Ullate and Barrios Manzano, 2019) that has gradually involved the municipal government to deal with health and security as the number of visitors exceeded the numbers of initial years and projections.

The outsider condition can happen also in different qualities and degrees. The Community Enterprise can be formed by full outsiders, but can also count with returned migrants, son or descendants of naturals, or full locals –born and residents in the village-. In Guitarvera festival we found a good balance as the CE counted with all types, ensuring legitimacy, but also external links facilitating interchanges with the outside (ideas, groups, visitors).

The SKI festival studied by Vestrum et al. is a case of D, outsiders trying to gain the municipal government support to organize the festival, but from the beginning it engaged local well-known actors moving towards A and towards B as they had a place in the organization. But as this was not enough to gain local public attraction it adapted and converged with the community agenda, places and events as including a sports base camp activity for youths and setting the outdoor opening concert around a traditional coffee meeting (Vestrum, 2017:20). In the same direction, being a Jazz festival, it also included Sami music and traditions.

When focusing in rural areas, a special attention must be given to some festivals, models of good practices. In these sense, we find important cases in Extremadura as they have been studied in the I International Conference on Musical Tourism and Festivals (Gómez-Ullate et al., 2017) and in the International Conference Musical Tourism: History, Geography and Didactics (Cremona, 20-22/11/2019). Festivals that in little villages have consolidated, engaging communities and giving the villages a unique opportunity to multiply their visitors for a 3 days a year. In these cases, culture and nature have a place in their agendas, cleaning and opening walking trails or doing trees planting collective actions. Environmental education and sensibilization play an important role in these festivals (e.g. El Festivalino in

Pescueza -148 inhabs.-, Magusto in Carbajo -200 inhabs.-, Guitarvera in Villanueva de la Vera -2021 inhabs-). These are good practices cases based on music festivals that through the activation of cultural powerful flows have a deep impact on the territories where they are organized. Producing major economic, social and cultural benefits to their populations. These festivals are promoted by associations called Community Enterprises that join local residents, newcomers and migrants together to creatively work on the cultural and natural heritage of these villages.

Nevertheless, the impact of the festival in the populations is relative, it does not stop depopulation. As Scott et al. have researched in Australia the "presence of the creative class, together with select rural amenity indicators, are powerful predictors of firm numbers but appear to have little influence over employment creation in rural Australia, (…) so the attraction of the creative class is unlikely to yield major economic development gains" (Scott et al., 2018).

Even though they do not make the population grow, they influence it in several ways:

- Avoiding businesses closure;
- Avoiding families' migration;
- Empowering social network, community and civic associations;
- Fostering public-private collaborations;
- Building on placemark and international image.

During the festival days, the villages receive visitors varying in numbers (20,000 in El Festivalino, 2,500 in El Magusto and 1,500 in Guitarvera aprox.), but as environmental issues are in the priorities of organizers, they have an impact as well on the local landscape as reforestation campaigns are fostered, trekking roads are cleaned and open to tourism, conferences are speaking about environmental local issues.

Guitarvera Festival, a Best Practices Model

Although it is the one attracting a more reduced number of visitors, we have considered the Guitarvera Festival in Villanueva de la Vera as a good practices case, since it has relevant cultural impact beyond the economics (Gómez-Ullate & al., 2019). This festival celebrated its 19th edition in 2020, with principles as self-sufficiency, sustainability, independence, community-based management, community involvement and animation, education, non-commercial. It has a strong impact in local development but an even higher cultural impact, for it is fostering and enhancing people of the region to perform, dance and play their musical heritage, in an intergenerational way, assuring its continuity and expansion.

A complete "fête de la musique", the cultural encounter as it is labeled by its founders, develops ethnography in traditional music and in local cultural heritage, foster the creation of music groups and other folk festivals as the "Toñá Piornalega". A community-focused festival, as the Boyle Arts Festival in Western Ireland (Mahon, M & Hyyrylainen, T, 2019), it values the role of the elders and devote important moments to them, and serves to fuel the transmission chain of tradition and culture.

We have done anthropological fieldwork and in-depth interviews to most relevant actors (organizers, local inhabitants, visitors, musicians and dancers, hostel and restaurant managers) in some of the past editions of Guitavera, gathering graphic information as the leaflets, films, recordings and pictures.

Figure 3. Villanueva de la Vera.
Source: Authors from Google Maps.

Villanueva de la Vera is a town in the province of Cáceres, one of the two provinces of the Autonomous Community of Extremadura, with roughly 2000 inhabitants that conserves its material and intangible cultural heritage in an exemplary manner and that also finds very interesting channels to manage it.

The Guitavera festival is one of these channels, a powerful collective and community instrument, a model of self-managed festival of social base and free of political or commercial interests, with a series of objectives, principles and three days programme that have changed little since its first edition. The Guitavera was born in 2002, inspired – as it is informed by one of its organizers in the *Encuentro de cuadrillas* in Barrandas, and small folk festivals organized by a national Radio station (Radio 3). In the last editions the nature and activities of the festival have changed little, focusing on a monograph subject and develop conferences, concerts, poetry and theatre performances, trekking walks, anthropological walks (with an ethnographer interviewing key informants on that year's issue) and simultaneous regional traditional music singing and dancing performance on 15 different places of the village. Monographs focus in important aspects of the cultural landscape and contemporary history of the village and region. The list of past editions include "The world of Children", Honey makers, Traditional Architecture, Weddings and marriages, Water, Popular music, Fire, Roads and lanes, Sewing and embroidery, Mills and wineries, Portals, Sweets and pastries, The swamp (Rosarito), Tobacco plant, Flowers and plants.

From their first graphic presentation and leaflets, we can extract the festival fundamentals:

We intend it to be a festive, participatory and joyful day that focuses on:

- *The figure of the "Guitarrero" (guitar player); what he has represented in the music and popular culture of our people.*
- *Traditions and customs transmitted by our elders.*

- *Crafts and tasks that were (and still are) every day in the lives of our people.*
- *We want to:*
- *Show Villanueva and its streets, corners, squares, etc... as a natural setting for group performances and activities.*
- *That the various cultural associations participate and give cultural life to the village.*
- *Give youth the opportunity to discover their folk, artistic and cultural roots.*
- *Recognize, value and thank our elders; they are the fundamental link in the chain in the transmission of our popular culture. We are, we feel, we enjoy... the way that "they" have transmitted it to us.*

The Guitavera has special influence on emigrants that reserve those dates to return to their hometown.

Returned migrants compose the festival`s steering committee and in interviews it was mentioned this way:

Who is more delighted with Guitavera are our peasants abroad. They come and encounter their roots, they feel identified and we touch a lot the sensitive fiber... [they say] «Well, I don't come in Easter and I come to the Guitavera».

All this is possible thanks to the altruism of the organizers who put all their nonprofit effort just "for the love of their village". The organizing committee formed by 10-15 independent persons from the village define themselves as "assemblearians in the way they work" (M., in interview 03/11/2019).

Festivals, Cultural Heritage and Identity

Ethnotourism as a form of experiental tourism should always count on music as an ideal vehicle to understand more deeply the cultural values, discourses and actions (Pomaina Pilamunga, 2015). However, there is complex and delicate relationship between cultural heritage and tourism and a risk to fall in different situations:

- To pervert the essential elements of culture when in its original place, but out of its context for staging culture for tourism purposes falling in what some authors call "staged authenticity" (Chhabra et al., 2003)
- To take out of the cultural and geographical context a cultural element to represent the village or the region.

The latest happens for example with Jarramplas, the festivity of Saint Sebastian in the village of Piornal, in which a masked character with the same name –Jarramplas- pursuit the population with a drum while receiving from the people a turnip rain as they try to take this masked devil out of town. An active cultural heritage safe guarder and promoter told us in an interview:

"I feel ambiguous sentiments about Jarramplas when I see it in a Tourism Exhibition [figure 4]. I know it is a powerful symbol for the village and for Extremadura, but it should not be taken out of the ritual celebration every year at Piornal." (L.P., in interview 05/11/2020)

Figure 4. Jarramplas at FITUR in 2012.
Source: Radio Interior. https://www.radiointerior.es/2012/01/21/noticia-22980/

As territory, cultural heritage and identity are unavoidably linked, identity should always be taken as a factor for rural development as it fosters and feeds tradition in a double direction as it is also reinforced by it.

Music Cities and Other Music-Based Development Projects

Beyond festivals and routes, the concept of Music Cities or villages give important clues and tips to rural development agents to find in musical heritage and music a powerful lever for dinamyzing their territories. The article deepens in the key elements and strategies that can be valued in that model.

A Music City, by its simplest definition, is a place with a vibrant music economy. There is a growing recognition among governments and other stakeholders that Music Cities can deliver significant economic, employment, cultural and social benefits (IFPI, 2015, p. 4). How each city defines success will vary. Some cities have set very ambitious goals for the delivery of economic, cultural and other benefits. For others, it comes down to creating a sustainable environment for music creation, for the sake of music, pure and simple. (IFPI, 2015, p. 10).

Table 1 shows what are the key elements for a music city. These can be attained in major music cities as Memphis, Nashville, New Orleans or London, but also in little villages.

It is important to understand the links between these key elements. Therefore, the originality of the music of a place, forming part of its musical heritage, can lead to the celebration of a festival what may also as well as attracting tourism, foster the creation of new groups and music scenarios.

A vibrant music economy drives value for cities in several important ways: job creation, economic growth, tourism development, city brand building and artistic growth. A strong music community has also been proven to attract other industrial investment, along with talented young workers who put a high value on quality of life, no matter their profession.

Strategies highlighted by experts to succeed in the management of a music place are shown in table 2.

These strategies may seem a bit far from rural municipalities capacities, but they are somehow organically undertaken by the village stakeholders as the community enterprises or independent engaged citizens. Music performance, research and education gather together in some agents as "folk-lovers" or Community musicians.

Table 1. Elements for a music city.

• Artists and musicians • A thriving music scene • Access to spaces and places • A receptive and engaged audience • Record labels and other music-related businesses • Multi-level government support for music • Broader city infrastructure • Music education • Music history and identity • Music tourism • Recognition of music as an economic driver • Strong community radio supporting local independent music • A distinct local sound or sounds

Source: International Federation of the Fonography Industry and Music Canadá (IFPI) (2015)

Table 2. Strategies for a music city.

• Developing a supportive policy framework • Establishing a city-run music office or hiring a music officer • Developing a Music Advisory Board • Engaging the broader music community • Ensuring access to spaces and places • Focusing on audience development • Creating a Music tourism plan

Source: International Federation of the Fonography Industry and Music Canadá (IFPI) (2015)

As important as Community Enterprises to run rural festivals are Community Musicians that is, "musicians committed to people, participation, context, diversity and equality of opportunity through which active music-making experiences happen" (Higgins, 2008). This is what we have seen in some practices in Spain dealing with the research, safeguarding and reproduction of musical heritage. At Festival Guitarvera, it was significant to see how the musicians and researchers that have compiled from elder informants local traditional songs would help them to sing these songs in a public act in the elderly residence when they failed to recall the lyrics.

A well known Spanish community musician, Joaquín Díaz, has devoted his whole life to compile and safeguard traditional Spanish music. With the economic support of the province administration, he founded the Joaquín Díaz Foundation in the very little village of Urueña in the province of Valladolid, counting less than 200 inhabitants, what later would become en "artists enclave" (Scott et al.) with the arrival of other musicians, artists and librarians. The village hosts two Museums of instruments, a museum of bells, a museum of the tale and some specialized bookstores. It was declared City of the Book in 2007 but it was music that boosted all.

Idanha-a-Nova, an UNESCO Creative City of Music

UNESCO has also favored the declaration of music cities in its program for creative cities. The first declared music city was Seville in 2006. A little Portuguese village from the region of Beira Baixa, Idanha-a-nova, has been also certified in 2015 as a music city by this program joining the list of UNESCO

Figure 5. Logo and website of Idanha-a-nova, creative music city.
Source: Municipal government of Idanha-a-Nova

creative music cities that includes Bologne, Glasgow, Gante, Brazzaville, Bogotá, Hammatsu, Mannheim, Hannover. A percussion instrument, the *adufe* played mostly by women, has served as a tractor element and icon for the branding of this music city (see figure 5).

The effort for the proposal and declaration of a music city is led by the municipal government, that from 2001 has undertaken as one of its short, medium and long term strategies to register the territory traditional music. The council has traditional music and dance groups linked to adufe ("ranchos folclóricos"), but also other drum groups (grupo de bombos), tunas or estudantinas, cross-national professional baroque music orchestras -Concerto Ibérico Orquestra Barroca (CIOB)-. International courses and masterclasses on ancient music have been organized during the last years. Since 2012 has organized and celebrated the festival "Fora do Lugar" ("out of place") a free entrance anual festival, that takes place in unusual music scenarios as churchs, oil and wine presses, historic houses, people houses, ancient agricole farms and other cultural spaces.

It also hosts and benefits of the presence of consolidated electronic music festivals as the Boom Festival, "a truly psychedelic global gathering of music, arts, culture & hands-on sustainability" (https://boomfestival.org/), that started in 1997 and now-a-days gather more than 25.000 people from more than 150 different countries.

A query to 140 local inhabitants from Idanha-a-Nova, about the Boom festival in 2008, a 15% did not know the festival but a 36% had already attended the festivals. From this percentage, a 77,8% has attended as visitors, a 18,5% for professional reasons and 3,7% has set up a stand in the festival. In a scale from 1 to 10, the festival was rated 7,79 in average, with only two responses below 5. 89,6% were agree

Table 3. Words synthetizing the Boom festival.

	Total Answers	Answers From Visitors	Answers From Non Visitors
Music	22,3%	22,2%	23,20%
Knowledge		14,8%	
Amusing	11,6%		
Culture	10,7%		
Different/Alternative	8,9%	14,80%	
Positive	8,0%	11,10%	
Drugs	8,0%		12,50%
Nature	7,1%		
Conviviality	7,1%		
Art	7,1%		10,70%

Source: Authors from Garcia, A., Nunes, S. (2009)

to host future Boom editions in Idanha-a-Nova. So, in general terms, the local inhabitants valorization of the festival was very positive. Asked for summarizing the festival in 2 words, we can see in table 3 de different responses given by those who attended and those who did not. (Garcia and Nunes, 2009)

We can see there are relevant differences in the perception and associations with the festival with drugs, but also with knowledge and with art. Local inhabitants' complains with the festival deal with the most usual negative impacts of rural festivals: polution and traffic jamming during the dates of celebration, alteration of prices, noise, alcohol and drug consumption.

SOLUTIONS AND RECOMMENDATIONS

From a cultural and social perspective, festivals favor creativity, the recovery of tangible and intangible heritage, diversify the cultural offer, improve the sociability and progress of civil society, reinforce identity through participation in a collective cultural event and in many cases, they recover public spaces as equipment to show the cultural initiative of the civil society. (Trejo Casero, 2017; Bonet, 2011). We would add that they motivate the creation of music groups, associations and other festivals, reinforcing the associative fabric around musical heritage, promoting musical vocations.

Music and cultural festivals and other heritage interventions as historical recreations, although considered by some authors as "staging culture" or "staging authenticity" (Mac-Cannell, 1976) reinforce identitarian links to the place and move migrants to come back to their hometowns at least once a year and sometimes definitely. Identity is as important as delicate for rural inhabitants. Any attempt to deal with crucial aspects of identity should have in consideration the local opinions and feelings. Under normal circumstances (and not in pandemic crisis) rural populations have shown to tolerate alterity more than expected and value all the beneficial aspects of music festivals no matter how alternative they are.

Community enterprises may have less or more support from the Administration but they can assume the key strategies for the making and managing a music city. From these strategies, the most common and effective are festivals based on –but not only- local music. Administrative support is more needed as a festival grows in number of visitors and consolidate.

Further than the measurable economic impacts that show high cost effectiveness rates, all these socio-cultural impacts and derivations are very difficult to measure but acknowledged by local inhabitants and visitors impacting relevantly the place. Long term strategies to value musical heritage may report future outcomes of different types as economic development, tourism attraction, cultural dynamization, educational reinforcement, international recognition.

FUTURE RESEARCH DIRECTIONS

Music tourism is a very emergent field of cultural tourism and tourism studies. Although festivals have already much more literature devoted to them from the social sciences. Other ways of music tourism as cultural routes or itineraries are much less studied yet. It is important to do longitudinal studies on the performance of urban and rural music cities to reinforce the management strategies for sustainable rural development.

The transversal character of music, present in almost all human aspects, makes of it a very useful resource for tourism and rural development. Even types of tourism that are not included among the cultural tourism as wildlife tourism or birdwatching can also be intertwined with music as it is shown in the documentary "The Birders: A Melodic Journey Through Northern Colombia" (Bleakney, 2019).

At present days, COVID-19 crisis has impacted deeply the music and cultural industry. Rural development based in music needs more than urban development presential conditions to take place. In fact, one of the distinctive factors of Guitarvera festival, deeply analyzed in this chapter is interactivity between public and performers. Although attempts have been done in the 2020 edition for a virtual celebration to avoid breaking its continuity, it has supposed but a subterfuge felt by organizers and public as not replacing the lively, bustling presential edition.

In fact, COVID-19 and the pandemic management has stopped the celebration of secular traditions as well as consolidated festivals. The psychological impact in their inhabitants of these dramatic breaks, should constitute a research line.

CONCLUSION

Festivals have proved to be a very important factor in rural development. Even though they do not solve the demographic challenges that European peripheral areas face, the help to mitigate some of them, attracting creative sector professionals, motivating migrants to come back, reinforcing rural dynamism and local identities.

Culture and cultural heritage can be and are fundamental tourism resources always that are legitimated by the populations that produce and safeguard that heritage. Legitimation of projects and initiatives may vary in time depending on the composition of the Community Enterprise leading those actions.

No matter the main musical focus of rural festivals, they should always try to incorporate local musicians and musical genres, and to hire local professionals. This will increase legitimacy and the possibilities to consolidate in the territory.

As well as festivals, there are other strategies to make music a lever of rural development. When these are varied and based on a rich musical heritage, a music city can be the best framework to create synergies and boost actions.

As intangible cultural heritage faces many different kinds of threats it is very important the top-bottom and bottom-top projects to safeguard it and put it into value. This is what the best practices cases here analyzed have done.

ACKNOWLEDGMENT

This research was supported by FEDER and Junta de Extremadura (ref. GR18186); the European Commission through the Erasmus+ programs [2019-1-ES01-KA204-065372].

REFERENCES

Argent, N., Tonts, M., Jones, R., & Holmes, J. (2013). A creativity-led rural renaissance? Amenity-led migration, the creative turn and the uneven development of rural Australia. *Applied Geography (Sevenoaks, England), 44*, 88–98. doi:10.1016/j.apgeog.2013.07.018

Bleakney, M. (Director). (2019). *The Birders: A Melodic Journey Through Northern Colombia* [Video file]. Retrieved from https://www.youtube.com/watch?v=T-jn1_thxK8

Bonet, L. (2011). Tipologías y modelos de gestión de festivals. In La gestión de festivales: miradas comparadas. Barcelona: Gescènic.

Chhabra, D., Healy, R., & Sills, E. (2003). Staged authenticity and heritage tourism. *Annals of Tourism Research, 30*(3), 702–719. doi:10.1016/S0160-7383(03)00044-6

Garcia, A., & Nunes, S. (2009). Boom festival: Um estudo sobre a percepção dos habitantes locais. *Egitânia Sciencia*, (4), 41–66.

García de León, M. A. (2017). Entre las dos orillas: paisajes musicales y turismo musical en Cuba. In *Turismo Musical y Festivales. Actas del I Congreso Internacional de Turismo Musical y Festivales*. Cáceres: Zenodo. 10.5281/zenodo.569803

Geertz, C. (1973). The interpretation of cultures. Basic Books, Inc.

Gómez-Ullate, M. (2009). *La comunidad soñada*. Plaza y Valdés. doi:10.5211/9788492751228

Gómez-Ullate, M., & Barrios Manzano, P. (2019). Festivales y desarrollo cultural en la provincia de Cáceres. Más allá del impacto económico. Trans 23 dossier: aproximaciones al estudio de las dinámicas e impactos de las celebraciones musicales en españa y portugal.

Gómez-Ullate, M., Barrios Manzano, P., & Gómez Pérez, J. (2017). Turismo Musical y Festivales. In *Actas del I Congreso Internacional de Turismo Musical y Festivales*. Cáceres: Zenodo. 10.5281/zenodo.569803

Higgins, L. (2008) The creative music workshop: event, facilitation, gift. *International Society for Music Education, 26*(4) 326–338. doi:10.1177/0255761408096074

International Federation of the Fonography Industry and Music Canadá (IFPI). (2015). The mastering of a music city. Key elements, effective strategies and why it's worth pursuing. *Music Canadá*. https://musiccanada.com/wp-content/uploads/2015/06/The-Mastering-of-a-Music-City.pdf

Kumar, N., Singh, G., & Parikh, T. S. (2011). Folk Music Goes Digital in India. *CHI 2011, Session: User Studies/Ethnography in Developing Regions*. 10.1145/1978942.1979151

MacCannell, D. (1976). *The tourist: A New Theory of the Leisure*. Class, MacMillan.

Mahon, M., & Hyyrylainen, T. (2019). Rural Arts Festivals as Contributors to Rural Development and Resilience. Sociologia Ruralis, 59(4), 612-635. doi:10.1111oru.12231

Naldi, L., Nilsson, P., Westlund, H., & Wixe, S. (2015). What is smart rural development? *Journal of Rural Studies*, *40*, 90–101. doi:10.1016/j.jrurstud.2015.06.006

Ng'ombe, J. L. (2000). *The communication model and folk media in development*. Sustainable Development Dimensions. http://www.fao.org/sd/CDdirect/CDan0029.htm

Pomaina Pimalunga, B. (2017). La musicología indígena y el turismo étnico: Estudio de caso Colta (Ecuador). In *Turismo Musical y Festivales. Actas del I Congreso Internacional de Turismo Musical y Festivales*. Cáceres: Zenodo. 10.5281/zenodo.569803

Scott, K., Rowe, F., & Pollock, V. (2018). Creating the good life? A wellbeing perspective on cultural value in rural development. *Journal of Rural Studies*, *59*(April), 173–182. doi:10.1016/j.jrurstud.2016.07.001

Stobart, H. (1994). Flourishing Horns and Enchanted Tubers: Music and Potatoes in Highland Bolivia. *British Journal of Ethnomusicology*, *3*(1), 35–48. doi:10.1080/09681229408567224

Trejo Casero, V. (2017). *Análisis de las repercusiones económicas y socio-culturales de un festival: El caso del festival de cultura popular "Guitarvera" en Villanueva de la Vera* (Trabajo Final de Máster, Universitat Oberta de Catalunya). Recuperado de: http://hdl.handle.net/10609/67005

UK Music. (2013). *Wish you were here. Music tourism's contribution to the UK economy*. UK Music. www.ukmusic.org/assets/general/LOWRESFORHOMEPRINTING.pdf

UNESCO. (2018). *Basic Texts of the 2003 Convention for the Safeguarding of the Intangible Cultural Heritage*. https://ich.unesco.org/doc/src/2003_Convention_Basic_Texts-_2018_version-EN.pdf

UNESCO. (2021). *Intangible Cultural Heritage*. https://ich.unesco.org/

Vestrum, I., Rasmussen, E., & Carter, S. (2017). How nascent community enterprises build legitimacy in internal and external environments. *Regional Studies*, *51*(11), 1721–1734. doi:10.1080/00343404.2016.1220675

World Encyclopedia of Puppetry Arts. (2016). *Bonecos de Santo Aleixo*. https://wepa.unima.org/es/bonecos-de-santo-aleixo/

ADDITIONAL READING

Vestrum, I., Rasmussen, E., & Carter, S. (2017). How nascent community enterprises build legitimacy in internal and external environments. *Regional Studies*, *51*(11), 1721–1734. doi:10.1080/00343404.2016.1220675

KEY TERMS AND DEFINITIONS

Community Enterprise: A type of business, company or enterprise with a director, management, employees created to run as a business with all assets controlled, managed, and directed by the community, with a board of directors elected from the community and all profits created contribute to the optimum health, human rights and shared prosperity of the community or sharing with local communities in the area to achieve shared goals and objectives.

Community Musician: A musician committed to, participation, context, diversity and equality of opportunity through which active music-making experiences happen.

Music Festival: An organized event, typically lasting several days, featuring performances by various musicians, singers, and groups.

Music Tourism: The act of visiting a city or town, to see a music festival or other music performances or cultural elements.

Musical Heritage: Musical works, artists who are living human treasures, musical instruments of high technical or historical value, manuscript scores and some editions of great compositions, original documents editions of great compositions, original documents that testify to the history of music, musical recordings of high music, musical recordings of high significance, and even certain types of players of these recordings, because of their technological and historical value.

Chapter 13
Sustainability of Cultural Heritage in Inland Territories:
The Case of World Heritage in Serbia

Milica Maksić Mulalić
Institute of Urban Planning Niš, Serbia

ABSTRACT

This chapter researches the possibilities for the sustainability of cultural heritage in inland territories, focusing on the case study of two archaeological sites in Serbia. Two plans, the plan for the archaeological site of Gamzigrad-Romuliana and the plan for the archaeological site of Caričin grad (Justiniana Prima), were analyzed from the aspect of sustainable development. These plans were prepared in compliance with the principles of preservation, protection, revitalization, and use of immovable cultural property for the purposes of science, education, presentation to the public and tourism. The conflicts between cultural heritage protection and development of tourism were analyzed. As a result of the analysis, recommendations for the sustainability of cultural heritage in inland territories are given in the chapter.

INTRODUCTION

The sustainability of World Heritage is a real challenge. Numerous conflicts have arisen in protected areas between the protection of cultural (and natural) assets on the one hand and development on the other. Construction in such areas is limited, and it has to be planned in accordance with the permitted capacities and protection regimes. In the process of preparing planning documents, it is necessary to systematize all these conflicts and find ways to harmonize them. This applies throughout the planning process and involves numerous actors.

Depending on their significance, according to the Law on Cultural Heritage, there are three categories of cultural heritage in Serbia: cultural heritage, cultural heritage of great importance and cultural heritage of exceptional importance (Zakon o kulturnim dobrima, 1994–2011). The environment surrounding immovable cultural property also enjoys protection, as well as the cultural property itself. Six cultural

DOI: 10.4018/978-1-7998-7339-6.ch013

heritage sites on the territory of the Republic of Serbia are inscribed on the World Heritage List, and another eleven are proposed for enrollment.

This chapter, which presents a case study of the archaeological sites Gamzigrad-Romuliana and Caričin grad, researches the possibilities regarding the sustainability of World Heritage in inland territories. The main research question is: how can conflicts between cultural heritage protection and the development of tourism be harmonized in inland territories?

Gamzigrad-Romuliana and Caričin grad were selected for the analysis because both sites are of international importance and are located in inland territories. Gamzigrad-Romuliana (inscribed on the World Heritage List) is located in Eastern Serbia, on the administrative territory of Zaječar. Caričin grad (Justiniana Prima) (tentative UNESCO list) is located in Southwest Serbia, in the municipality of Lebane. Both sites have plans which were prepared in compliance with the principles of preservation, protection, revitalization and use of immovable cultural property for the purposes of science, education, presentation to the public and tourism.

Firstly, the key points of the institutional and planning frameworks for both sites are presented, in order to understand the broader institutional environment. Secondly, the detailed regulation plans for Gamzigrad-Romuliana and Caričin grad are analyzed from the aspect of sustainable development. Special attention is given to the integral treatment of immovable cultural heritage within its surrounding area, and to protection and revitalization based on integral and continuous spatial and urban planning. The conflicts between cultural heritage protection and the development of tourism are analyzed. By researching planning solutions and the mechanisms for coordinating actors and institutions included in the process, general recommendations for improving the sustainability of cultural heritage in inland territories are given.

BACKGROUND

In terms of covering the dimensions of sustainability, despite an increasing number of references in cultural heritage as a key for sustainable urban development, there is a lack of interdisciplinary understanding of how urban dynamics interact with cultural heritage (Guzmán et al., 2017). At the practical level, operational gaps emphasize a need to focus even more on developing incorporation and planning tools (Mubaideen & Kurdi, 2017).

Being registered on the UNESCO list comes with a great responsibility, both nationally and locally, to protect and preserve the World Heritage site for future generations. Indeed, legislation, policies and strategies should ensure protection of the outstanding universal value of World Heritage properties, promote active participation by the community and all stakeholders, and support the preservation of natural and cultural heritage (UNESCO WHC, 2016).

According to Li et al. (2020a), international cultural heritage management develops an inclusive and integrated approach primarily through a bottom-up process of decision making. This process seeks to collaborate with and empower local communities in the entire process of cultural heritage management (Li et al., 2020a). Local administration should promote new management approaches of world heritage cities, such as a process-oriented approach, which is sufficiently flexible, organizationally decentralized, favouring a multi-stakeholder approach (Petrić et al., 2020).

In order to increase tourism attractiveness of UNESCO World Heritage sites, policy makers and hospitality managers should secure a better allocation of local resources and expertise to implement successful sustainable tourism development plans (Canale et al., 2019). Governing body should adopt specific strategies to engage local actors in decision-making and idea-generation processes, which could help to integrate resources within the system, co-create value and handle conflicts when the goals are not mutually aligned (D'Arco et al., 2021).

Many cities worldwide are developing various documents, mechanisms and institutional arrangements in order to protect, restore or manage cultural heritage sites. Examples of world practice are different. The study of three towns (Stavanger, Mosjøen and Risør) has shown that the conservation plan itself was not the most important factor to the relative success of the conservation of these areas. Preservation plans were combined with management tools adopted locally and practised with flexibility to allow for the changes necessary to modern living (Nyseth & Sognnæs, 2013).

According to Stunpo (2011), in the case of the Syracuse archaeological site in Italy, a detailed urban plan was developed for the historic core, which defines the policies and criteria of intervention and functions compatible with the historic urban fabric. A central management office was established that allowed experts to guide the management process and its strategic coordination. Financing procedures were simplified, allowing the private sector significant participation (Stunpo, 2011).

In Oaxaca de Juárez, Mexico, there was early involvement of public and private actors, as well as civil society, in the preservation and restoration of heritage sites. Several plans for preservation and management were developed, including a specific management plan for the area (Plan de Manejo para el Centro Histyrico de la Ciudad de Oaxaca) (Quartesan & Romis, 2011). In the Old Town of Aleppo, a comprehensive protection and development plan was developed through a participatory process, along with special construction rules and guidelines on conservation and construction standards. A list of facilities that can be used to attract new businesses was prepared and the development of tourism promoted. The Old Town Directorate was established, with the aim of developing, planning, coordinating and implementing works (Spiekermann, 2011).

In Lijiang, stakeholder groups have already been engaged in the decision-making of local heritage management, such as discussing and approving new policies and institutional regulations. The representatives of both native and migrant residents were invited to governmental meetings, together with other stakeholders, such as experts and local administrators (Li et al., 2020b). In the Seongbuk-dong district of Seoul, efforts have been made to strengthen community participation, including the introduction of a "Community Planning Group" policy and to make new institutional opportunities for community participation from the planning stage (Seo, 2020).

In Kasthamandap, the local community within the monument's vicinity, educated with conservation facts and procedures, raised questions to the authorities about the correct procedures for conservation. The voice has been raised in favor of community engagement and awareness, quality of work, transparency in the process and against the unhealthy procurement (Joshi et al., 2021). Inspired by the Historic Urban Landscape approach, the City of Ballarat in Australia facilitated a large conversation with the community called Ballarat Imagine, to be better informed on their interests and needs. This participatory engagement project successfully produced well established procedures and a community vision of local conservation and development (Li et al., 2021).

At Bannack State Park in North America, collaborative efforts between the Montana Office of Public Instruction (OPI) and Montana Fish Wildlife and Parks (FWP) increased visibility of indigenous histories at state park sites through primary school education (Briwa, 2021). In the case of the Big Wild Goose Pagoda Area in Xi', the special agency, the Qujiang Management Committee, as a quasi-governmental branch of the local government, generated great benefits for developers and the local government, while it adversely affected local communities (Zhao et al., 2021).

There is a growing number of innovative tools for community-led heritage management, tourism planning and inclusive governance for the sustainable development of heritage in rural contexts (Gallou & Fouseki, 2019). In Óbidos (Central Region, Portugal), while seeking to reverse the process of social and economic decline, the elected mayor initiated a development strategy to convert the rural-based economy into one based on culture and creativity, supported by a strong marketing strategy: "Creative Óbidos". Two municipal companies were created to manage local initiatives, and diverse public-private partnerships were promoted with higher education institutions, training companies, and business associations (Caldeira & Tomaz, 2015). In order to manage the properties in the historic center of Český Krumlov (South Bohemia, Czech Republic) that are registered by UNESCO, the local administration established a company, the Český Krumlov Development Fund, and its own department of tourism, Destination Management, in order to ensure the implementation of the City Council's strategy (Caldeira & Tomaz, 2015).

In order to overcome COVID-19 pandemic problems, new solutions for cultural heritage tourism are suggested. Frey and Briviba (2021) propose that the heavily visited historical sites are to be replicated in a suitable nearby location. Advanced digital technology, such as augmented and virtual reality, holograms, and digital twins are to be used to make the cultural sites attractive to all sorts of tourists (Frey & Briviba, 2021).

Cultural heritage is not affirmed as a development resource in Serbia and the protection of cultural heritage is still treated sectorally. There is still no national strategy, while the legal solutions from 1994 have been surpassed. A specific problem is the status of cultural heritage on the UNESCO World Heritage list (Zakon o Prostornom planu Republike Srbije od 2010. do 2020. godine, 2010).

Cultural heritage in rural areas of Serbia has enormous potential as a development resource. Rural areas cover approximately 85% of the total territory of Serbia and about 55% of the population lives in these areas (Ministarstvo poljoprivrede, šumarstva i vodoprivrede, 2009). However, Igić et al. (2015) point out that these sites are facing huge problems and that many of them are endangered. Uneven valorization of cultural heritage and in some cases the absence of its protection, as well as the lack of modern technologies, are some of the problems.

Since 2016, Serbian tourism has been recording a constant growth of all indicators, with changes in their structure. In terms of competitiveness, it was ranked in 2019 by the World Economic Forum as 83rd among 140 countries in the world. However, insufficient and inadequate valorization of natural and anthropogenic tourist potentials, insufficiently affirmed concept of sustainable tourism, insufficient development of complementary activities, are some of the limitations for tourism development (Prostorni plan Republike Srbije od 2021. do 2035. godine).

The protection of cultural heritage and tourism development are recognized as topics in spatial and urban planning in Serbia. The arrangement of cultural heritage sites in Serbia is mostly carried out through spatial plans for special purpose areas and detailed regulation plans. Spatial plans for special purpose areas are adopted for areas that require a special regime of organization, arrangement, use and protection of space, and especially for: an area with natural, cultural, historical or ambient values. A detailed regulation plan contains in particular detailed purpose of the land, the rules of arrangement and rules of construction by units and zones (Zakon o planiranju i izgradnji, 2009-2020).

Figure 1. Gamzigrad-Romuliana (1) and Caričin grad (2) on the map of Serbia

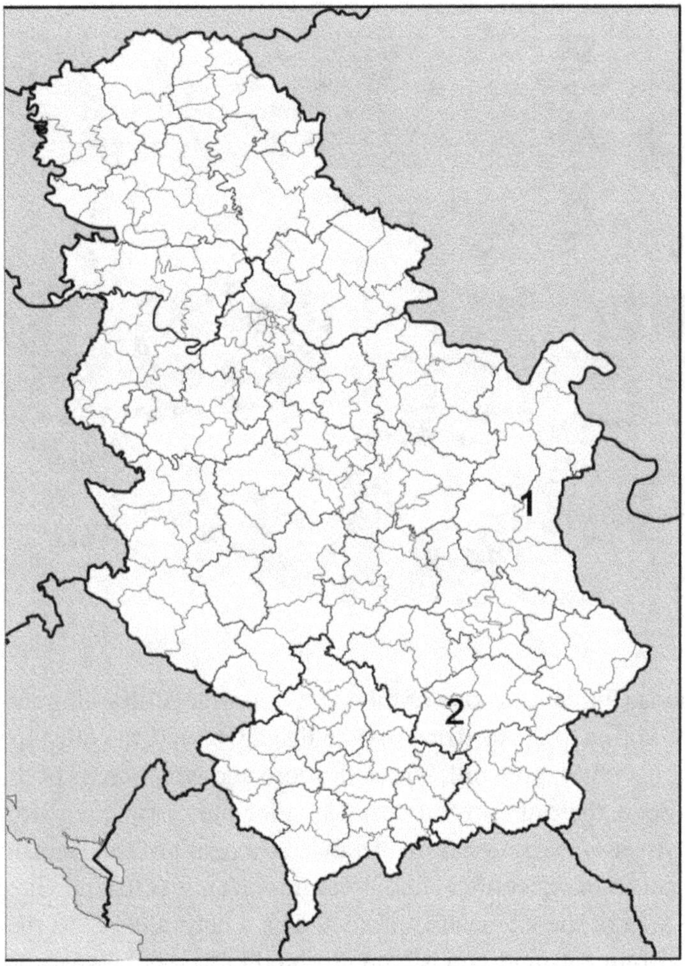

GAMZIGRAD-ROMULIANA AND CARIČIN GRAD (JUSTINIANA PRIMA) SITES

The archaeological site of Gamzigrad-Romuliana is located in Eastern Serbia, on the administrative territory of Zaječar, south of the settlement of Gamzigrad (Figure 1). There are 42 settlements on the territory of the City of Zaječar, but apart from the city itself, all the others are rural. According to the census data from 2011, the administrative territory of City of Zaječar, with its 42 settlements, had 59,461 inhabitants, and Gamzigrad, the nearest settlement to Romuliana, had 683 inhabitants (Republički zavod za statistiku, 2014).

According to the volume and structure of projected investments for the City of Zaječar, 32% are focused on developing the economy (small and medium enterprises), 21% on developing infrastructure, 18% on developing tourism, 16% on developing agriculture and 13% on improving human resources. The business climate in Zaječar is relatively unfavorable. The unemployment rate is still above the national average, and the level of salaries is still lower than the national average. There has also been a decrease in the number of people in employment. Lack of funds for local development programs and sustainable tourism are one of the threats for tourism development (Prostorni plan teritorije grada Zaječara, 2012).

Figure 2. The archaeological site of Gamzigrad-Romuliana
(author's archive)

Gamzigrad-Romuliana (Figure 2) is one of the most important cultural monuments from the ancient period in the world. In addition to the fortified royal palace, which is the most important site, on its eastern side is a sacral complex on Magura hill, and in the core and wider areas of the archaeological site, as well as in its protection zone, there are several dozen smaller sites, from prehistory to the Middle Ages. It was declared a cultural asset of exceptional importance as a result of the Decision on the determination of immovable cultural goods of exceptional and great importance (Official Gazette of the SRS, 14/79). The central part of the area of the site is also protected as a natural asset in the category of "protected immovable cultural environments and goods", determined by a special study by the Republic Institute for Nature Protection and the Republic Institute for the Protection of Cultural Monuments.

Since 2007, it has been inscribed on the UNESCO list with the name "Gamzigrad-Romuliana, Palace of Galerius". According to UNESCO, the complex meets the following criteria: Criterion (III) - The fortifications, palace and memorial complex are a unique testimony of the Roman architectural tradition imbued with the ideological program of the Second Tetrarchy and the Gallery as its builder, and Criterion (IV) - the group of buildings that makes up the architectural complex is unique in that it intertwines the ceremonial and memorial program (UNESCO).

The late Roman palace and memorial complex of Gamzigrad-Romuliana were built by the Roman emperor Gaius Valerius Galerius Maximian, the successor of Diocletian in the Second Tetrarchy, at the end of the 3rd and the beginning of the 4th century. This was confirmed by the discovery of the sculpted head of the royal figure during the excavation of the bathroom. The palace is known as Felix Romuliana, after the emperor's mother.

Archaeological works have uncovered the remains of an older and a younger fortification. The palace complex is surrounded by a unique defense system: a double rampart was formed from the remains of the older fortification, which were incorporated into the younger one. Both fortifications were built in a time interval of about ten years. The communication that connected the eastern and western gate inside the ramparts divided the space into two parts for different purposes. The northern half encompasses the imperial complex palace with a small temple and a monumental altar, while in the southern half there are buildings for public use (a large temple with two crypts of rectangular base, and thermal baths) and facilities for servicing the palace (horeum and single-nave building with portico).

The basic problem with regard to the functioning of the site is the inadequate traffic access and dual function of the main access road to the archaeological site, which is also used as an access road for agricultural machines. The sacral complex on Magura Hill is available only to pedestrians and off-road vehicles, and it has no designated parking space. In addition, the site has no auxiliary facilities (reception center for visitors, adequate accommodation facilities, etc.). The existing building that accommodates researchers is located within the fortifications of the Imperial Palace and is in a poor structural condition. There are no bicycle or excursion trails in the complex. There are also 35 kV and 10 kV overhead power lines that pass through the site.

The archaeological site of Caričin grad (Justiniana Prima) is located in Southwest Serbia, in the municipality of Lebane, one of the poorest municipalities in Serbia (Figure 1). Out of 38 rural settlements in the municipality of Lebane, 7 have less than 100 inhabitants and 19 have between 100 and 400; there are 8 settlements from 400 to 1000 inhabitants and only 3 settlements have over 1000 inhabitants. According to the Census data from 2011, the total population of the municipality of Lebane in 39 settlements was 22,000 inhabitants, and in the nearest settlements to Caričin grad - Štulac and Svinjarica, there were 279 and 111 inhabitants respectively (Republički zavod za statistiku, 2014).

The municipality of Lebane is an economically devastated area. Economic backwardness and poverty are particularly pronounced in the marginal hilly and mountainous parts of the area. In addition to the negative natural increase and the negative migration rate, there is an unfavorable age structure of the population. In the current economy of the municipality of Lebane, the participation of agricultural activities is dominant, and the participation of crafts and small and medium enterprises is far smaller. There are almost no tourist activities in the municipality (Prostorni plan opštine Lebane 2009-2024, 2011).

Caričin grad (Justiniana Prima) complex (Figure 3) is one of the best-preserved early Byzantine urban units, with a city center, large and spacious suburbs, surrounding fortifications, unique defensive lines around the city, basilicas outside the city, a dam and lake, an aqueduct, a craft center between two river valleys, a necropolis, and a completely preserved historical landscape of the ancient city. It was determined as a cultural asset by Decision no. 257/49 from 12.02.1949 (Official Gazette of the Republic of Serbia, no. 54/IV): the site Caričina – Gradište, as well as all the terrain between the Svinjarička and Caričinska rivers located around Gradište. It was categorized as immovable cultural goods of exceptional importance for the Republic of Serbia on 07.04.1979 (Official Gazette of the SRS no. 14/79), and since 2010, it has been listed on the UNESCO preliminary list.

Although without direct material evidence, the site has been defined as the early Byzantine city of Justinian Prima, which was built around 535 by Emperor Justinian I near his birthplace, Taurision. According to archaeological findings, the fortified city, in terms of its urban structure and architecture, represents a combination of late antique and early Christian architecture. The city consisted of an urban core in the ramparts (Acropolis, Upper Town and Lower Town) and a wider suburb. Research carried out in the city has brought to light four rings of ramparts, eight basilicas, numerous public buildings, thermal baths, a central square, wide streets and porticos, an aqueduct, a large cistern, a water tower, a water and sewage network, a dam with a lake, craft furnaces, and a network of fortifications in its immediate vicinity.

Figure 3. The archaeological site of Caričin grad
(author's archive)

Outside the city, situated on gentle slopes, there is a wide suburb, with an area of about 20 ha, protected by a palisade. The city area also includes a craft center, located at the foot of the city, on the banks of two rivers. In addition to this, there are also the remains of a dam-bridge on the river Caričina, rising above which, on a cliff, is a smaller rectangular fortification. Three more fortresses in the immediate vicinity should be added to this fortification, which, all together, formed a single system of defense for the city.

The remains of the Byzantine city buildings are at different levels of preservation. The building structures discovered and researched long ago that have not been preserved or restored are collapsing and are covered by uncultivated vegetation. The site is used only minimally for educational purposes (with the exception of archaeological student internships), and is particularly poorly used by local preschools and schools. Within the zone of protection regime I, near the site itself, there are facilities to accommodate researchers, which are in very poor structural condition.

The main limitation refers to the functioning of the archaeological site itself. Caričin grad has not yet been fully explored, and what has been explored is not adequately protected. The site has no regulated access, and no adequate access road or parking space; neither does it have any adequate accompanying facilities. The access road passes partly through the archaeological site. Existing 10 kV and 0.4 kV overhead transmission lines pass through the site.

Key Points With Regard to Institutional and Planning Frameworks

Both sites have adopted plans as part of the formal planning system. The Spatial plan for the area of Romuliana-Gamzigrad archaeological site was adopted by the Government of Serbia in 2004. This plan differentiated the area of the archaeological site into protection regime zones I, II.1, II.2 and III and gave general guidance for the organization and protection of the land (Prostorni plan područja arheološkog nalazišta Romulijana-Gamzigrad, 2004). Development of the Detailed regulation plan for the "first level of protection for the archaeological site Romuliana-Gamzigrad with the expansion area" was initiated on the basis of a Decision of the City of Zaječar in 2014 with the goal to organize the archaeological site, to build accompanying facilities, and to determine areas for public and other uses (Plan detaljne regulacije "prostora zone I stepena zaštite arheološkog nalazišta Romulijana-Gamzigrad sa proširenjem", 2018). The area of the detailed regulation plan fully covered protection regime zone I, as well as part of zone II.1.

The Spatial plan of the Republic of Serbia from 2010 to 2020 classified Caričin grad as a priority cultural area that should enjoy special treatment (Zakon o Prostornom planu Republike Srbije od 2010. do 2020. godine, 2010). The Spatial plan for the special purpose area of Radan Mountain, which was prepared for the natural asset of Radan Mountain, paid special attention to Caričin grad. It differentiated the area of the archaeological site into zones of protection regimes I, II.1 and II.2 and envisaged the development of a detailed regulation plan for protection zones I and II.1 (Prostorni plan područja posebne namene Radan planine, 2014).

The development of the Detailed regulation plan for the archaeological site of Caričin grad was initiated in 2015 by the municipality of Lebane. The plan completely covers zones with protection regimes I and II.1 over an area of 134 ha. The largest part of the land covered by the plan - 108.93 ha (81.6% of the area covered by the plan) is currently privately owned. The Republic of Serbia owns 24.58 ha (18.41% of the total area of the plan). The municipality of Lebane is registered in the cadastre as the user of 16.28 ha, and the National Museum as the user of 8.3 ha (Plan detaljne regulacije arheološkog nalazišta Caričin grad, 2018).

In addition to documents that are part of the formal planning system, the Management plan for the archaeological site of Caričin grad was prepared in 2014. The document was created as one of the seven management plans in seven countries of Southeast Europe, within the project Ljubljana Process II: Rehabilitation of Our Common Heritage, coordinated by the Regional Working Group for Culture and Society of the Regional Cooperation Council, which was supported by the European Commission and the Council of Europe program. The management plan paid special attention to the development of tourism through the creation of thematic tourist routes and their implementation in the tourist offer, as well as the organization of study visits for employees in tourism and culture and pedestrian and bicycle trails (Menadžment plan arheološkog lokaliteta Caričin grad (Justiniana Prima), 2014).

Conflicts among actors in the process of preparing the detailed regulation plan for Romuliana-Gamzigrad were efficiently resolved by the Commission for Plans of the City of Zaječar and its members – representatives from public enterprises, including Roads of Serbia, and representatives from the local self-government, who coordinated the plan. Meetings were supported by the Institute for Urban Planning of the City of Niš (the institution which elaborated the detailed regulation plan), with cooperation from the Republic Institute for the Protection of Cultural Heritage, and the National Museum, which manages the archaeological site. As Maksić et al. (2015) pointed out, during the preparation of the detailed regulation plan, cooperation with other professions was accomplished as an important criterion in the field of archaeological heritage protection.

As the development of the detailed regulation plan for Caričin grad was financed by European Progress, a meeting was held at an early stage in the preparation of the plan, organized by European Progress, in which the key actors were present (Republic Institute for the Protection of Cultural Monuments, Institute for the Protection of Cultural Monuments of Niš, Archaeological Institute Belgrade, Institute of urban planning of the City of Niš (the institution that prepared plan), local government representatives, local citizens). It should be stressed that this meeting was not typical of standard planning practices and was not mandatory according to the law on planning and construction. However, the meeting was extremely important for solving key problems in the early phase of developing the plan and defining the final planning solutions. At this meeting, key issues in the development of the plan were resolved. They refer to the precise definition of protection zones, harmonization of the needs of the inhabitants of the area and protection of the cultural property, defining the local road that crosses the site (planned road without a road curtain), and defining the location for the tourism zone (municipal requirements for development in relation to the protection of cultural heritage).

The following section presents and analyzes the harmonization of conflicts between cultural heritage and construction in the function of tourism, by means of the planning solutions.

Harmonization of Conflicts

Both plans have strict measures within the zone of protection regime I. In zone I of Gamzigrad-Romuliana, only construction that supports immovable cultural property can be carried out. Within zone 1 of Caričin grad, only construction works that relate exclusively to the cultural property and the preservation and presentation of the archaeological site can be performed, that is, the conservation and restoration of the architectural remains.

Specific types of activities and construction are allowed in zone 1. Thus, in the case of Gamzigrad-Romuliana, within the fortified palace, there is a possibility of adapting and introducing new contents of the towers that are part of the younger and older fortifications, in order to improve the function and presentation of the complex (souvenir shops, exhibition spaces, work spaces for research and education, etc.). The construction of protective structures above certain objects in the archaeological site is planned: above the mausoleum of Galerius and Romula on Magura Hill and above part of the palace. Depending on the condition of the mosaics and floors, it is also possible to place protective open protective structures above other buildings. In Caričin grad, when needed, it is possible to build protective structures for individual facilities within the fortified city. In order to improve the functioning and presentation the complex, it is possible to introduce contents such as information boards and markers for the direction of movement of visitors.

The plan for Gamzigrad-Romuliana limits the maximum number of visitors permitted in the zone of protection regime I to 300 (with an increase of a maximum of 50 tolerated at peak times). Both plans contain precise locations for tourist complexes, permitted facilities within this areas, and precise building capacities. Within the western entry point of Gamzigrad-Romuliana, the construction of volume-independent facilities is planned (Figure 4). The plan defines a visitor center that will be oriented towards the archaeological site and will contain an accessible space for gathering visitors; an accommodation facility for visitors, with a planned capacity of up to 10 double rooms and 2 suites, with terraces; a research center; an accommodation facility for researchers, which should be in the immediate vicinity of the research center, with accommodation units in the form of ten double rooms. At the point of the eastern parking lot, the construction of a facility to receive visitors is also planned, with a maximum gross construction area of 431m^2.

The plan for Caričin grad defines a complex that includes a visitor center and parking space for 33 passenger vehicles and 7 buses is planned at the entrance point, on the location next to the existing approach to the archaeological site, outside the zone of protection regime 1 (Figure 5). The visitor center complex will contain a souvenir shop, space for a guard and guide service, a multipurpose hall for the presentation of sites and occasional exhibitions, as well as accommodation facilities for researchers with a maximum of 10 double rooms. The area covered by the center should not exceed 500m^2.

Figure 4. The detailed regulation plan for Gamzigrad-Romuliana
(author's archive)

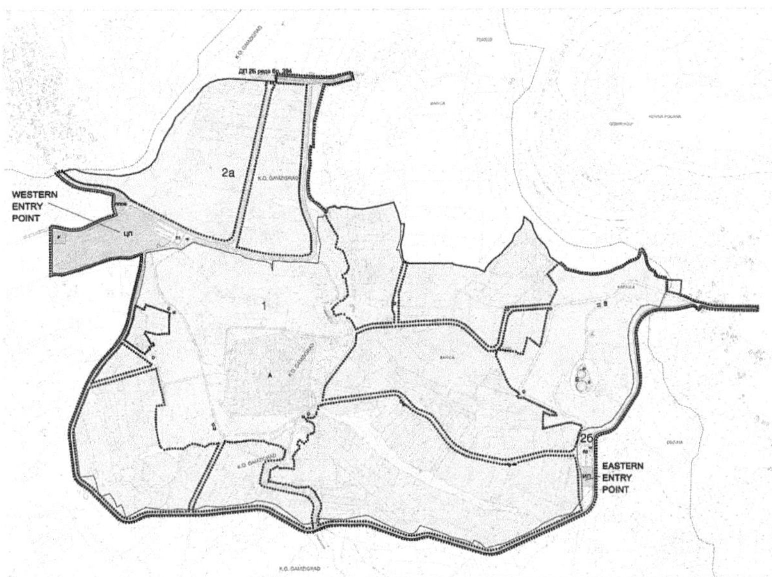

As tourism in the municipality of Lebane is not sufficiently developed, the municipality has demanded the definition of an appropriate location for the development of tourism within the scope of the plan. The plan defines the precise location, permitted facilities and building capacities. Construction that will develop and improve the tourist offer is planned in the area of Donja Caričina, north of the archaeological site (Figure 5). A tourist center with cabins, an auto-camping area with an appropriate number of parking spaces, souvenir shops and retail spaces, playgrounds for children and public toilets will be formed in that area. The tourist center's facilities should be at ground level and be visually integrated into the surroundings. Within this zone, in addition to the tourist complex, it is possible to organize ethno-contents that are related to the cultural good in terms of tourism: larger restaurants with local food and accommodation or households with similar offers, and spaces for summer schools, workshops or art colonies.

Both plans envisage the construction of hiking trails mostly on the routes of the existing field roads and trails, through the entire area of the plan. They are planned to access all of the localities of the archaeological site. The construction of bicycle paths is also planned through the entire area covered by the plans.

Infrastructure solutions that follow the development of tourism are harmonized with the protection zones. In Gamzigrad-Romuliana, all types of access road traffic will be stopped at the entrances to the zone of the I degree of protection, at the parking lots-terminals (within the zone with the transitional protection regime of II.1). Construction of a road with a special regime is planned in protection zone I, with turning points near the western parking lot and the Magura site, for communication within zone I, in order to present the main archaeological sites. It is intended for integrated pedestrian, bicycle and car traffic in the form of tourist vehicles. It can also be used as necessary for road traffic in exceptional cases (as access for communal and fire-fighting vehicles, as well as other similar intervention activities).

In Caričin grad, all types of access road traffic will be stopped at the entrance to protection zone II.1. From the existing state road, a road under a special regime is planned, without the possibility of

Figure 5. The detailed regulation plan for Caričin grad
(author's archive)

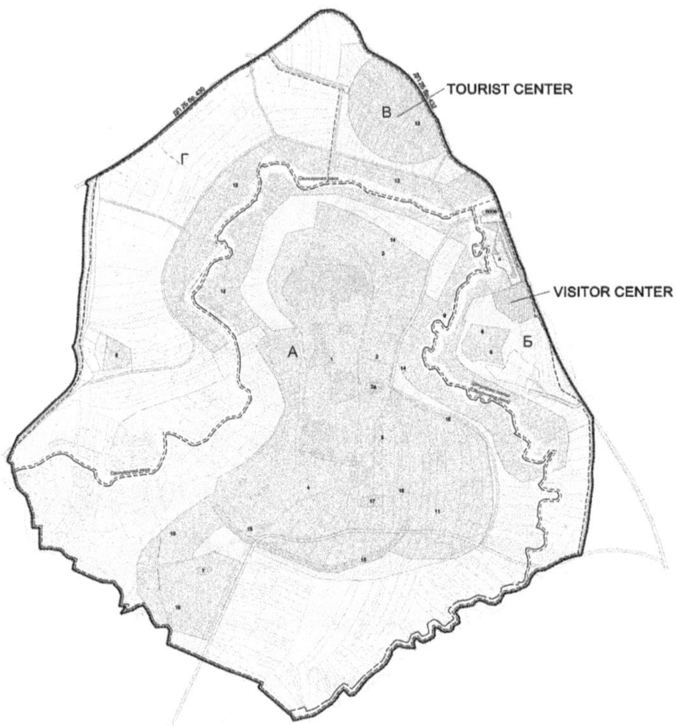

car access. Road access is allowed only in certain cases, such as access to emergency vehicles or transportation of materials relating to conservation works or archaeological research. The road curtain in the development must not be monolithic and should be easily removed. The road will lead only to the main entrance to the site.

Given the specificity of the planning area and the need to preserve the cultural and natural landscape, the existing 35 kV and 10 kV overhead power lines will be routed underground in Gamzigrad-Romuliana. For the transmission lines with voltage levels of 35 kV and 10 kV, a protection zone (corridor) is provided until their termination. In Caričin grad, it is necessary to lay all the existing 10 kV and 0.4 kV overhead transmission lines in the ground by means of cables. The newly planned 10 kV cable lines will be connected to the existing network in the southeastern section and can be laid through the perimeter roads of the plan, which raises the protection and level of security of the power supply for the planned consumers within the scope of the plan.

In Gamzigrad-Romuliana, it is forbidden to discharge any wastewater into the watercourse of Dragan's creek. The construction of a sewerage network and a wastewater treatment plant is planned for the purpose of wastewater drainage from the visitor center. The minimum requirement for sanitary facilities and the planned eastern parking lot is the construction of impermeable septic tanks. In Caričin grad, used water will be collected through the sewerage network along the bicycle/walking paths and the planned infrastructure corridor, and taken to the planned wastewater treatment plant located along the southern bank of Svinjarički potok River, immediately after the inflow of Štulacki potok River (Caričinske reke). From the complex for visitors, the used water will also be taken to the plant by the planned sewerage pipes under the road.

Table 1. The conflict harmonization measures

Gamzigrad-Romuliana	Caričin Grad
• Allowing only construction that supports immovable cultural property. • Limiting the maximum number of visitors in zone I. • Defining strict locations of entrance points. • Limiting the capacity for the visitor center. • Defining strict measures for construction. • Stopping all types of traffic at the entrances to zone I. • Construction of a road with a special regime. • Construction of hiking trails and bicycle paths. • Underground routing of existing 35 kV and 10 kV overhead power lines. • Construction of a sewerage network and a wastewater treatment plant. • Natural arrangement of the banks of the watercourses.	• Allowing only construction that supports the preservation and presentation of the archaeological site. • Defining an appropriate location for the development of tourism. • Defining strict measures and uses for the tourist center's facilities. • Stopping all types of traffic at the entrance to zone II.1. • Construction of a road with a special regime. • Construction of hiking trails and bicycle paths. • Underground routing of existing 10 kV and 0.4 kV overhead transmission lines. • Construction of a sewerage network and a wastewater treatment plant. • Natural arrangement of the banks of the watercourses.

In both plans, the arrangement of the banks of the watercourses and their possible regulation will be carried out in such a way that they correspond to the natural configuration of the terrain as much as possible. All regulatory interventions in the riverbed can be reduced to so-called natural regulation, which makes it possible not only to preserve the existing biodiversity, but also to increase the diversity of biocenoses, especially phytocenoses, in the coastal zone.

The conflict harmonization measures are summarized in Table 1.

SOLUTIONS AND RECOMMENDATIONS

As a result of the analysis of these two plans, certain recommendations for the sustainability of cultural heritage in inland territories are proposed. Conflicts between cultural heritage protection and the development of tourism, can be harmonized, by applying appropriate criteria related to degrees of protection to the arrangement and construction of cultural heritage sites. Opportunities for involvement of the local population in the process of protection and promotion of cultural heritage should be encouraged.

The rules for the use of space and construction and the capacities of the planned activities must be defined in accordance with the protection zones of the archaeological sites. In the case of the Gamzigrad-Romuliana and Caričin grad sites, the protection zones were defined by higher order plans - spatial plan for the special purpose areas. While for the Gamzigrad-Romuliana site the zones have been completely retained, for Caričin grad the zones were modified during the procedure for preparing the detailed regulation plan, in accordance with the conditions of the Republic Institute for the Protection of Cultural Monuments, since in the meantime, new archaeological facts emerged.

In areas of cultural heritage sites in inland territories, special attention must be paid to the development of local communities and the definition of sustainable types of tourism. Neither plan forbids tourism, but appropriate forms for their promotion have been found. The construction of a limited number of accommodation facilities is allowed in both plans (in the case of Gamzigrad-Romuliana within the visitor center complex, and in the case of Caričin grad, within the special area for tourism development with cabins). In Gamzigrad-Romuliana, within the fortified palace, there is a possibility of adapting and introducing new functions to the towers, in order to present the complex more effectively (souvenir shops, exhibition spaces, work spaces for research and education, etc.). Both plans encourage the development of a tourist product relating to special interests (thematic routes: hiking, cycling), as well as educational types of tourism.

Infrastructural development follows the specificities of the planning areas. Similar solutions are given in both plans (traffic roads planned outside the area of protection zone 1, roads under a special regime through zone 1, the displacement of electric overhead power lines under the ground to not disturb the views, the natural regulation of the banks of watercourses).

The spatial plans for special purpose areas that were prepared for both sites are an excellent framework for defining the planning solutions. The holder of their preparation in Serbia is the ministry in charge of construction, and their expert control is performed by the national level commission formed by the competent ministry. In addition, the detailed regulation plans for Gamzigrad-Romuliana and Caričin grad were subjected to the consent of the relevant government ministry, in order to verify the planning decisions and harmonize them with higher-level plans. In this way, the development and construction of these important areas is controlled at the national level. In addition, a management plan was prepared for Caričin grad, which UNESCO recommends for its localities.

Different mechanisms at the local level were developed to harmonize the views of the actors involved. In the case of Gamzigrad-Romuliana, the Commission for Plans of the City of Zaječar proved to be an excellent mechanism, while in the case of Caričin grad, informal workshops (which are not required by law) proved to be an excellent forum. It is especially important that forums are organized at an early stage of preparing a planning document.

FUTURE RESEARCH DIRECTIONS

Spatial and urban planning is only one of the instruments for directing development and construction in the areas of cultural heritage sites. In order to improve the cultural heritage management and management tools in Serbia, diverse world practices are of special importance. Cities all over the world are developing various instruments to protect, restore and manage cultural heritage sites. In terms of the legal framework, such instruments can be conventions, laws, regulations, or recommendations, and in terms of the planning framework, spatial and urban plans, management plans (UNESCO recommendations), integrated development plans, action plans, special construction rules, sustainable tourism development plans, etc. World practices also show various institutional arrangements developed to manage cultural heritage sites (Central Management Office, The Old Town Directorate, Český Krumlov Development Fund, Destination Management, Qujiang Management Committee). The subject of future research could be a detailed analysis of the possibilities and limitations of introducing the experiences of different world practices in Serbia.

CONCLUSION

The sustainability of cultural heritage in rural areas is a complex task, as opportunities for tourism are limited and have to be planned in accordance with permitted capacities and protection regimes. A more complicated task concerns UNESCO World heritage sites, since policies and strategies should ensure the protection of objects of outstanding universal value.

This chapter has shown that continuous harmonization of economic, environmental, and social aspects of sustainable development can be achieved through the process of developing planning documents and the mechanisms for coordinating actors. In the aspect of institutional cooperation, the experiences in the development of planning documentation in Serbia for immovable cultural properties show that continuous cooperation with institutions in charge of cultural heritage protection is necessary throughout the entire process of their development. It implies, inter alia, that clear requirements regarding the definition of protection zones should be submitted to the institution preparing the plan in a timely manner. Of particular importance is the involvement of all actors in the early stages of a plan's development (institutions for protection of cultural heritage, representatives of local governments, residents of the planning area, interested investors, etc.).

In the case of cultural heritage in Serbia, the system of spatial and urban planning provides mechanisms for the harmonization of conflicts in the area of cultural heritage, and mechanisms for coordinating actors. Although in Serbia in the system of spatial and urban planning, significant attention is paid to cultural heritage through the development of spatial plans for special purpose areas and detailed regulation plans, it is necessary to develop other documentation, such as: development plans, action plans, and above all, management plans for sites on the UNESCO List, recommended by UNESCO and binding under the Law on the Spatial Plan of the Republic of Serbia from 2010 to 2020.

The treatment of cultural heritage in the planning system continuously needs improvements. One of the ways of making improvements is, as Maksić et al. (2018) found, through the development of new heritage management models and new mechanisms for active involvement of the public, private and civil sectors, in the early phases of formulating policies and plans. Another way, as seen by Maksić (2019), is the continuous education of planners in the field of spatial planning.

An additional problem in the management of cultural heritage in rural areas is the limited financial and institutional support by institutions that manage the sites. In the development of new models of site management, special attention should be paid to this problem. One model used elsewhere in the world is the development of special institutions that will manage the site. Financial measures, combined with the use of various types and sources of funding (not only public sector funds) should be established, including greater use of foreign funds (UNESCO, the European Community) for sites on the World Heritage List.

REFERENCES

Briwa, R. (2021). The changing terrain of heritage at a Montana state park. *Journal of Historical Geography*, *71*, 17–27. doi:10.1016/j.jhg.2020.06.007

Caldeira, E., & Tomaz, N. (2015). Culture in development strategies of small and medium-sized European cities: A comparative analysis. In S. Hristova, M. Dragićević Šešić, & N. Duxbury (Eds.), *Culture and Sustainability in European Cities: Imagining Europolis* (pp. 86–100). Routledge.

Canale, R. R., De Simone, E., Di Maio, A., & Parenti, B. (2019). UNESCO World Heritage sites and tourism attractiveness: The case of Italian provinces. *Land Use Policy*, *85*, 114–120. doi:10.1016/j.landusepol.2019.03.037

D'Arco, M., Presti, L. L., Marino, V., & Maggiore, G. (2021). Is sustainable tourism a goal that came true? The Italian experience of the Cilento and Vallo di Diano National Park. *Land Use Policy*, *101*, 105198. doi:10.1016/j.landusepol.2020.105198

Frey, B. S., & Briviba, A. (2021). A policy proposal to deal with excessive cultural tourism. *European Planning Studies*, 1–18.

Gallou, E., & Fouseki, K. (2019). Applying social impact assessment (SIA) principles in assessing contribution of cultural heritage to social sustainability in rural landscapes. *Journal of Cultural Heritage Management and Sustainable Development*, *9*(3), 352–375. doi:10.1108/JCHMSD-05-2018-0037

Guzmán, P. C., Roders, A. P., & Colenbrander, B. J. F. (2017). Measuring links between cultural heritage management and sustainable urban development: An overview of global monitoring tools. *Cities (London, England)*, *60*, 192–201. doi:10.1016/j.cities.2016.09.005

Igić, M., Maksić, M., Cekić, N., & Vasov, M. (2015). Problems of Cultural Heritage Protection in Rural Areas in Serbia. In *Proceedings of the III International Conference "Modern Technologies in Cultural Heritage"* (pp. 36-44). TU Sofia.

Joshi, R., Tamrakar, A., & Magaiya, B. (2021). Community-based participatory approach in cultural heritage reconstruction: A case study of Kasthamandap. *Progress in Disaster Science*, *10*, 100153. doi:10.1016/j.pdisas.2021.100153

Li, J., Krishnamurthy, S., Roders, A. P., & van Wesemael, P. (2020a). Community participation in cultural heritage management: A systematic literature review comparing Chinese and international practices. *Cities (London, England)*, *96*, 102476. doi:10.1016/j.cities.2019.102476

Li, J., Krishnamurthy, S., Roders, A. P., & van Wesemael, P. (2020b). Informing or consulting? Exploring community participation within urban heritage management in China. *Habitat International*, *105*, 102268. doi:10.1016/j.habitatint.2020.102268

Li, J., Krishnamurthy, S., Roders, A. P., & van Wesemael, P. (2021). Imagine the Old Town of Lijiang: Contextualising community participation for urban heritage management in China. *Habitat International*, *108*, 102321. doi:10.1016/j.habitatint.2021.102321

Maksić, M. (2019). Governance Capacity of the City of Niš, Serbia, in the Spatial Planning Process. *Journal of Planning Education and Research*. Advance online publication. doi:10.1177/0739456X19844564

Maksić, M., Cekić, N., Vasov, M., & Igić, M. (2015). Problems of Archaeological Site of Felix Romuliana in Serbia. In *Proceedings of the III International Conference "Modern Technologies in Cultural Heritage"* (pp 22-28). TU Sofia.

Maksić, M., Dobričić, M., & Trkulja, S. (2018). Institutional limitations in the management of UNESCO cultural heritage in Serbia: The case of Gamzigrad-Romuliana archaeological site. *Land Use Policy*, *78*, 195–206. doi:10.1016/j.landusepol.2018.06.055

Menadžment plan arheološkog lokaliteta Caričin grad (Justiniana Prima). (2014). [Management plan for the archaeological site of Caričin grad (Justiniana Prima)]. Academic Press.

Ministarstvo poljoprivrede, šumarstva i vodoprivrede. (2009). *Plan strategije ruralnog razvoja 2009–2013* [Strategic plan for rural development 2009-2013]. Academic Press.

Mubaideen, S., & Al Kurdi, N. (2017). Heritage conservation and urban development: A supporting management model for the effective incorporation of archaeological sites in the planning process. *Journal of Cultural Heritage, 28*, 117–128. doi:10.1016/j.culher.2017.05.007

Nyseth, T., & Sognnæs, J. (2013). Preservation of old towns in Norway: Heritage discourses, community processes and the new cultural economy. *Cities (London, England), 31*, 69–75. doi:10.1016/j.cities.2012.05.014

Petrić, L., Hell, M., & van der Borg, J. (2020). Process orientation of the world heritage city management system. *Journal of Cultural Heritage, 46*, 259–267. doi:10.1016/j.culher.2020.07.009

Plan detaljne regulacije arheološkog nalazišta Caričin grad. (2018). *Službeni glasnik grada Leskovca, 14.*

Plan detaljne regulacije "prostora zone I stepena zaštite arheološkog nalazišta Romulijana-Gamzigrad sa proširenjem. (2018). *Službeni list grada Zaječara, 14.*

Prostorni plan opštine Lebane 2009-2024. (2011). *Sl. glasnik grada Leskovca, 13.*

Prostorni plan područja arheološkog nalazišta Romulijana-Gamzigrad. (2004). *Službeni glasnik RS, 131.*

Prostorni plan područja posebne namene Radan planine. (2014). *Sl. glasnik RS, 110.*

Prostorni plan Republike Srbije od 2021. do 2035. godine (nacrt) [Law on Spatial Plan of the Republic of Serbia from 2021 to 2035 (draft)] (in Serban).

Prostorni plan teritorije grada Zaječara (Sl. list grada Zaječara, 15/12) [Spatial plan of the territory of City of Zaječar) (Official Gazette of city of Zaječar, 15/12)] (in Serban).

Quartesan, A., & Romis, M. (2011). Oaxaca de Juбrez, Mexico. In E. Rojas & F. Lanzafame (Eds.), *City Development: Experiences in the Preservation of Ten World Heritage Sites* (pp. 13–59). Inter-American Development Bank.

Republički zavod za statistiku (2014). *Popis stanovništva, domaćinstava i stanova 2011. u Republici Srbiji: Stanovništvo* [2011 Census of Population, Households and Dwellings in the Republic of Serbia: Population]. Academic Press.

Seo, U. S. (2020). Urban regeneration governance, community organizing, and artists' commitment: A case study of Seongbuk-dong in Seoul. *City. Cultura e Scuola, 21*, 100328. doi:10.1016/j.ccs.2019.100328

Spiekermann, M. (2011). Aleppo, Syria. In E. Rojas & F. Lanzafame (Eds.), *City Development: Experiences in the preservation of ten World Heritage Sites* (pp. 309–329). Inter-American Development Bank.

Stunpo, S. (2011). Siracusa, Italy. In E. Rojas & F. Lanzafame (Eds.), *City Development: Experiences in the preservation of ten World Heritage Sites* (pp. 255–283). Inter-American Development Bank.

UNESCO WHC. (2016). *Operational Guidelines for the Implementation of the World Heritage Convention*. UNESCO. Retrieved: December 09, 2020 from http://whc.unesco.org/en/list/1253

Zakon o kulturnim dobrima (Službeni glasnik RS, 71/94, 52/2011 - dr. zakoni i 99/2011 - dr. zakon) [Law on Cultural Heritage ("Official Gazette of RS", 71/94, 52/ 2011 – other laws and 99/2011 – other laws)] (in Serban).

Zakon o planiranju i izgradnji (Službeni glasnik RS, 72/2009, 81/2009 - ispr., 64/ 2010 - odluka US, 24/2011, 121/2012, 42/2013 - odluka US, 50/2013 - odluka US, 98/2013 - odluka US, 132/2014, 145/2014, 83/2018, 31/2019, 37/2019 i 9/2020) [Law on Planning and Construction (Official Gazette of RS, 72/2009, 81/2009 - April, 64/2010 - Decision CC, 24/ 2011, 121/2012, 42/2013 - Decision CC, 50/2013 - Decision CC, 98/2013 – Decision CC, 132/2014, 145/2014, 83/2018, 31/2019, 37/2019 and 9/2020)] (in Serban).

Zakon o Prostornom planu Republike Srbije od 2010. do 2020. godine (Službeni glasnik RS, 88/2010) [Law on Spatial Plan of the Republic of Serbia from 2010 to 2020 (Official Gazette of RS, 88/2010)] (in Serban).

Zhao, Y., Ponzini, D., & Zhang, R. (2020). The policy networks of heritage-led development in Chinese historic cities: The case of Xi'an's Big Wild Goose Pagoda area. *Habitat International*, *96*, 102106. doi:10.1016/j.habitatint.2019.102106

KEY TERMS AND DEFINITIONS

Caričin Grad (Justiniana Prima): Archaeological site on UNESCO's tentative list, located in Southwest Serbia in the municipality of Lebane.

Conflicts: Opposing actions of incompatible activities.

Gamzigrad-Romuliana: One of the most important cultural monuments from the ancient period in the world, archaeological site on UNESCO's list, located in Eastern Serbia, on the territory of the City of Zaječar.

Governance: A set of mechanisms that shape social activities either by the administration, the private sector, or the civil society sector.

Institutions: A significant practice, relationship, or organization (formal or informal).

Mechanism: A process, technique, or system for achieving a result.

Spatial and Urban Planning: Methods by which the public sector influences the distribution of activities in space.

Sustainability: Balancing economic, environmental, and social factors in equal harmony.

UNESCO: Specialized agency of the UN that promotes international collaboration in education, science, and culture.

World Heritage: Site designated by UNESCO for having cultural, historical, scientific, or other form of significance

Chapter 14
Enogastronomic Tourism in Times of Pandemic

Mario Ossorio
University of Campania "Luigi Vanvitelli", Italy

ABSTRACT

The chapter investigates the relevance of events in enogastronomic tourism during the pandemic. Over last years, typical food and beverage have increasingly grown their relevance within the tourism industry. The evolution of the concept of enogastronomic tourism and the economic consequences on the rural territories are illustrated. In addition, the different categories of enogastronomic tourists are examined, emphasizing the relevance of experience within the industry. In the second section, the role of the events and the experiences connected to them, in the wine industry, are analysed. Lastly, the chapter describes la Strada del Vino dei Campi Flegrei, isola d'Ischia e Vesuvio, a territory near the City of Naples, where well-known Italian wines are produced and many wine-centered events are arranged, strengthening the attractiveness of that rural area.

INTRODUCTION

Nowadays, tourism is one of the most important economic industries. It generates a multiplier effect economically and can directly and indirectly benefit many sub-industries. Tourism has a leading role in the development of societies, and it represents a source of income for those offering services because people spend to satisfy their needs when they are visiting places during the holidays (Sormaz et al., 2016). Over the last decades, people's needs require a higher and higher level of goods and services, which increases production, investment, and income in societies. Accordingly, wine and food tourism has become a distinct industry in tourism and not just a supporting factor when tourists are travelling (Everett, 2016). Although there is increasing relevance for local food consumption, scant attention has been given to exploring typical food and drink consumption in the context of hospitality and tourism (Kim et al., 2009). In fact, leisure travellers try to experience many food and drink events, both when food is considered to be a main part of the holiday and when it represents a supplementary element (Stone and

DOI: 10.4018/978-1-7998-7339-6.ch014

Migacz, 2016). Evaluating the typical products entails a wider view than just commercializing them in order to generate rural development in an integrated and sustainable way (Slee, 1993).

The main aim of this chapter is to highlight the role of events in enogastronomic tourism as a means to strengthen the attractiveness of a destination during the pandemic era.

The chapter is articulated into three sections. The first section illustrates the origin of the modern concept of enogastronomic tourism. The socio-cultural changes in the last years of XX centuries stimulated new needs in the customers' minds. These needs are based on a slow life, a rediscovery of genuine food, and the appreciation for settings different from ordinary urban ones, such as a rural context. Emerging needs contribute to the definition of enogastronomic tourists who are attracted by typical cuisine but also by other interests connected to the local food and beverages, such as gazing at the landscape or learning about the production processes of a typical food. This chapter describes the role of the customers' experiences connected to enogastronomy and the possibilities offered by a rural context. Lastly, the socio-economic advantages of enogastronomic tourism for a territory are illustrated.

Section 2 illustrates the motivations behind the selection of an enogastronomic destination. It points out the different importance that each category of travel attributes to food and beverages during a trip and the increasingly relevant role of experience in enogastronomic tourism. Also, the section highlights the potential of wine tourism in the context of the experience economy. More specifically, the model 4E of Pine II and Gilmore (1998) is applied to wine tourism, illustrating the kinds of events which a traveller may participate in and the experiences associated with the events.

Section 3 illustrates Le Strade del Vino dei Campi Flegrei, Isola d'Ischia e Vesuvio which represents an effort to create synergies between the several wine producers and wine centred event organizers in order to coordinate and strengthen the wine industry of those areas that are rich in history, culture, and traditions and have volcanic soils which provide grapes and wines with delicate and unique flavours. The main scope of the Strade del Vino is wider than valuing the local wine. It intends to treasure typical craftsmanship, local food, archaeological sites, and the natural beauty of the territory.

ENOGASTRONOMIC TOURISM: THE EVOLUTION OF AN INDUSTRY

Enogastronomic tourism has its roots in old times. In the XIX century, some tour guides described in a very detailed way gastronomic specialities that tourists might find and the tastes and flavours in different Italian cities and towns. (Croce and Perri, 2018). Those guides contained so much information about the typical food in each region that they created some disorientation in a tourist's mind regarding the itinerary to follow. Anyway, over the last few years, a significant evolution of the enogastronomic tourism concept has been occurring, following the relevant social and economic change processes.

Many factors have contributed to creating the actual concept of enogastronomic tourism. First of all is the greater and the more widespread wealth of mankind in advanced and emerging economies. This has permitted people to satisfy new and more sophisticated needs. Experiencing cultural and local food permits one to achieve self-realization, the fifth stage of Maslow's hierarchy of needs (Guzel and Apaydin, 2016), making food tourism a driver in choosing destinations (Tikkanen, 2017). The improvement of communication and transportation has reduced the distance between countries and permitted people to more easily discover new destinations and new foods typical of the regions. Furthermore, the enogastronomic industry has become more attractive because of an increasing awareness of the importance of health and genuine foods and of the pleasure of spending time around a good dish (Guruge,

2020). Besides, gastronomic tourism matches the actual propensity to adopt a behaviour respectful of the environment. It is eco-sustainable and takes care of the quality, the origin, and the traceability of food products (Antonioli and Mottironi, 2013). Lastly, at the end of the twentieth century, global food is widespread throughout the occidental countries. This is due to both a fashion phenomenon and the needs of a fast paced lifestyle which has pushed people toward a standardization of behaviours (including food preferences), putting aside rural and local cuisines. Anyway, in the same years, an opposite phenomenon has been arising that exalts a slow life and values the typical food products of some territories. Also, it reflects the desire of many people to discover their own cultures, such as their domestic rural towns.

Those socio-economic changes lead to a new enogastronomic tourism concept that contains, but is not limited to, the consumption of local speciality food, which represents just a part of a wider range of products and services. In fact, many enogastronomic products feature symbolic traits. They offer an emotional feeling because they directly recall a certain territory and its traditions and culture which have long roots and permit a transfer people's behaviours to the next generations (Asero and Patti, 2009). The land of origin of food or wine specialities represents not only a simple logistic context but becomes a central element of a destinations' offering. Under this perspective, the consumption of local food and beverages represents a relevant facet such as exploring local culture and territories.

Enogastronomic tourism represents a social phenomenon that combines culture, the social environment, history, and people (Garibaldi and Pozzi, 2018), inglobing a culture in foods and beverages that are peculiar to a geographical area. In fact, food and drink encourage travellers to evaluate an abstract culture through tangible means (Lin and Mao, 2015). Because of its strong territorial implications, food can be considered to be a point of union between the authenticity of a territory and tourists who are characterized by an increasing desire for genuine and involving experiences that are closely connected to the peculiarities of the territory visited (Nocifora et al., 2011; de Salvo et al., 2013).

Going deeper, gastronomic identity is determined when two factors match each other–namely, the environment (climate and geography) and culture (religion, history, ethnic diversity, traditions, values, and beliefs) (Seyitoglu and Ivanov, 2020). It embraces the materials, techniques, flavours, recipes, and manners that belong to a specific region (Harrington, 2005).

Scholars have proposed different approaches to the definition of gastronomic tourism (Pavlidis and Markantonatou, 2020). Sharples and Hall (2004) consider gastronomic tourism to be a travel experience in an area with distinct gastronomy. It embodies visits to food producers, gastronomic and culinary events, rural markets, and other activities strictly related to food. In this approach, gastronomy rises as the main driver for selecting a tourist destination.

Smith and Xiao (2008) emphasize the experience gained during gastronomic travel, where the tourist learns, values, and experiences typical local gastronomy products. Therefore, even if experience is not the first scope of the travel, it is considered to be very relevant in choosing the travel destination.

Long (2004) notes that the gastronomic tourist is motivated by discovering new flavours, and food represents an extraordinary means to appreciate new cultures and lifestyles.

The abovementioned definitions emphasize the relevance of the experience for the travellers. Each tourist is not satisfied with just tasting typical foods and beverages. He or she intends to feel involved in the typical context that he or she is visiting.

Gastronomic tourists look for a unique set of attractions and activities that enable them to make their vacation distinctive. During their gastronomic trips, they seek a "noteworthy experience" to talk about when they come back to their origin country. This kind of experience is possible if a tourist discovers the traditions, culture, and values and has direct contact with the local community (Antonioli Corigliano,

1999). In addition, typical enogastronomic food and wine products have a relevant feature, they can be brought back home and be consumed after the trip. The possibility to shift their consumption after the holidays extends the experience of the travel. This reminds the traveller of the feelings and the pleasure of the vacation and becomes an efficient testimonial of the country visited (Asero and Patti, 2009).

With regard to the traveller's noteworthy experiences, enogastronomic tourism involves visiting special places related to some distinctive foods, food manufacturers, or eating festivals together with consuming a typical dish, learning about the production processes of a typical local food, or eating a typical dish from the hands of a very famous chef, as well as observing the different phases in the preparation of a certain dish (Hall et al. 2003; Hall and Mitchell, 2005).

Under these circumstances, enogastronomic tourism can be fully experienced in a rural context. Rural areas provide the ideal scenario for combining wine tasting with gastronomic products and many other activities, such as learning about cooking, horse-riding, hunting, bird watching, photography, etc. (Martinengo and Savoja, 1999; Antonioli Corigliano, 2000).

Gastronomy in all its expressions entails the feeling of identification between food and beverages and the territory, and it involves both tourists and residents (Sims, 2009). If, on one hand, enogastronomy symbolizes the lifestyle of a unique geographical area, since its core values are connected to its own culture and traditions (Kilic et al., 2017), on the other hand, it speeds up the development of tourism in that area.

Estimating typical products plays a relevant role in the development of rural areas in a sustainable way with several overall socio-economic advantages during an economic crisis period. In fact, peripheral rural areas, not yet invested in a modernisation process, may adopt a conservative rural development model that, on one side, protects the elements constituting it (Sortino and Chang, 2008) and, on the other side, shows competences and capabilities accumulated locally in order to turn local tangible and intangible resources into high-value typical foods and beverages (de Salvo et al., 2013). This new paradigm preserves local value and therefore satisfies post-modern tourist consumption necessities. Anyway, it requires cooperation between different territorial actors because the social relations networks and the associative tendencies confer to the product a collective dimension (Beeton, 2006, de Salvo et al., 2013). Cultural routes, markets, events, and festivals are just some of many forms through which local gastronomic heritage may become a tourist attraction and promote the territorial development of its inhabitants (Privitera et al., 2018)

More specifically, with regard to the typical products featured in a certain territory, the main advantages are (de Salvo et al., 2013) the following: 1) higher profits for the agricultural firms located in the rural areas which, in turn, create benefits to the local economic network, 2) the creation of qualified employment, 3) the regeneration of the traditional activities, 4) the development of gastronomic tourism which can support the economic sustainability of the surrounding territories.

ENOGASTRONOMIC TOURISM IN ITALY AND THE EFFECT OF PANDEMIC

Over the last years before the Covid-19 pandemic, enogastronomic tourism in Italy had significantly increased. Museums, churches, and archaeological sites were the main tourist places until some decades ago. For many travellers, putting together a gastronomic interest with another cultural activity in Italy has become the main scope of travel in the country. Between 2013 and 2017, tourists increased their consumption in the Italian enogastronomic sector by 70%, from 131 million euros to 223 million euros (Rapporto ENIT, 2018). Half of the revenues of the Italian enogastronomic industry are generated by tourists coming from the United States, United Kingdom, Austria, Switzerland, and France. Many travellers bind together enogastronomic experiences with other cultural activities, pointing out that food and beverage sectors generate advantages for other cultural industries. On average, each tourist that is motivated to choose Italy for its typical food and beverages spends 149,90 euro per a day. This data is relevant if compared to the consumption of other cultural motives underlying the choice to visit Italy. A tourist selecting Italy spends about 128,70 euro per a day for a visit to an art city, 122,90 euro for a sport vacation, or 109,30 euro for a mountain holiday. Enogastronomic interest is the main scope of travel in Italy for 22,3% of the Italian tourists and 22,9% of foreign tourists. The Rapporto sul turismo enogastronomico italiano (Garibaldi, 2019) points out that in 2018 the experience of putting together food, wine, and night events displayed a greater growth rate of bookings in Italy. The report highlights that an enongastronomic tourist in Italy is eclectic with a propensity to have a wide set of experiences, confirming theoretical studies on the behaviour of the traveller interested in typical food and beverages.

The Covid-19 pandemic unexpectedly stopped the increasing growth of Italian enogastronomic tourism. From March to May 2020, the government in Italy adopted laws aimed to contain the spread of the coronavirus. This limited or interrupted many economic activities, including the tourism sector. From January to September 2020, the occupancy in Italian hotels was reduced by 50.9% compared to the previous year (Istat). Foreign tourists in Italy in summertime represented 40% of the previous year. During the summer of 2020, the legislative chaos and uncertainty connected to a potential new wave of coronavirus lead to a summertime completely different from the previous ones and stimulated "proximity tourism" that entailed the rediscovery of the places nearby and represented a relevant opportunity for many small villages (*borghi*) where physical distancing is more feasible.

During the lockdown of March-May 2020 in Italy, a relevant increase in online wine and food tasting occurred, especially among Millennials and X Generation (GFK Sinottica). In the summertime, less restrictive national laws stimulated people to enjoy enogastronomic open-air events. The great success of some enogastronomic events, such as Calici di Stelle e Cantine Aperte in Vendemmia, witnessed renewed demand for this industry after the winter peak of the sanitary crisis, and the showed the relevance of this sector for the Italian economy. In November 2020, a report of the Wine Tourism International think thank highlighted that, compared to pre-pandemic period, guided tours and wine tasting in cellars decreased by 24% and 22%, respectively; while wine tasting in vineyards and outdoor activities increased by 6% and 9%, respectively. Many enogastronomic experts predict that a firms' competitive advantage within this industry will be increasingly based on great differentiation of products/services, active participation of the tourists in a certain enogastronomic event, a direct relationship with customers, a relevant emphasis on experiences, and being sustainable and on-line.

GASTRONOMIC TOURIST MOTIVATIONS AND THE ROLE OF EXPERIENCE IN THE WINE INDUSTRY

Studies on tourism emphasize that the choice of a destination is articulated in two main factors: the need to escape and the need to seek (Dann, 1981). People travel because they are pushed by internal factors or attracted by the features of a destination (Devesa-Fernández et al., 2009). Over the last year, gastronomy has represented the main motivation in the selection of a destination. Due to the interconnection between food, territory, tradition and culture, a perfect symbiosis occurs between gastronomy and tourism (Berbel-Pieda et al., 2019).

Many studies on gastronomic tourism have focused on the motivations and typologies of tourists towards gastronomy (Seyitoğlu and Ivanov, 2020). Hjalager (2004) identified four categories of gastronomic tourists. Diversionary and recreational travellers have greater propensity for familiar food; therefore, they are not much interested in local foodstuffs, and they don't consider food as a peak experience. On the other side, existential and experiential travellers have a stronger interest in local food which is perceived as a peak experience.

Boyne et al. (2003) propose a classification of gastronomic tourists that takes into account their effort to seek food-related information during their travel. Using this variable, the authors differentiate travellers that actively seek information on food when they travel to a destination from passive travellers that may be interested in information on food only when they encounter it and from uninterested travellers that may join in food related activities, anyway.

Local food may be actively searched by authenticity seekers. It is consumed but not actively searched by moderates, and it is not preferred by comfort seekers that search for familiar food in a familiar environment.

Using the categories of Özdemir and Seyitoğlu (2017), the relevance of food during travel has also been analysed by Björk and Kauppinen-Räisänen (2016). They identified the following categories of gastronomic tourists: 1) experiencers who find essential food and food-related activities; 2) enjoyers also appreciate gastronomy during their vacation, although their main target is not gastronomic activities; 3) survivors who consider the local food as a means to satisfy their physiological needs.

Through the different typologies offered by previous studies, Quan and Wang (2004) emphasized that local food may be considered to be a peak experience if it represents the main motivation of the traveller, or it is a supporting experience if food represents just a need. Under these circumstances, travellers appreciating local food will be particularly attracted by gastronomic events because these permit them to have a high value multi-sensorial experience. On the other hand, gastronomic events are able to boost the attractiveness of a place, generating economic benefits for the community.

From the different motivations pointed out in the literature on enogastronomic tourism emerges the relevance of the experience associated with the consumption of food and beverages. Over last decades, experience has increasingly become a factor which firms leverage to obtain a durable competitive advantage. "*As services, like goods before them, increasingly become commoditized - think of long-distanced telephone services sold solely on price - experiences have emerged as the next step in what we call the progression of economic value*" (Pine II and Gilmore, 1998). Firms intending to differentiate themselves in the experience economy use services as the stage and products as the props in order to create memorable experiences. While previous economic offerings, such as commodities, goods or services, are not personal, experience lives only in the mind of an individual and cannot be the same experience of another individual. Each experience is unique because it is generated by the interaction between the staged event and the state of mind of the individual (Pine II and Gilmore, 1998).

Rifkin (2000) points out the relevance of global tourism in the experience economy. Since the second part of XX century, tourism has become one of the most prominent industries in the world due to the cultural experiences offered to travellers. In particular, wine tourism has emerged as an industry that is able to offer unique experiences. It may be defined as "*visitations to vineyards, wineries, wine festivals and wine shows for which grape wine tasting and/or experiencing the attributes of a grape wine region are the prime motivation factors for visitors*" (Hall et al., 2000, p. 3). The set of the possible experiences within this industry may be classified using the 4E of the experience model (Pine II and Gilmore, 1999). It considers 2 dimensions: customer participation and the connection. The first dimension indicates how

Figure 1. Typical wine tourist experiences within Pine II and Gilmore (1999) model
(Source: Quadri-Felitti & Fiore, 2012)

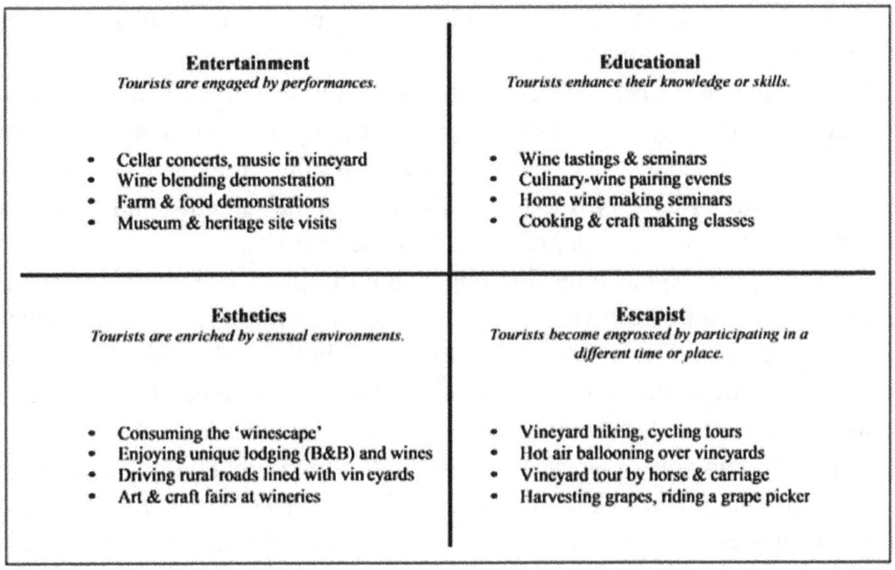

much the customer is able to influence the performance, if at all. At the end of the spectrum lies active participation which refers to customers that play key roles in generating the performance or the event. At the opposite end of the spectrum lies passive participation, where the "guest" joins the event as an observer or listener.

The connection dimension which unites customers on the basis of performance has absorption at one end and immersion at the other end. Observing a landscape immersed in the sounds and smells of nature is more immersing than watching a documentary on TV. One of the main motives that stimulates people to attend an event-specific place is to engage in activities that they have not had the possibility to experience in their daily life, and they may have access to it only as traveller or event-goer (Getz, 2008).

Matching the two dimensions, four categories of experiences are identified (Figure 1). Entertainment entails that the customer is more passively than actively involved in the creation of the performance. Customers tend to be external to the event rather than immersed in the action. Themed festivals centred on wine provide attendees with entertainment activities, combining, for example, wine and jazz or wine and art (Williams and Kelly, 2001; Yuan et al., 2005). Many studies point out the linkages between art, music and cultural heritage (Charters, 2006; Williams and Kelly, 2001; Yuan et al., 2005) because *wine is positioned as a luxury item that includes a lifestyle inclusive of art and culture* (Quadri-Felitti and Fiore, 2012). Wineries and vineyards represent an ideal scenario where cultural activities may be realized in order to entertain the attendee, strengthening the attractiveness of a destination.

Educational is one of the main drivers of wine tourists (Charters and Ali-Knight, 2000; Williams and Kelly, 2001). Educational events involve more active participation, but the customers' connection is more likely one of absorption than of immersion. Learning is a very perceived need of consumers in literature on wine tourism (Fountain and Charters, 2010). Personal development, growth, and learning are very relevant factors in the wine tourism industry for travellers who consider the level of sensations to be important (Galloway et al., 2008). Wine tasting and wine-food events at restaurants, winemaking seminaries, and culinary classes by chefs and farmers are the common educational activities that are featured by the active and absorption dimensions.

Tours of vineyards are good examples of events that satisfy both educational and entertainment needs (Getz and Carlsen, 2008). In fact, tourists learn about viticulture and are involved in it by organizers.

On one hand, customers in the aesthetic experiences are immersed in a unique environment. On the other hand, they are able to enjoy it passively. Views of the world's wine regions, the landscape and townscapes, and rural roads lined with vineyards and orchards offer joy and pleasure that is very different from the everyday urban lifestyle (Urry, 1995). Similarly, rural inns and bed and breakfasts characterize typical accommodations that are much appreciated by wine tourists.

Escapist experiences give the customer the opportunity to participate actively in an event and, simultaneously, to be immersed in the environment. *The escapist experience is highly immersive wherein consumers are engrossed in a different time or place* (Quadri-Felitti and Fiore, 2012, p. 8). On one hand, tourists in this experience are immersed in a rural, natural setting and enjoy the wine region's vineyard. On the other hand, they may actively participate in wine country recreation, such as harvesting or crashing the grapes or hot air ballooning over vineyards. Some studies highlight that just a part of winery tourists have purchasing and enjoying wine as the main scope of their travel (Mitchell and Hall, 2006). Indeed, winery tourists are interested in living a multi-sensorial experience that winery destinations are able to offer. The escapist dimension is dependent on the number of activities the tourist may be engrossed in at the wine destination.

A wine destination is a place where the tourist may enjoy, learn, be amused, and absorbed.

WINE EVENTS IN LA STRADA DEI VINI DEI CAMPI FLEGREI, ISOLA D'ISCHIA E DEL VESUVIO

The contagion risk associated with Covid-19 stopped non-essential economic activities in many countries during worst period of the pandemic. In many countries, the social distancing imposed by national governments in order to limit the spread of the coronavirus imposed limitations to attending events in a closed space. In Italy, some of the industries compromised by those restricting rules were tourism, leisure, and entertainment, including restaurants, pubs and coffee shops.

It is worthwhile noting that wine tourism represents a kind of industry that may guarantee a high level of safety from virus spread for several reasons. First, many of the activities connected to wine tourism may be arranged in external spaces where activities may be experienced without threatening the physical distancing required. Second, specifically in Italy, the structure of the wine tourism industry is fragmented, implying that wine events are offered by a great number of wine firms. Consequently, on one hand, this means that the demand connected to wine tourism industry is not concentrated in few wine firms, on the other hand, many wine producers that at now just vinify and sell wine may potentially extend their activities, including the organization of events associated with the production and consumption of wine, in order to satisfy the increasing wine tourism demand that requires a high-standard of safety.

In addition, since May 2020 wine firms adopted a specific protocol, particularly with regard to visiting the cellar. It was written by an international committee of wine tourism experts and integrated with national and regional laws. The protocol is divided into the following eleven sections: 1) Communication before a visit (e.g., informing visitors of rules with respect to safety in the cellar and not inviting those with Covid-19 symptoms to visit cellar) 2) Inside the cellar (e.g., checking body temperature at the entrance, verifying the use of a face mask, favouring the use of hand-sanitizers and/or soap), 3) Reception (e.g., indicating the safe social distance to use during the visit, favouring contactless payment, providing visitors with digital invoices), 4) Tasting (e.g., controlling the respect for physical distance, forbidding the sharing of food and beverages), 5) Cellar tour (e.g., favouring small groups of visitors, sanitizing spaces following tours, favouring the permanence of visitors in open-air spaces, establishing a one way itinerary), 6) Wine shop (e.g., establishing the maximum number of visitors per room, favouring digital consultation relative to wines), 7) End of visit (e.g., avoiding the use of a guestbook, using a digital questionnaire to measure customer satisfaction, inviting visitors to share cellar experiences through digital content), 8) Spaces (e.g., forbidding the use of elevators, constantly making soap available in bathrooms, positioning hand-sanitizer distributors at every stop of the tour), 9) Restaurant (e.g., maintaining the appropriate distance between tables, using digital menus), 10) Team (e.g., providing personnel with the necessary protective equipment, washing work clothes daily), 11) Contracts with tour operators (e.g., asking for the documents required by the government).

Therefore, the safety needs inducted by the pandemic may stimulate the identification of new opportunities for wine producers. The organization of recreational activities associated with the wine tourism, including the offering of experiences such as gazing at the landscape or visiting a museum near the vineyard, are not just easy to repeat events but unique events that may provide firms with the opportunity to translate a weakness into a strength.

Wine associated with music, paintings, archaeological sites, museums, harvesting and graping, and wonderful landscape are just some of unique experiences offered by wine firms located in Strada del Vino dei Campi Flegrei, isola d'Ischia e Vesuvio (Figure 2) that inglobes some rural geographical areas near the city of Naples. History, traditions, and the ability of producers to create high-quality wines make it very intriguing to be analysed. In addition, many typical products in this zone are protected by Italian and international laws that define their identity and geographical origin. This protection guarantees the high quality standards and uniqueness of a product. More specifically, in the Strada del Vino mentioned above, there are typical products protected by the DOP (Denominazione di Origine Protetta – Protected Designation of Origin) mark which is attributed to products whose quality characteristics depend essentially or exclusively on the production area and the traditional production techniques.

Over last years before pandemic crisis, events were gradually increased following a rising demand. This increased the heterogeneity of local wine tourism offerings that, in turn, makes the destination more attractive. The naturalistic, archaeological, and architectonic beauty of that area may be carefully treasured and, together with wine offerings, exert a multiplier effect on local economic development.

As Scaccheri (2010) points out, the essential factors of Strada del Vino are as follows:

1. The wine product has to represent not just something to drink, with certain characteristics, but also recall the quality connected to the specific production methodologies and the environment from which it was produced. Wine has to contain tangible and intangible aspects. Its consumption should stimulate a multi-sensorial experience that not only embodies smell and taste but also the culture, landscape, beauty, traditions, and the history of the territory. Therefore, wine should represent an asset that bonds the set of tourist attractions contained in the wide setting where la Strada del Vino is located.
2. The territory is represented by the geographical and physical area, its socio-economic development model, values, traditions, culture, history, and the technologies developed and adopted in that context.

Figure 2. Wine growing zones in Campania
(Source: www.enospace.it)

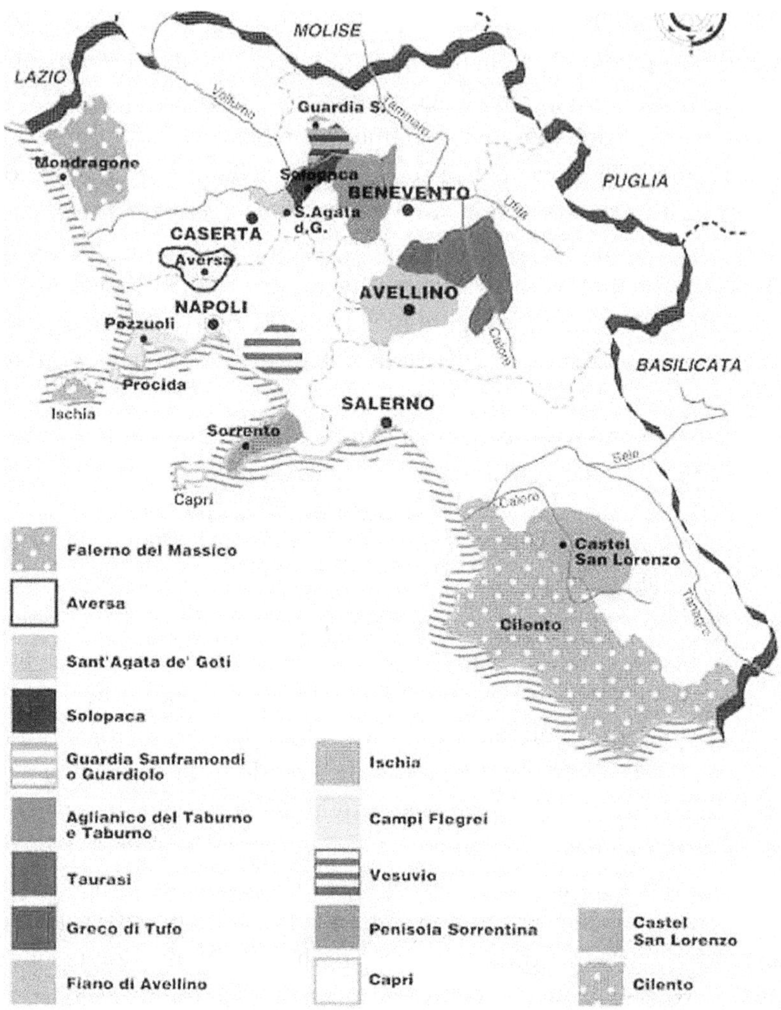

3. The environment consists of the people's sensitivity toward socio-environmental responsibility and consequently by the laws and rules aimed to preserve the quality of the product and the environmental context. The quality and genuineness of the products are traits more and more appreciated by the evolved customers.
4. The actors involved are wine producers (organized through independent firms, associations, cooperatives, etc.) distributors, tour operators, B&B's, hotels, restaurants, farmhouses, and all the actors joined in the wine sector and local institutions aimed to promote the territory.

Strada del Vino Campi Flegrei is a territory rich in culture, beauty, ancient tastes, smells, cooking recipes and the setting of the famous *Banchetti luculliani* in the ancient Roman age. For many centuries, this territory - very glorified by ancient Romans - has been preserving ancient recipes rooted in fish dishes and mixing the sophistication of oriental-Greek cuisine with the simplicity of the Roman one. In the itinerary, that begins in Pozzuoli, the bulk of ancient Roman remains, the tourist may gaze at the amphitheatre, archaeological parks, temples, legendary places such as the l'antro della Sibilla Cumana, Averno lake, and the Acropoli protected by the Sibilla Myth and surrounded by grapevines. Not only historical places, but also the natural attractiveness may be admired, such as the unique landscapes, the volcanic phenomena of bradyseism, solfatara, and thermes. In fact, Campi Flegrei is a vast area of volcanic origin. The soil is rich in ashter, lapillus, tuff, and other microelements that confer unique flavours to the grapes and wines. The word "Flegreo" derives from the Greek word flègo "burn", "fire", referring to fumaroles that can be noticed in that area. Those naturalistic scenarios provide tourists with a sensational experience that guarantees an immersion in the environment. One of the most peculiar towns of Campi Flegrei is Cuma. It was founded in the nineth century B.C. by ancient Greeks from Eubea and brings back many traditions, such as growing the Falanghina grape which makes a delicate white wine. Therefore, the grapes of Campi Flegrei have very ancient roots, described from the ancient Roman Age by writers such as Catone, Giovenale, Stazio, and Plinio who emphasized the peculiar traits of these grapes.

In this itinerary, many events were arranged and confirmed during the summer and autumn months of the pandemic period, because they leveraged on the wide and external spaces that avoided overcrowding. Jazz in the Vineyard, for example, was arranged with respect for the social distancing laws. It is an event that offers a multisensorial experience, where the participant may admire the extraordinary landscape of the Pozzuoli gulf, the Capri Island and Monte di Procida, and the unique colours of the Mediterranean sunset, enjoy the smells and the unique flavours of the vineyard, listen to relaxing jazz music, gaze at an artist painting a picture using wine, and taste typical food and the Piedirosso and Falanghina wines.

In the Vesuvius area of La Strada dei Vini, escapist experiences are offered by wine producers. The festival of the Harvest on Vesuvius is one of the main events that permits travellers to live as a vigneron for a day, visiting an ancient Munazei, a structure where wine was produced in ancient times, and lunching as a farmer. The vineyard is collocated in an area located at 400 meters above the sea and the landscape embodies the colours of Mediterranean sea and the majesty of the Vulcan, in the National Park of Vesuvius (Figure 3). The atmosphere created for this event leads the traveller in a unique dimension, making him or her an active participant in the several activities arranged. After providing participants with a kit for spending the day in the vineyard and typical clothes, they are involved in harvesting the grapes with vineyard staff and some experts. Other activities, such as grape pressing, are alternated with wine and food tasting and visiting the ancient fascinating cellars carved in stone where participants may learn the whole winemaking process.

Figure 3. Casa Setaro vineyard
(Source: www.ilMattino.it)

CONCLUSION

During last decade before the pandemic, enogastronomic tourism grew at a very high-rate. On one hand, some destinations exploited the growth of demand to strengthen their offerings, increasing the variety of services, attractions, and experiences. On the other hand, peripheral rural areas seized the consumers' interests to adopt a socio-economic development model that is able to turn local tangible and intangible resources into high-value typical food and beverages (de Salvo et al., 2013).

The Covid-19 pandemic forced many national governments to issue laws and rules aimed to limit overcrowding in order to limit the contagion. Consequently, many economic activities exercised in close spaces have been compelled to interrupt their services. The gastronomy industry has been hard hit by the Covid-19 crisis. In Italy, restaurants, bars, and coffee shops interrupted their activities for many weeks, or they limited them just to take-away service. Anyway, many firms operating in the enogastronomic industry leverage the organization of events in order to offer travellers not only local food and beverages but also experiences, such as attending a typical event or being immersed in a unique environment that permits tourists to live multi-sensorial events that are not repeatable in ordinary day by day life. Events organized in open spaces, such as farms, agricultural estates, or vineyards may represent a great opportunity for firms operating in the enogastronomic industry because social distancing and personal safety would be guaranteed.

La Strada del Vino dei Campi Flegrei, Ischia e Vesuvius presents very interesting examples of the experiences offered by the wine tourism industry in the pandemic period. Wine producers arrange events that permit travellers to choose from a wide range of experiences, in unique environments, with the possibility of actively participating in activities of the vineyard while respecting the safety of participants.

REFERENCES

Antonioli Corigliano, M. (1999). Strade del vino ed enoturismo. Distretti turistici e vie di comunicazione. Milan: Franco Angeli.

Antonioli Corigliano, M. (2000). I Distretti Turistici e le aggregazioni fra attori per lo sviluppo del prodotto destinazione. In Turismo: una tappa per la ricerca. Bologna: CNR, Progetto strategico "Turismo e Sviluppo Economico".

Asero, V. & Patti, S. (2009). Prodotti enogastronomici e territorio: La proposta dell'enoturismo. *XVI Rapporto sul turismo italiano.*

Beeton, S. (2006). *Community development through tourism.* Landlinks Press. doi:10.1071/9780643093881

Berbel-Pineda, J. M., Palacios-Florencio, B., Ramírez-Hurtado, J. M., & Santos-Roldán, L. (2019). Gastronomic experience as a factor of motivation in the tourist movements. *International Journal of Gastronomy and Food Science, 18*, 18. doi:10.1016/j.ijgfs.2019.100171

Björk, P., & Kauppinen-Räisänen, H. (2016). Local food – a source for destination attraction. *International Journal of Contemporary Hospitality, 26*(2), 177–194. doi:10.1108/IJCHM-05-2014-0214

Boyne, S., Hall, D., & Williams, F. (2003). Policy, Support and Promotion for Food-Related Tourism Initiatives. *Journal of Travel & Tourism Marketing, 14*(3-4), 14. doi:10.1300/J073v14n03_08

Charters, S. (2006). *Wine & Society: The Social and Cultural Context of a Drink*. Elsevier Butterworth-Heinemann. doi:10.4324/9780080458038

Charters, S., & Ali-Knight, J. (2000). Wine tourism—A thirst for knowledge? *International Journal of Wine Marketing*, *12*(3), 70–80. doi:10.1108/eb008715

Corigliano, M., & Mottironi, C. (2013). *Planning and Management of European Rural Peripheral Territories Through Multifunctionality: The Case of Gastronomy Routes in Trends in European Tourism Planning and Organisation*. Academic Press.

Croce, E., & Perri, G. (2018). *Il turismo enogastronomico. Progettare, gestire, vivere l'integrazione tra cibo, viaggio, territorio*. Franco Angeli.

Dann, G. M. S. (1981). Tourism motivation: An appraisal. *Annals of Tourism Research*, 8.

de Salvo, P., Hernández Mogollón, J. M., Clemente, E. D., & Calzati, V. (2013). Territory, tourism and local products. The extra virgin oil's enhancement and promotion: A benchmarking Italy-Spain. *Tourism and Hospitality Management*, *19*(1), 23–34. doi:10.20867/thm.19.1.2

Devesa-Fernández, M., Laguna-García, M., & Palacios-Picos, A. (2009). A structural model for the influence of visit motivation on the satisfaction with tourist visits. *Revista de Psicología del Trabajo y de las Organizaciones*, *24*(2).

ENIT. (2018). *Rapporto Annuale dell'Agenzia Nazionale del Turismo*. ENIT.

Everett, S. (2016). *Food and Drink Tourism: Principles and Practice*. Sage Publications.

Fountain, J., & Charters, S. (2010). Generation Y as wine tourists: their expectations and experiences at the winery-cellar door. In P. Benckendorff, G. Moscardo, & D. Pendergast (Eds.), *Tourism and Generation Y*. CAB International.

Galloway, G., Mitchell, R., Getz, D., Crouch, G., & Ong, B. (2008). Sensation seeking and the prediction of attitudes and behaviours of wine tourists. *Tourism Management*, *29*(5), 950–966. doi:10.1016/j.tourman.2007.11.006

Garibaldi, R. (2019). *Rapporto sul Turismo Enogastronomico in Italia*. Academic Press.

Garibaldi, R., & Pozzi, A. (2018). Creating tourism experiences combining food and culture: An analysis among Italian producers. *Tourism Review*, *73*(2), 73. doi:10.1108/TR-06-2017-0097

Getz, D., & Carlsen, J. (2008). Wine tourism among Generations X and Y. *Tourism (Zagreb)*, *56*(3).

Guruge, M. C. B. (2020). Conceptual Review on Gastronomy Tourism. *International Journal of Scientific and Research Publications*, *10*(2).

Guzel, B., & Apaydin, M. (2016). Gastronomy tourism, motivation and destinations. In O. St. Kliment (Ed.), *Global issues and trends in tourism*. University Press.

Hall, C. M., Mitchell, R., & Sharples, L. (2003). *Consuming Places: The Role of Food, Wine and Tourism on Regional Development. Food Tourism around the World: Development, Management and Markets*. Butterworth-Heinemann.

Hall, C. M., Sharples, E., Cambourne, B., & Macionis, N. (Eds.). (2000). *Wine Tourism Around the World: Development, Management and Markets*. Butterworth-Heinemann.

Hall, M., & Mitchell, R. (2005). Gastronomic tourism: comparing food and wine tourism experiences. In Niche Tourism, Contemporary Issues, Trends and Cases. Academic Press.

Harrington, R. J. (2005). Defining Gastronomic Identity. *Journal of Culinary Science & Technology*, *4*(2-3), 129–152. doi:10.1300/J385v04n02_10

Hjalager, A. M. (2004). What do tourists eat and why? Towards a sociology of gastronomy and tourism. *Tourism (Zagreb)*, *52*(2).

Kilic, G., Yucedag, N., & Aytekin, E. (2017). An Evaluation Towards Gastronomy Tourism. *Developments in Social Sciences, 551*.

Kim, Y. G., Eves, A., & Scarles, C. (2009). Building a model of local food consumption on trips and holidays: A grounded theory approach. *International Journal of Hospitality Management*, *28*(3), 3. doi:10.1016/j.ijhm.2008.11.005

Lin, L., & Mao, P. (2015). Food for memories and culture – A content analysis study of food specialties and souvenirs. *Journal of Hospitality and Tourism Management*, *22*, 22. doi:10.1016/j.jhtm.2014.12.001

Long, L. (2004). *Culinary Tourism: Exploring the Other through Food*. The University Press of Kentucky.

Martinengo, M. C., & Savoja, L. (1999). *Il turismo dell'ambiente*. Guerini Studio.

Mitchell, R., & Hall, C. M. (2006). Wine tourism research: The state of play. *Tourism Review International*, *9*(4), 307–332. doi:10.3727/154427206776330535

Nocifora, E., de Salvo, P., & Calzati, V. (2011). *Territori lenti e turismo di qualità, prospettive innovative per lo sviluppo di un turismo sostenibile*. Franco Angeli.

Özdemir, B. & Seyitoğlu, F. (2017). A conceptual study of gastronomical quests of tourists: authenticity or safety and comfort? *Tourism Management Perspective, 23*.

Pavlidis, G., & Markantonatou, S. (2020). Gastronomic tourism in Greece and beyond: A thorough review. *International Journal of Gastronomy and Food Science*, *21*, 21. doi:10.1016/j.ijgfs.2020.100229 PMID:32834883

Pine, I. I. J., & Gilmore, J. (1998). *Welcome to the experience economy*. Harvard Business Review, July-August.

Pine, I. I. J., & Gilmore, J. (1999). *The Experience Economy*. Harvard Business School Press.

Privitera, D., Nedelcu, A., & Nicula, V. (2018). Gastronomic and food tourism as an economic local resource: Cases studies from Romania and Italy. *Geo Journal of Tourism and Geosites*, *2018*, 1.

Quadri-Felitti, D., & Fiore, A. M. (2012). Experience economy constructs as a framework for understanding wine tourism. *Journal of Vacation Marketing*, *18*(1), 3–15. doi:10.1177/1356766711432222

Quan, S., & Wang, N. (2008). Towards a structural model of the tourist experience: An illustration from food experiences in tourism. *Tourism Management*, *25*(3), 297–305. doi:10.1016/S0261-5177(03)00130-4

Rifkin, J. (2000). *L'era dell'accesso. La rivoluzione della new economy*. Oscar Mondatori.

Scaccheri, A. (2010). *Vino e territorio. Tipicità del vino e gusti del consumatore*. Franco Angeli.

Seyitoğlu, F., & Ivanov, S. (2020). A conceptual study of the strategic role of gastronomy in tourism destinations. *International Journal of Gastronomy and Food Science*, *21*, 21. doi:10.1016/j.ijgfs.2020.100230

Sharples, L., & Hall, C. M. (2004). The consumption of experiences or the experience of consumption? An introduction to the tourism of taste. In *Food Tourism Around the World*. Routledge.

Sims, R. (2009). Food, place and authenticity: Local food and the sustainable tourism experience. *Journal of Sustainability, 3*.

Slee, B. (1993), Endogenous development: a concept in search of a theory. *Options Mediterraneenns, A*(23).

Smith, S. L., & Xiao, H. (2008). Culinary tourism supply chains: A preliminary examination. *Journal of Travel Research*, *46*(3), 289–299. doi:10.1177/0047287506303981

Sormaz, U., Akmese, H., Gunes, E., & Aras, S. (2016). Gastronomy in tourism. *Procedia Economics and Finance*, 39.

Sortino, A., & Chang, M. (2008). Pattern endogeni di sviluppo dell'agricoltura dicotomizzata: basi teoriche per una nuova politica agricola comunitaria. In *Acqua, agricoltura e ambiente nei nuovi scenari di politica comunitaria*. Franco Angeli.

Stone, M. J., & Migacz, S. (2016). *The American culinary traveler: Profiles, behaviors, and attitudes*. World Food Travel Association.

Tikkanen, I. (2007). Maslow's hierarchy and food tourism in Finland: Five cases. *British Food Journal*, *109*(9), 721–734. doi:10.1108/00070700710780698

Urry, J. (1995). *Consuming Places*. Routledge.

Yuan, J., Cai, L. A., Morrison, A. M., & Linton, S. (2005). An analysis of wine festival attendees' motivations: a synergy of wine, travel and special events? *Journal of Vacation Marketing 11*(1).

KEY TERMS AND DEFINITIONS

Economy Experience: According to Pine II and Gilmore (1998), it is the next economy following the agrarian economy, the industrial economy, and the service economy. Each experience is unique because it is generated by the interaction between the staged event and the state of mind of the individual.

Escapist Experiences: They give the customer the opportunity to participate actively in an event and, simultaneously, to be immersed in the environment.

Compilation of References

Aasestrela (2021). *UFCD 0244 – Execução de Peças de Olaria.* https://www.aasestrela.com/site/index.php?option=com_content&view=category&layout=blog&id=1&Itemid=46

Adrianto, L., & Matsuda, Y. (2004). Study on Assessing Economic Vulnerability of Small Island Regions. *Environment, Development and Sustainability, 6*(3), 317–336. doi:10.1023/B:ENVI.0000029902.39214.d0

AFAR: The Experiential Travel Guide. (n.d.). Https://www.afar.com

Agência de Turismo de Portugal. (2017). Estratégia 2027 - Liderar o Turismo do Futuro. Author.

Agência Nacional de Águas (ANA). (2005). *O Turismo e o Lazer e sua Interface com o Setor de Recursos Hídricos.* ANA.

Agência para o Desenvolvimento e Coesão. (2021). *Plano Global de Avaliação do Portugal 2020.* Agência para o Desenvolvimento e Coesão I.P.

Agudelo, D., Bretón-López, J., & Buela-Casal, G. (2003). Análisis bibliométrico de las revistas de Psicología Clínica editadas en castellano. *Psicothema, 15*, 507–516.

Ahmet, S. (2012). *Vize sakin şehir, Kırklareli için turizmde yeni bir marka ve kalkınma modeli cittaslow sakin şehir Vize.* Trakya Kalkınma Ajansı Kırklareli Yatırım Destek Ofisi.

Akpınarlı, H., & Arslan, P. (2017). Dokumacılığın Yerel Kalkınma Projeleri İle Yaşatılması Örneği: Üzümlü Dastar Dokumaları. In A. Aytaç (Ed.), *IX. Uluslararası Türk Sanatı, Tarihi ve Folklörü Kongresi.* Anka Basım.

Akşit, N. (2014). Yaratıcı (Kreatif) Turizm. *Uluslararası Sosyal Araştırmalar Dergisi, 7*(31), 786–795.

Aláez, F. L. (1993). Bibliometría: (análisis bivariante). Madrid: EYPASA.

Albuquerque, H. (2004). *Turismo sustentável nas zonas costeiras: O caso das praias de Mira.* Dissertação de Mestrado, Universidade de Aveiro, Portugal.

Aldeias de Montanha. (2021). *Aldeias de Montanha.* https://www.aldeiasdemontanha.pt/pt/aldeias/

Aldeias do Xisto. (2018). *Aldeias.* https://aldeiasdoxisto.pt

Aldeias Históricas de Portugal. (2021). *Planear viagem.* https://aldeiashistoricasdeportugal.com/planear-viagem

Alsos, G. A., Eide, D., & Madsen, E. L. (2014). Innovation in tourism industries. In G. A. Alsos, D. Eide, & E. L. Madson (Eds.), *Handbook of research on innovation in tourism industries* (pp. 1–26). Edward Elgar.

Altınay, M., & İstanbullu Dinçer, F. (2017). Geleneksel El Sanatlarının Yaratıcı Turízm Kapsamında Değerlendírílmesí. *Journal of Recreation and Tourism Research, 4*(1), 343–352.

Amaral, S. & Ferreira, M.T. (2010) *Caracterização da Pesca Recreativa nas águas interiores em Portugal Continental - Pesca Lúdica e Desportiva Pesca Desportiva em Albufeiras do Centro e Sul de Portuga.l Contribuição para a redução da eutrofização*. Instituto Superior de Agronomia Autoridade Florestal Nacional.

Amazonas, I. (2014). *Gestão ambiental na hotelaria: Tecnologias e práticas sustentáveis aplicadas nos hotéis de João Pessoa*. Dissertação de Mestrado, Universidade Federal da Paraíba, Brazil.

Andrade, B. (2006). *Turismo e sustentabilidade no município de Florianópolis: Uma aplicação do método da pegada ecológica*. Dissertação de Mestrado, Universidade Federal de Santa Catarina, Brazil.

Antonioli Corigliano, M. (1999). Strade del vino ed enoturismo. Distretti turistici e vie di comunicazione. Milan: Franco Angeli.

Antonioli Corigliano, M. (2000). I Distretti Turistici e le aggregazioni fra attori per lo sviluppo del prodotto destinazione. In Turismo: una tappa per la ricerca. Bologna: CNR, Progetto strategico "Turismo e Sviluppo Economico".

Antunes, L. (2016). *Governança territorial, inovação e turismo sustentável: Uma análise comparative das redes Casas Brancas e Aldeias de Xisto*. Dissertação de Mestrado, Universidade de Coimbra, Portugal.

Arencibia, A., & González, C. (2013). *La Cooperación Transfronteiriza*. Amarú Ediciones.

Argent, N., Tonts, M., Jones, R., & Holmes, J. (2013). A creativity-led rural renaissance? Amenity-led migration, the creative turn and the uneven development of rural Australia. *Applied Geography (Sevenoaks, England)*, *44*, 88–98. doi:10.1016/j.apgeog.2013.07.018

Arlinghaus, R. (2008). *Code of Practice for Recreational Fisheries*. EIFAC 251h session, Antalya, Turquia.

Asero, V. & Patti, S. (2009). Prodotti enogastronomici e territorio: La proposta dell'enoturismo. *XVI Rapporto sul turismo italiano*.

Azevedo, F. (2017). *Estudo para a Comissão REGI - A situação económica, social e territorial dos Açores (Portugal)*. European Commission.

Bac, D. P. (2012). From Slow Food to Slow Tourism. Department of Economics, University of Oradea, Faculty of Economics.

Bahar, O., Özdemir, Ö., & Akyürek, S. (2020). Yeşilüzümlü Kuzugöbeği Mantar ve Dastar Festivalinin Kırsal Kalkınma ve Kültürel Değerler Üzerindeki Rolü. *Sosyal ve Beşeri Bilimler Araştırmaları Dergisi*, *21*(46), 131–159.

Baldacchino, G. (2006). Warm versus cold water island tourism: A review of policy implications. *Island Studies Journal*, *1*(2), 183–200.

Balletto, G., Milesi, A., Ladu, M., & Borruso, G. (2020). A dashboard for supporting slow tourism in green infrastructures. A methodological proposal in Sardinia (Italy). *Sustainability (Switzerland)*, *12*(9), 3579. Advance online publication. doi:10.3390u12093579

Barcelona Creative Tourism. (2005). *Barcelona Creative Tourism*. Retrieved from https://www.barcelonacreativa.info/

Bătinaú, R., & Sorocovschi, V. (2012). *The Hydrographic Network – Mobilisation Vector of Pollutants in Roúia Montana Area, Roúia Montană in Universal History*. Cluj University Press.

Beeton, S. (2006). *Community Development through Tourism*. Landlinks Press. doi:10.1071/9780643093881

Bekar, A., Kocatürk, E., & Sürücü, Ç. (2017). Gastronomi Festivallerinin Algılanan Destinasyon İmajına ve Tekrar Ziyaret Etme Niyetine Etkisi. *Journal of Tourism and Gastronomy Studies*, *5*(2), 28–36. doi:10.21325/jotags.2017.109

BelediyesiS. (2021). http://seferihisar.bel.tr/seferihisarda-9-tohum-takas-senligi-basliyor

Benavides, C., Guzmán, V., & Quintana, C. (2011). The evolution of family firm literature as a research discipline. *Cuadernos de Economía y Dirección de la Empresa*, *14*, 78–90.

Berbel-Pineda, J. M., Palacios-Florencio, B., Ramírez-Hurtado, J. M., & Santos-Roldán, L. (2019). Gastronomic experience as a factor of motivation in the tourist movements. *International Journal of Gastronomy and Food Science*, *18*, 18. doi:10.1016/j.ijgfs.2019.100171

Bernardie-Tahir, N. (2011). *L'usage de l'île*. Petra.

Bialk-wolf, A., Pechlaner, H., & Nordhorn, C. (2013). The role of culture in building regional innovation systems and its impact on business tourism – The case of the Nuremberg Metropolitan Region. Poznan University of Economics Review, 13(4), 111–129.

Biggs, R., Schlüter, M., & Schoon, M. (2015). *Principles for Building Resilience: Sustaining Ecosystem Services in Social-Ecological Systems*. Cambridge University Press. doi:10.1017/CBO9781316014240

Binkhorst, E. (2007). Creativity in tourism experiences: The case of Sitges. In G. Richards & J. Wilson (Eds.), *Tourism Creativity and Development* (pp. 125–144).

Björk, P., & Kauppinen-Räisänen, H. (2016). Local food – a source for destination attraction. *International Journal of Contemporary Hospitality*, *26*(2), 177–194. doi:10.1108/IJCHM-05-2014-0214

Blapp, M., & Mitas, O. (2018). Creative tourism in Balinese rural communities. *Current Issues in Tourism*, *21*(11), 1285–1311. doi:10.1080/13683500.2017.1358701

Bleakney, M. (Director). (2019). *The Birders: A Melodic Journey Through Northern Colombia* [Video file]. Retrieved from https://www.youtube.com/watch?v=T-jn1_thxK8

Bochechas, J. (2000). Pesca nas águas interiores. Floresas de Portugal, 217-221.

Bonet, L. (2011). Tipologías y modelos de gestión de festivals. In La gestión de festivales: miradas comparadas. Barcelona: Gescènic.

Booyens, I., & Rogerson, C. M. (2015). Creative Tourism in Cape Town: An Innovation Perspective. *Urban Forum*, *26*(4), 405–424.

Bosworth, G., Rizzo, F., Marquardt, D., Strijker, D., Haartsen, T., & Thuesen, A. (2016). Identifying social innovations in European local rural development initiatives. *Innovation*, *29*(4), 442–461. doi:10.1080/13511610.2016.1176555

Boyne, S., Hall, D., & Williams, F. (2003). Policy, Support and Promotion for Food-Related Tourism Initiatives. *Journal of Travel & Tourism Marketing*, *14*(3-4), 14. doi:10.1300/J073v14n03_08

Brainerd, S. M. (2010). *European Charter on Angling and Biodiversity. Presentation of NINA2 and ESUSG3 specialist on 9 April 2010*.

Briguglio, L., & Kisanga, E. J. (Eds.). (2004). *Vulnerability and resilience of small states*. Commonwealth Secretariat and Islands and Small States Institute of the University of Malta.

Briguglio, L., Persaud, B., & Stern, R. (2006). *Toward an outward-oriented development strategy for small states: Issues, opportunities, and resilience building* (W. B. G. I. M. Fund, Ed.). World Bank Group / International Monetary Fund.

Briwa, R. (2021). The changing terrain of heritage at a Montana state park. *Journal of Historical Geography*, *71*, 17–27. doi:10.1016/j.jhg.2020.06.007

Brouder, P. (2012). Creative outposts: Tourism's place in rural innovation. *Tourism Planning & Development, 9*(4), 383–396. doi:10.1080/21568316.2012.726254

Bruner, R. F. (2004). *Think Globally Act Locally*. Understanding Strategy.

Bucci, A., Sacco, P. L., & Segre, G. (2014). Smart endogenous growth: Cultural capital and the creative use of skills. *International Journal of Manpower, 35*(1/2), 33–55. doi:10.1108/IJM-08-2013-0193

Butler, R. (1986). Literature as an influence in shaping the image of tourist destinations: A review and case study. In J. Marsh (Ed.), Canadian studies of parks, recreation and foreign lands (pp. 111–132). Department of Geography, Trent University.

Butler, R. (2000). Literary tourism. In J. Jafari (Ed.), *Encyclopedia of tourism* (p. 360). Routledge.

Butler, R. W. (1980). The concept of a tourist area cycle of evolution: Implications for management of resources. *The Canadian Geographer. Geographe Canadien, 24*(1), 5–12. doi:10.1111/j.1541-0064.1980.tb00970.x

Butler, R. W. (2006). Contrasting coldwater and warmwater island tourism destinations. In G. Baldacchino (Ed.), *Extreme tourism: Lessons from the world's cold water islands* (pp. 247–257). Elsevier.

Buttimer, A., & Seamon, D. (Eds.). (1980). *The Human Experience of Space and Place*. Routledge.

Byrd, E. T. (2007). Stakeholders in Sustainable Tourism Development and their Roles. *Tourism Review, 62*(2), 6–13. doi:10.1108/16605370780000309

Cabero Diéguez, V., & Caramelo, S. (2001). The Evolution of the Spanish-Portuguese Border and the Process of European Integration. Anais… Congress of the European Regional Science Association, 41.

Cabero Diéguez, V., Campesino Fernández, A. J., & López Trigal, L. (1995). El conocimiento de las franjas fronterizas. Aportación de los geógrafos españoles. *Boletín de la Asociación de Geógrafos Españoles*, (21-22), 93–106.

Cacilhas, N. (2020). Roteiro literário Levantado do Chão. *Dos Algarves: A Multidisciplinary e-Journal, 37*, 106-122. . doi:10.18089/DAMeJ.2020.37.7

Caffyn, A. (2012). Advocating and implementing slow tourism. *Tourism Recreation Research, 37*(1), 77–80. doi:10.1080/02508281.2012.11081690

Çağlı, I. B. (2012). *Türkiye'de yerel kültürün turizm odaklı kalkınmadaki rolü: Gastronomi turizmi örneği* (Doctoral dissertation). Institute of Science And Technology.

Caldeira, E., & Tomaz, N. (2015). Culture in development strategies of small and medium-sized European cities: A comparative analysis. In S. Hristova, M. Dragićević Šešić, & N. Duxbury (Eds.), *Culture and Sustainability in European Cities: Imagining Europolis* (pp. 86–100). Routledge.

Calderón, F. J. (2015). Reviewing the Spanish-Portuguese border: Conflict,interaction and cross-border cooperation. *Estudios Fronterizos, nueva época, 16*(31).

Calderón, F. J. (2017). Fronteras intraeuropeas, desactivación fronteriza, cooperación transfronteriza e instituciones: El caso de La Raya ibérica [Borders within Europe, border deactivation, cross-border cooperation and institutions: The Iberian Raya case]. *Estudios Fronterizos (México), 18*(36), 78–101. doi:10.21670/ref.2017.36.a04

Camarero, L., & Oliva, J. (2016). Understanding Rural Change: Mobilities, Diversities and Hybridizations. *Social Studies, 13*(2), 93–112. doi:10.5817/SOC2016-2-93

Campesino Fernández, A.-J. (2016). Paisajes del agua y turismo fluvial en la Raya ibérica. In Paisaje, cultura territorial y vivencia de la geografía. Libro homenaje al profesor Alfredo Morales Gil. Universidad de Alicante. doi:10.14198/LibroHomenajeAlfredoMorales2016-05

Campesino Fernández, A-J. & Jurado, J. M. (Eds.). (2014). *Turismo de Frontera (III).* Productos turísticos de la Raya ibérica. Edición de la III Jornada de Turismo de Frontera, Huelva, 3-4 de abril, 2014, Servicio de publicaciones, Universidad de Huelva.

Çanakkale Tarım ve Orman Genel müdürlüğü. (2021). www.canakkale.tarimorman.gov.tr

Canale, R. R., De Simone, E., Di Maio, A., & Parenti, B. (2019). UNESCO World Heritage sites and tourism attractiveness: The case of Italian provinces. *Land Use Policy, 85*, 114–120. doi:10.1016/j.landusepol.2019.03.037

Cànoves, G., Blanco-Romero, A., Prat, J. M., & Villarino, M. (Eds.). (2017). Turismo de interior en España. Productos y dinámicas territoriales [Inland tourism in Spain. Products and territorial dynamics.]. València, Spain: Universitat de València, Servei de Publicacions, Colección Desarrollo Territorial.

Çapar, G., & Yenipınar, U. (2016). Somut olmayan kültürel miras kaynağı olarak yöresel yiyeceklerin turizm. *Journal of Tourism and Gastronomy Studies, 4*, 100–115.

Carrizo-Sainero, G. (2000). Toward a concept of bibliometrics. *Journal of Spanish Research on Information Science, 1*(2), 59–86.

Carvalho, M. S., de Lima, J. I. S., & Kastenholz, E. (2014). Criatividade Cultural - que oportunidade para destinos rurais? *PASOS Revista de Turismo y Patrimonio Cultural, 12*(3), 635–648. doi:10.25145/j.pasos.2014.12.047

Carvalho, N. (2012). *Ecoturismo - Estudo de caso: Zmar-Eco camping resort. Projeto final de licenciatura.* Universidade Atlântica.

Carvalho, R., Ferreira, A. M., & Figueira, L. M. (2016). Cultural and Creative tourism in Portugal. *PASOS. Revista de Turismo y Patrimonio Cultural, 14*(5), 1075–1082. doi:10.25145/j.pasos.2016.14.071

Castro, E., Loureiro, F., Gomes, H., & Vieira, G. (2018). O Património Geológico do Geopark Estrela e a sua Valorização. In Geopatrimónio - Geoconhecimento, Geoconservação e Geoturismo. CEGOT.

Castro, E., Vieira, G., Loureiro, F., Patrocínio, F., Firmino, G., Gomes, H. & Fernandes, M. (2017). Managing the Estrela Aspiring Geopark: A framework for promoting geoheritage conservation. *Conference Ciência 2017.*

Castro, E., Vieira, G., Loureiro, F., Patrocínio, F., Firmino, G., Gomes, H., & Fernandes, M. (2017). Managing the Estrela Aspiring Geopark: A framework for promoting geoheritage conservation. *Conference Ciência 2017.*

Cejudo, E., & Navarro, F. (Eds.). (2020). *Neoendogenous Development in European Rural Areas: Results and Lessons.* Springer International Publishing. doi:10.1007/978-3-030-33463-5

Chão do Rio. (2018). *Chão do Rio.* https://www.chaodorio.pt/pt-pt

Charters, S. (2006). *Wine & Society: The Social and Cultural Context of a Drink.* Elsevier Butterworth-Heinemann. doi:10.4324/9780080458038

Charters, S., & Ali-Knight, J. (2000). Wine tourism—A thirst for knowledge? *International Journal of Wine Marketing, 12*(3), 70–80. doi:10.1108/eb008715

Chhabra, D., Healy, R., & Sills, E. (2003). Staged authenticity and heritage tourism. *Annals of Tourism Research, 30*(3), 702–719. doi:10.1016/S0160-7383(03)00044-6

CIMBSE-Comunidade Intermunicipal das Beiras e Serra da Estrela. (2021). *Quem Somos*. http://cimbse.pt/apresentacao/quem-somos

Cittaslow. (2021). *Uluslararası Yavaş Şehir Ağı*. https://www.cittaslow.org/content/organization

Cittaslowturkiye. (2021). *Yavaş Şehir Türkiye Ağı*. https://cittaslowturkiye.org/

Coelho, I. (2009). *Análise das consequências do crescimento do turismo no município de Porto Seguro a partir da década de 90 sob o ponto de vista do desenvolvimento sustentável*. Projeto de Bacharelato, Universidade Federal da Bahia.

Colletis-Wahl, K., & Pecqueur, B. (2001). Territories, development and specific resources: What analytical framework? *Regional Studies*, *35*(5), 449–459. doi:10.1080/713693825

Conselho da Europa. (2021). https://www.consilium.europa.eu/pt/policies/green-deal/

Consultores de Polítcas Públicas (CPC). (2014). *Análise da situação económica, social e ambiental da zona transfronteiriça de espanha e portugal*. Doc. de Trabalho para a programação 2014-2020, Programa de Cooperação Transfronteiriça Espanha Portugal.

Cooper, A. J. (2012). *Essentials of tourism*. Pearson.

Corbelle-Rico, E., Crecente-Maseda, R., & Santé-Riveira, I. (2012). Multi-scale assessment and spatial modelling of agricultural land abandonment in a European peripheral region: Galicia (Spain), 1956–2004. *Land Use Policy*, *29*(3), 493–501. doi:10.1016/j.landusepol.2011.08.008

Corigliano, M., & Mottironi, C. (2013). *Planning and Management of European Rural Peripheral Territories Through Multifunctionality: The Case of Gastronomy Routes in Trends in European Tourism Planning and Organisation*. Academic Press.

Costa, R., Costa, C., Brandão, M., & Breda, Z. (2014). *Turismo nos Países Lusófonos: Conhecimento, Estratégia e Territórios* (Vol. 1). Escolar Editora.

Courel Mountains Global Geopark. (2021). *500 millones de años caminando* [500 million years walking]. Retrieved January 19, 2021, from https://www.courelmountains.es/

Couret, C. (2012). Barcelona Creative Tourism. *Journal of Tourism Consumption and Practice*, *4*(2), 132–134.

Cowx, I. G., Arlinghaus, R., & Cooke, S. J. (2010). Harmonis-ing recreational fisheries and conservation objectives foraquatic biodiversity in inland waters. *Journal of Fish Biology*, *76*(9), 2194–2215. doi:10.1111/j.1095-8649.2010.02686.x PMID:20557659

Cox de Moura-Leite, R. M. (2009). ¿Cómo es la responsabilidad social empresarial en las PYMES? Administrando en entornos inciertos = managingin uncertain environment. Escuela Superior de Gestión Comercial y Marketing, ESIC.

Crane, D. (1969). Social structure in a group of scientists: A test of the "invisible college" hypothesis. *American Sociological Review*, *34*(3), 335–352. doi:10.2307/2092499

Croce, E., & Perri, G. (2018). *Il turismo enogastronomico. Progettare, gestire, vivere l'integrazione tra cibo, viaggio, territorio*. Franco Angeli.

Cunha, L. (2011). Autenticidade e Inovação: factores de renovação dos destinos turísticos maduros. Cogitur Journal of Tourism Studies, 4(4), 9-28.

Cunha, L. (2003). *Introdução ao Turismo* (2nd ed.). Editorial Verbo.

Çürük, G. (2017). Yeşil Üzümlü Beldesinde Geleneksel Dastar Dokumacılığı. *Milli Folklor*, *29*(115), 170–183.

D'Arco, M., Presti, L. L., Marino, V., & Maggiore, G. (2021). Is sustainable tourism a goal that came true? The Italian experience of the Cilento and Vallo di Diano National Park. *Land Use Policy*, *101*, 105198. doi:10.1016/j.landusepol.2020.105198

d'Italia, A. B. A. (2018). *L'ASSOCIAZIONE*. https://www.borghiautenticiditalia.it/lassociazione

Dall'Aglio, S. (2011). *ECONSTAT strategies for the travel industry*. Slow Tourism Seminar, Bled, Slovenia.

Dann, G. M. S. (1981). Tourism motivation: An appraisal. *Annals of Tourism Research*, 8.

Davallon, J. (2016). Penser le patrimoine selon une perspective communicationnelle [Thinking about heritage as a communicative fact]. *Science and Society*, *99*(99), 15–29. doi:10.4000ds.5257

Dax, T. (2019). Development of mountainous regions: smart specialization approaches as a means to overcoming peripheralization. In I. Kristensen, A. Alexandre Dubois, & J. Teräs (Eds.), *Strategic Approaches to Regional Development. Smart Experimentationin Less-Favoured* (pp. 52–67). Routledge.

De Bakker, F. G., Groenewegen, P., & Den Hond, F. (2005). A bibliometric analysis of 30 years of research and theory on corporate social responsibility and corporate social performance. *Business & Society*, *44*(3), 283–317. doi:10.1177/0007650305278086

de Salvo, P., Hernández Mogollón, J. M., Clemente, E. D., & Calzati, V. (2013). Territory, tourism and local products. The extra virgin oil's enhancement and promotion: A benchmarking Italy-Spain. *Tourism and Hospitality Management*, *19*(1), 23–34. doi:10.20867/thm.19.1.2

Declaração de Arouca, (2011). Congresso Internacional de Geoturismo. In *Geotourism in Action*. Arouca.

Decoville, A., & Durand, F. (2018). Exploring cross-border integration in Europe: How do populations cross borders and perceive their neighbours? *European Urban and Regional Studies*, 1–24. doi:10.1177/0969776418756934

Demirarslan, S., & Demirarslan, O. (2010). İnşaat Sektöründe Cam ve Cam Kökenli Malzeme Kullanımı. *CAMGERAN 2010 Uluslararası Katılımlı Uygulamalı Cam Sempozyumu,* 125–130.

Deniz, T. (2017). Hızlı dünyada sürdürülebilir mekânlar: Sakin kentler. *İnsan ve Toplum Bilimleri Araştırmaları Dergisi, 6*(3), 1399-1412.

Dessein, J., Soini, K., Fairclough, G., & Horlings, L. (2015). *Culture in, for and as sustainable development: Conclusions from the COST action IS1007 investigating cultural sustainability.* Jyväskylä, Finland: University of Jyväskylä. Retrieved from https://www.cost.eu/media/publications/Culture-in-for-and-as-Sustainable-Development-Conclusions-from-the-COST-Action-IS1007-Investigating-Cultural-Sustainability

Devesa-Fernández, M., Laguna-García, M., & Palacios-Picos, A. (2009). A structural model for the influence of visit motivation on the satisfaction with tourist visits. *Revista de Psicología del Trabajo y de las Organizaciones*, *24*(2).

Di Meo, G. (2008). *Processus de patrimonialisation et construction des territoires* [Processes of heritagization and construction of the territoires]. Retrieved December 15, 2020, from https://halshs.archives-ouvertes.fr/halshs-00281934

Dias, J. (2016). Environmental sustainability measurement in the Travel & Tourism Competitiveness Index: An empirical analysis of its reliability. *Ecological Indicators*, *73*, 589–596. doi:10.1016/j.ecolind.2016.10.008

Dickinson, J., & Lumsdon, L. (2010). *Slow Travel and Tourism*. Earthscan. doi:10.4324/9781849776493

Dissart, J. C., & Seigneuret, N. (Eds.). (2020). *Local Resources, Territorial Development and Well-being*. Edward Elgar Publishing. doi:10.4337/9781789908619

Ditton, R. B., Holland, S. M., & Anderson, D. K. (2002). Recreational fishing as tourism. Fisheries. *Afsjournals, 27*(3), 17-24. http://www.rbff.org/uploads/Research_section/fishing_as_tourism.pdf

DN – Diário de Notícias. (2018). *Ciclo "12 em Rede - Aldeias em Festa" anima aldeias históricas até dezembro*. https://www.dn.pt/portugal/ciclo-12-em-rede---aldeias-em-festa-anima-aldeias-historias-ate-dezembro-9256131.html

Domínguez García, L., Swagemakers, P., & Fernández, X. S. (2010). Mountain Food Production: Promising niches for rural development in Galicia, Spain. In *Proceedings of the 9th European IFSA Symposium: Building sustainable rural futures* (pp. 1607-1616). Vienna: Universität für Bodenkultur.

Dominguez, L., Aliste, J., Ibáñez Martinez, M. L., Natário, M., & Fernandes, G. (2013). *Estudio Socioeconomico de la Frontera entre Portugal y España*. Edita Riet.

Duxbury, N., Bakas, F. E., & Pato de Carvalho, C. (2019). Why is research–practice collaboration so challenging to achieve? A creative tourism experiment. *Tourism Geographies*, 1–26.

Eagles, F. J. P., McCool, F. S. & Haynes, C. D. (2002). *Sustainable Tourism in Protected Areas: Guidelines for Planning and Management*. The United Nations Environment Programme, World Tourism Organization, International Union for Conservation of Nature and Natural Resources, World Commission on Protected Areas (WCPA) Best Practice Protected Area Guidelines Series No. 8.

Ellender, B., Weyl, O., & Winke, R. H. (2009). Who uses the fishery resources in South Africa's largest impoundment? Characterising subsistence and recreational sectors on Lake Gariep. *Water S.A., 35*(5), 677–684. doi:10.4314/wsa.v35i5.49194

Embratur (Brasil). (1996). *Guia Brasileiro de Pesca*. Almeida Prado & Conceição.

ENIT. (2018). *Rapporto Annuale dell'Agenzia Nazionale del Turismo*. ENIT.

EQF - Eça De Queiroz Foundation. (2018). *Relatório e Contas*. Fundação Eça de Queiroz

EQF - Eça De Queiroz Foundation. (2020). *Plano de Actividades e Orçamento 2021*. Fundação Eça de Queiroz.

EQF - Eça De Queiroz Foundation. (2021). *Web of the Foundation*. https://feq.pt/

EQF - Eça De Queiroz Foundation. (n.d.). *Estatutos da Fundação Eça de Queiroz*. Fundação Eça de Queiroz.

Ergüven, M. H. (2011). Cittaslow – yaşamaya değer şehirlerin uluslararası birliği: Vize örneği. *Organizasyon ve Yönetim Bilimleri Dergisi, 3*(2), 201–210.

Ertaş, Y., & Gezmen-Karadağ, M. (2013). Sağlıklı beslenmede Türk mutfak kültürünün yeri. *Gümüşhane Üniversitesi Sağlık Bilimleri Dergisi, 2*(1), 117–136.

Eskiyörük, D. (2016). Antik Çağda Zeytin ve Zeytinyağı: Kilikya Bölgesi Örneği [Olive and Olive Oil in the Ancient Age: Cilicia Region Example]. *Journal of Tourism and Gastronomy Studies, 4*(1), 228–228.

ESPON. (2007). Cross-Border Cooperation. Cross-Thematic Study of INTERREG and ESPON activities. Luxembourg: INTERACT and ESPON.

ESPON. (2017). Shrinking rural regions in Europe Towards smart and innovative approaches to regional development challenges in depopulating rural regions. Luxemburg: Monitoring Committee.

Etikan, S. (2007). Kaya Köy Halıları. *Milli Folklor, Uluslararası Kültür Araştırmaları Dergisi, 10*(19), 83–90.

Etikan, S. (2016). Fethiye Alaçul Dokumaları. *Journal of International Social Research, 9*(42), 967–977. doi:10.17719/jisr.20164216206

European Commission (2015). *Europe without borders. The Schengen area.* doi:10.2837/78709

European Commission. (2017). *My Region, My Europe, Our Future. Seventh report on economic, social and territorial cohesion*. European Commission.

European Commission. (2019). *Country Report Portugal 2019. Including an In-Depth Review on the prevention and correction of macroeconomic imbalances*. European Commission.

European Committee of the Regions. (2020). *EU strategy for Rural Revival* (NAT-VII/006). The EU's Assembly of Regional and Local Representatives.

European Parliament. (2018). EU Agenda for Rural, Mountainous and Remote Areas (2018/2720 RSP). *Official Journal of the European Union, C*, 11.

Everett, S. (2016). *Food and Drink Tourism: Principles and Practice*. Sage Publications.

FAO. (2008). Codigo de conducta de la pesca recreativa. *Version en espanhol aceptada y presentada por Confederacion Espanola de Pesca Maritima de Recreo Responsable*. http://www.fao.org/3/a-i0363s.pdf

Farsani, N.T., Coelho, C., & Costa, C. (2011). Geotourism and Geoparks as Novel Strategies for Socio-economic Development in Rural Areas. *International Journal of Tourism Research, 13*, 68–81. doi:10.1002/jtr.800

Farsani, N. T., Celeste Coelho, C., Carlos Costa, C., & Carlos Neto, C. (2011). *Geoparks and Geotourism. New Approaches to Sustainability for the 21st Century*. Universal Publishers.

Fath, J. & Hunya, G. (2001). Cross-Border economic co-operation on present and future EU borders, a summary of findings, Enlargement. *Research Bulletin*, 2.

Feio, A., & Guedes, M. C. (2013). Architecture, tourism and sustainable development for the Douro region. *Renewable Energy, 49*, 72–76. doi:10.1016/j.renene.2012.01.063

Féret, S., Berchoux, T., Requier, M., & Abdelhakim, T. (2020). Framework providing definitions, operational typology and review of EU strategies for rural areas. *Working Document from SHERPA: Sustainable Hub to Engage into Rural Policies with Actors*. Retrieved February 16, 2021, from www.rural-interfaces.eu

Fernandes, G. (2015). Geografia, Cultura e Desenvolvimento Turístico em Territórios de Montanha. Das Perspetivas às Logicas de Atuação e Governação Territorial, Coleção Instituto Politécnico da Guarda, Guarda.

Fernandes, G. (2019). Rural depopulation, social resilience and context costs in the border municipalities of central Portugal. Dichotomies of social reorganization vs absence of public policies. *Economía Agraria y Recursos Naturales - Agricultural and Resource Economics, 19*(1), 121-149.

Fernandes, G. (2014). *Regiões de fronteira e desenvolvimento turístico: desafios, estratégias e utopias para a BIN-SAL 1492. In A JANGADA DE PEDRA – Geografias Ibero-Afro-Americanas*. Universidade do Minho.

Fernandes, G. (2020). Ecocultural Heritage, Tourism, Resilience and Landscapes Dynamics in Border Mountains of the Iberia Central Range. In A. Rocha, A. Abreu, J. V. de Carvalho, D. Liberato, E. Alén González, & P. Liberato (Eds.), *Advances in Tourism, Technology and Smart Systems, 171* (pp. 489–504). Springer. doi:10.1007/978-981-15-2024-2_43

Fernandes, G., Castro, E., Firmino, G., & Ferreira, A. (2016). Geoparks and regional and local development: aspiring Estrela Geopark. *Abstracts European Geoparks Conference*. 10.13140/RG.2.2.28770.94405

Fernandes, G., Castro, E., & Tracana, R. B. (2021). Tourism as a Strategy for Geo-Education and Cultural Valorization of Territories: Promoting Ecocultural Values and Geoconservation in Estrela Geopark, Portugal. In F. Brandão, Z. Breda, R. Costa, & C. Costa (Eds.), *The Role of Tourism in Achieving Sustainable Development Goals* (pp. 117–137). IGI Global. doi:10.4018/978-1-7998-5691-7.ch007

Fernandes, G., & Natário, M. (2020). Entrepreneurship in Border Territories: Context Costs, Resilience, and Logics of Action of the Companies and Entrepreneurs – Study of the Border Municipalities of Beira Interior in Portugal and Salamanca in Spain. In J. Dantas & L. Carvalho (Eds.), *Handbook of Research on Approaches to Alternative Entrepreneurship Opportunities* (pp. 129–152). IGI Global. doi:10.4018/978-1-7998-1981-3.ch007

Fernandes, G., Natario, M., & Braga, A. (2014). *Entrepreneurship and structural constraints to development of the bordering municipalities of beira interior norte. In Proceedings Renaissance of 11w Regions of Southern Europe.* APDR.

Fernandes, G., Natario, M., & Braga, A. (2016). Empreendedorismo e dinamismos socioeconómicos em territórios de fronteira. Uma abordagem à região transfronteiriça da bin-sal (beira interior norte-salamanca). In *Las fronteras en la investigación peninsular: temáticas y enfoques contemporáneos*. Publicaciones Universidade de Santiago de Compostela.

Fernandes, G., Natário, M., Daniel, A., & Braga, A. (2012). Evolução demográfica em municípios de baixa densidade no interior centro de Portugal. In *Respuestas de la Geografía Ibérica a la crisis actual. XIII Coloquio Ibérico de Geografía*. AGE e APG.

Fernández, R. (1998). Análisis bibliométrico de la Producción Científica. Revista de Divulgación Científica y Tecnológica de la Asociación Ciencia Hoy, 8(44).

Ferreira, J., Silva, S., Seixas, L., Lopes, A., Fortes, C., Reis, M., Poseiro, P., Capitão, R., Malvarez, G., & Tenedório, J. (2014). *The resource valuation and the challenge of coastal ecosystem-based management implementation*. Retrieved from http://repositorio.lnec.pt:8080/handle/123456789/1007022

Ferreira, P. F. (2014). *Slow cities (Cittaslow): Os espaços urbanos do movimento slow*. Academic Press.

Ferreira. M. T. (2009). *Estudo estratégico para a gestão das pescas continentais*. PAMAF medida 4 – ied, acção 4.4: estudos estratégicos. ICNF.

Ferreira, M. T., & Guerreiro De Brito, A. (2009). Águas interiores supericiais. In H. Pereira, T. Domingos, L. Vicente, & V. Proença (Eds.), *Ecossistemas e Bem-Estar Humano em Portugal*. Fundação da Faculdade de Ciências da U. L. e Escolar Editora.

Fethiye Chamber Of Commerce and Industry. (2020). *Fethiye Farming*. Retrieved from https://www.fto.org.tr/index_eng.aspx

Fethiye District Governorship Building. (2020). *Fethiye*. Retrieved from http://www.fethiye.gov.tr/

Fethiye Municipality. (2020). *Fethiye*. Retrieved from https://fethiye.bel.tr/

Flores, M.J., Muñoz, V., & Cabañero, M. (2004). *Análisis bibliométrico comparativo entre tres disciplinas del ámbito de la salud*. Comunicación presentada en la XV Jornadas de Interrelación de Enfermería, La Cañada, 5 y 6 Mayo.

Fondazione Slow Food. (2018). *Fondazione Slow Food - Salva la biodiversità, salva il pianeta*. https://www.fondazioneslowfood.com/it/

Fountain, J., & Charters, S. (2010). Generation Y as wine tourists: their expectations and experiences at the winery-cellar door. In P. Benckendorff, G. Moscardo, & D. Pendergast (Eds.), *Tourism and Generation Y*. CAB International.

Fraj-Andrés, E., López-Pérez, M. E., Melero-Polo, I., & Vázquez-Carrasco, R. (2012). Company image and corporate social responsibility: Reflecting with SMEs' managers. *Marketing Intelligence & Planning*, *30*(2), 266–280. doi:10.1108/02634501211212019

Fredheim, H., & Khalaf, M. (2016). The significance of values: Heritage value typologies re-examined. *International Journal of Heritage Studies*, *22*(6), 466–481. doi:10.1080/13527258.2016.1171247

Frey, B. S., & Briviba, A. (2021). A policy proposal to deal with excessive cultural tourism. *European Planning Studies*, 1–18.

Frost, W., & Laing, J. (2013). Communicating persuasive messages through slow food festivals. *Journal of Vacation Marketing*, *19*(1), 67–74. doi:10.1177/1356766712461403

Fullagar, S., Wilson, E., & Markwell, K. (2012). Starting slow: Thinking through slow mobilities and experiences. In S. Fullagar, K. Markwell & E. Wilson (Eds.), Slow Tourism Experiences and Mobilities (pp. 1-10). Channel View Publications.

Fullagar, S., Markwell, K., & Wilson, E. (2012). *Slow Tourism. Experiences and Mobilities*. Channel View Publications. doi:10.21832/9781845412821

Galán Ladero, M., Galera Casquet, C., Valero Amaro, V., & Díaz Méndez, M. (2006). Cause-related marketing in Spain. A consumer behaviour perspective. *35th EMAC Conference Proceedings*.

Gallou, E., & Fouseki, K. (2019). Applying social impact assessment (SIA) principles in assessing contribution of cultural heritage to social sustainability in rural landscapes. *Journal of Cultural Heritage Management and Sustainable Development*, *9*(3), 352–375. doi:10.1108/JCHMSD-05-2018-0037

Galloway, G., Mitchell, R., Getz, D., Crouch, G., & Ong, B. (2008). Sensation seeking and the prediction of attitudes and behaviours of wine tourists. *Tourism Management*, *29*(5), 950–966. doi:10.1016/j.tourman.2007.11.006

García de León, M. A. (2017). Entre las dos orillas: paisajes musicales y turismo musical en Cuba. In *Turismo Musical y Festivales. Actas del I Congreso Internacional de Turismo Musical y Festivales*. Cáceres: Zenodo. 10.5281/zenodo.569803

Garcia, A., & Nunes, S. (2009). Boom festival: Um estudo sobre a percepção dos habitantes locais. *Egitânia Sciencia*, (4), 41–66.

García-Rosell, J. C., Haanpää, M., Kylänen, M., & Markuksela, V. (2007). From firms to extended markets: A cultural approach to tourism product development. *Tourism: An International Interdisciplinary Journal*, *55*(4), 445–459.

Garibaldi, R. (2019). *Rapporto sul Turismo Enogastronomico in Italia*. Academic Press.

Garibaldi, R., & Pozzi, A. (2018). Creating tourism experiences combining food and culture: An analysis among Italian producers. *Tourism Review*, *73*(2), 73. doi:10.1108/TR-06-2017-0097

Geertz, C. (1973). The interpretation of cultures. Basic Books, Inc.

Gelbman, A., & Timothy, D. J. (2010). From hostile boundaries to tourist attractions. *Current Issues in Tourism*, *13*(3), 239–259. doi:10.1080/13683500903033278

Gerzeninsesi. (2017). (https://www.gerzeninsesi.com/guncel/gerzenin-parklarina-el-yapimi-ahsap-tabelalar-asildi-h6945.html

Getz, D., & Carlsen, J. (2008). Wine tourism among Generations X and Y. *Tourism (Zagreb)*, *56*(3).

Giampiccoli, A., & Kalis, J. H. (2012). Community-based tourism and local culture: the case of the amaMpondo. *Revista de Turismo y Patrimonio Cultural*, 173-188.

Godinho, A. M. (2017). Prefácio. In A Viagem que Ainda Ninguém Fez. Por 49 Aldeias – Maravilha de Portugal. Edição 7 Maravilhas.

Golja, T., & Krstinić, M. N. (2010). Corporate social responsibility in tourism - the most popular tourism destinations in Croatia: Comparative analysis. *Management, 15*(2), 107-121.

Gomes, H., Fernandes, M., Castro, E., & Vieira, G. (2017). The (Serra da) Estrela Aspiring Geopark (Portugal): preserving geoheritage, while promoting science and its links to local communities. In *Geophysical Research Abstracts* Vol. 19. EGU General Assembly.

Gomes, H., Loureiro, F., Cezar, L., Castro, E., & Vieira, G. (2019). Interpreting Late Pleistocene paleoenvironments through the geosites of the Estrela Geopark (Central Portugal). In *International Meeting on Paleoclimate: change and adaptation – book of abstracts.* Instituto Terra e Memória - Centro de Geociências da Universidade de Coimbra.

Gómez-Ullate, M., & Barrios Manzano, P. (2019). Festivales y desarrollo cultural en la provincia de Cáceres. Más allá del impacto económico. Trans 23 dossier: aproximaciones al estudio de las dinámicas e impactos de las celebraciones musicales en españa y portugal.

Gómez-Ullate, M. (2009). *La comunidad soñada.* Plaza y Valdés. doi:10.5211/9788492751228

Gómez-Ullate, M., Rieutort, L., Kamara, A., Santos, A. S., Pirra, A., & Gotay, M. (2020). Demographic Challenges in Rural Europe and Cases of Resilience based on Cultural Heritage Management. *European Countryside, 12*(3), 408–431. doi:10.2478/euco-2020-0022

Gonçalves, A., Barbosa, A., Maia, A., Martinho, A., Domingues, D., Alexandrino, P., d'Eça, P., Vieira, S., & Varandas, S. (2008). *A pesca nas águas interiores do Entre Douro e Minho.* Associação Regional do Norte da Pesca Desportiva. Jódique- Artes Gráficas, Lda.

Gordin, V., & Matetskaya, M. (2012). Creative tourism in Saint Petersburg: The state of the art. *Journal of Tourism Consumption and Practice, 4*(2), 55–77.

Gössling, S., Scott, D., & Hall, C. (2013). Challenges of tourism in a low-carbon economy. *Wiley Interdisciplinary Reviews: Climate Change, 4*(6), 525–538. doi:10.1002/wcc.243

Gössling, S., & Wall, G. (2007). Island tourism. In G. Baldacchino (Ed.), *A world of islands* (pp. 429–454). Agenda Academic e Institute of Island Studies.

Governo da República Portuguesa, X. I. I. (2020). *Guia Fiscal do Interior.* https://www.portugal.gov.pt/pt/gc22/comunicacao/documento?i=guia-fiscal-do-interior

Gray, M. (2019). Geodiversity, geoheritage and geoconservation for society. *International Journal of Geoheritage and Parks, 7*(4), 226–236. doi:10.1016/j.ijgeop.2019.11.001

Grecu, F., & Iosif, D. (2014). *La notion du Géosite et son pertinence dans une étude sur la région des Gorges du Danube en Roumanie.* Analele Universităţii Bucureşti: Seria Geografie LXIII.

Grove, K. (2018). *Resilience.* Routledge. doi:10.4324/9781315661407

Grydehøj, A., & Kelman, I. (2020). Reflections on conspicuous sustainability: Creating Small Island Dependent States (SIDS) through Ostentatious Development Assistance (ODA)? *Geoforum, 116*, 90–97. doi:10.1016/j.geoforum.2020.08.004

Guia da Cidade. (2008). *Na Rota do Artesanato da Serra.* https://www.guiadacidade.pt/pt/art/na-rota-do-artesanato-da-serra-19679-05

Guia de Viagem Foge Comigo. (2014). *Aldeias Históricas de Portugal.* Foge Comigo.

Guiver, J., & McGrath, P. (2016). Slow tourism: Exploring the discourses. *Dos Algarves: A Multidisciplinary e-Journal, 27*, 11-34. DOI: doi:10.18089/DAMeJ.2016.27.1

Guiver, J., McGrath, P., & Torkington, K. (2016). Special Issue Slow Tourism [Editorial]. *Dos Algarves: A Multidisciplinary e-Journal, 27*, 5-10. . doi:10.18089/DAMeJ.2016.27.0

Gunn, C. A., & Var, T. (2002). *Tourism Planning: Basics, Concepts and Cases* (4th ed.). Routledge.

Guruge, M. C. B. (2020). Conceptual Review on Gastronomy Tourism. *International Journal of Scientific and Research Publications, 10*(2).

Guzel, B., & Apaydin, M. (2016). Gastronomy tourism, motivation and destinations. In O. St. Kliment (Ed.), *Global issues and trends in tourism*. University Press.

Guzmán, P. C., Roders, A. P., & Colenbrander, B. J. F. (2017). Measuring links between cultural heritage management and sustainable urban development: An overview of global monitoring tools. *Cities (London, England), 60*, 192–201. doi:10.1016/j.cities.2016.09.005

Hackston, D., & Milne, M. J. (1996). Some determinants of social and environmental disclosures in New Zealand companies. *Accounting, Auditing & Accountability Journal, 9*(1), 77–108. doi:10.1108/09513579610109987

Hall, M., & Mitchell, R. (2005). Gastronomic tourism: comparing food and wine tourism experiences. In Niche Tourism, Contemporary Issues, Trends and Cases. Academic Press.

Hall, C. M. (2010). Island destinations: a natural laboratory for tourism: introduction. *Asia Pacific Journal of Tourism Research, 15*(3), 245–249. doi:10.1080/10941665.2010.503613

Hall, C. M., Harrison, D., Weaver, D., & Wall, G. (2013). Vanishing Peripheries: Does Tourism Consume Places? *Tourism Recreation Research, 38*(1), 71–92. doi:10.1080/02508281.2013.11081730

Hall, C. M., Mitchell, R., & Sharples, L. (2003). *Consuming Places: The Role of Food, Wine and Tourism on Regional Development. Food Tourism around the World: Development, Management and Markets*. Butterworth-Heinemann.

Hall, C. M., Sharples, E., Cambourne, B., & Macionis, N. (Eds.). (2000). *Wine Tourism Around the World: Development, Management and Markets*. Butterworth-Heinemann.

Hampton, M. P. (2002). Offshore pariahs? Small Island economies, tax havens, and the re-configuration of global finance. *Elsevier Science Ltd., 30*(9), 1657–1673.

Harrington, R. J. (2005). Defining Gastronomic Identity. *Journal of Culinary Science & Technology, 4*(2-3), 129–152. doi:10.1300/J385v04n02_10

Hartman, S. (2018). Resilient tourism destinations? Governance implications of bringing theories of resilience and adaptive capacity to tourism practice. In Destination Resilience – Challenges and Opportunities for Destination Management and Governance. Routledge. doi:10.4324/9780203701904-5

Hayaloğlu, A. A. (2008). Türkiye'nin peynirleri-Genel bir perspektif. *Türkiye, 10*, 21–23.

Heitmann, S. (2011). *Tourist Behaviour and Tourism Motivation. P. Robinson, S. Heitmann, & P. Dieke in, Research Themes for Tourism*. CAB International.

Hendrix, H. (2014). Literature and tourism: Explorations, reflections, and challenges. In Lit&Tour: Ensaios sobre Literatura e Turismo. Vila Nova de Famalicão: Húmus.

Henriques, M. H., & Brilha, J. (2017). UNESCO Global Geoparks: A strategy towards global understanding and sustainability. *Episodes, 40*(4), 349–355. doi:10.18814/epiiugs/2017/v40i4/017036

Herbert, D. (2001). Literary places, tourism and the heritage experience. Annals of Tourism Research, 28(2), 312-333. doi:10.1016/S0160-7383(00)00048-7

Hernández Borge, J. (2011). Despoblación y envejecimiento: Galicia hacia el año 2020 [Depopulation and ageing: Galicia towards the year 2020]. *Revista Galega de Economía, 20*, 121–140.

Hernández-Ramírez, J. (2017). Turismo en la frontera: Patrimonialización y cooperación transfronteriza en una periferia de la Unión Europea. *Etnográfica (Lisboa), 21*(2), 385–409. Advance online publication. doi:10.4000/etnografica.4940

Herrera Madueño, J., Larran Jorge, M., & Lechuga Sancho, M. P. (2014). Evolución de la literatura sobre la responsabilidad social en pymes como disciplina científica. *Revista Europea de Dirección y Economía de la Empresa, 24*(2015), 117–128.

Hidalgo, A., Dominguez, J., & Benzinho, C. (2014). *Costes de Contexto Tranfronterizos en el ámbito Empreserial- Territorio BIN-SAL, Diputacion de Salmanca*. OAEDER.

Higginbottom, K. (2004). *Wildlife Tourism: Impacts, management and planning*. Cooperative Research Centre for Sustainable Tourism.

Higgins, L. (2008) The creative music workshop: event, facilitation, gift. *International Society for Music Education, 26*(4) 326–338. doi:10.1177/0255761408096074

Hjalager, A. M. (2004). What do tourists eat and why? Towards a sociology of gastronomy and tourism. *Tourism (Zagreb), 52*(2).

Hoitsy, G., Woynarovich, A., & Moth-Poulsen, T. (2012). *Guide to trout farming based angling tourism*. FAO.

Hoppen, A., Brown, L., & Fyall, A. (2014). Literary tourism: Opportunities and challenges for the marketing and branding of destinations? *Journal of Destination Marketing & Management, 3*(1), 37–47. doi:10.1016/j.jdmm.2013.12.009

Hortelano, L. A. (2015). El Patrimonio territorial como activo turístico en la «raya» de Castilla y León con Portugal. *Cuadernos de Turismo*, (36), 247–268. doi:10.6018/turismo.36.230981

Hose, T. A. (2000). European "Geotourism" – geological interpretation and geoconservation promotion for tourists. In Geological Heritage: its conservation and management. Academic Press.

Hose, T. A. (1995). Selling the Story of Britain's Stone. *Environmental Interpretation, 2*, 16–17.

Hose, T. A. (2005). Geoturismo europeo. Interpretacíon geológica y promoción de la conservación geológica para turistas. In D. Barretino, W. A. P. Winbledon, & E. Gallego (Eds.), *Patrimonio geológico: conservación y gestión*. Istituto Tecnológico Geomineiro de España.

Hose, T. A. (2012). 3G's for modern geotourism. *Geoheritage, 4*(1-2), 7–24. doi:10.100712371-011-0052-y

Hotel Association of Turkey (TÜROB). (2020). *2019 Muğla Tourist Statistics*. Access from http://www.turob.com/tr/istatistikler/mayis-2020-mugla-turizm-istatistikleri

Howard, B. (2014). Brazil leads world in reducing carbon emissions by slashing deforestation. *National Geographic*. Retrieved from https://www.nationalgeographic.com/science/article/140605-brazil-deforestation-carbon-emissions-environment

Hung, W. L., Lee, Y. J., & Huang, P. H. (2016). Creative experiences, memorability and revisit intention in creative tourism. *Current Issues in Tourism, 19*(8), 763–770. doi:10.1080/13683500.2013.877422

IBAMA (Brasil). (2010). *Sensibilização da Comunidade para o Turismo de Pesca*. Programa Nacional do Desenvolvimento da Pesca Amadora.

IBGE – Instituto Brasileiro de Geografia e Estatística. (2017). *Estatísticas – população.* Retrieved from https://www.ibge.gov.br/estatisticas-novoportal/sociais/populacao.html

ICNF. (2020). *Edital – Zona de Pesca Reservada das Lagoas da Serra da Estrela 2020.* ICNF.

Igić, M., Maksić, M., Cekić, N., & Vasov, M. (2015). Problems of Cultural Heritage Protection in Rural Areas in Serbia. In *Proceedings of the III International Conference "Modern Technologies in Cultural Heritage"* (pp. 36-44). TU Sofia.

Ilbery, B. (2014). *The Geography of Rural Change.* Routledge. doi:10.4324/9781315842608

INE – Statistics Portugal. (2019). *Annual Provisional Resident Population Estimates.* https://www.ine.pt/xportal/xmain?xpid=INE&xpgid=ine_indicadores&contecto=pi&indOcorrCod=0008273&selTab=tab0&xlang=pt

INE. (2020). *Estatísticas do turismo 2019.* Lisboa: Instituto Nacional de Estatística, I. P.

Instituto Galego de Estatística. (2020). *Statistical Open Data of Galicia.* Retrieved January 15, 2021, from http://www.ige.eu/web/index.jsp?paxina=001&idioma=es

International Federation of the Fonography Industry and Music Canadá (IFPI). (2015). The mastering of a music city. Key elements, effective strategies and why it's worth pursuing. *Music Canadá.* https://musiccanada.com/wp-content/uploads/2015/06/The-Mastering-of-a-Music-City.pdf

IPDT. (2016). *Plano estratégico e de marketing do turismo dos Açores.* IPDT - Instituto de Turismo.

Iqbal, M. (1993). *International Trade in NonWood Forest Products: An Overview. Working Paper Misc/93/11.* Rome, Italy: Food and Agriculture Organization of the United Nations. Sec. 7.1. http://www.fao.org /documents/show_cdr.asp?url_file=/docrep/ x5326e/x5326 e00.htm

Ivanova, P. (2013). Creativity and sustainable tourism development. *Икономика 21, 3*(2), 108–120.

Jang, H. J., & Jung, T. (2015). *Go slow and curvy: Understanding the philosophy of the Cittaslow slowcity phenomenon.* Springer.

Jarábková, J., & Hamada, M. (2012). Creativity and Rural Tourism. *Creative and Knowledge Society*, *2*(2), 5–15. doi:10.2478/v10212-011-0022-4

Jelinčić, D. A., & Žuvela, A. (2012). Facing the Challenge? Creative Tourism in Croatia. *Journal of Tourism Consumption and Practice*, *4*(2), 78–90.

Jenkins, R. (1994). Rethinking ethnicity: Identity, categorization and power. *Ethnic and Racial Studies*, *17*(2), 197–223. doi:10.1080/01419870.1994.9993821

Joshi, R., Tamrakar, A., & Magaiya, B. (2021). Community-based participatory approach in cultural heritage reconstruction: A case study of Kasthamandap. *Progress in Disaster Science*, *10*, 100153. doi:10.1016/j.pdisas.2021.100153

Jurado, J. M., & Pazos-García, F. J. (2016). Población y turismo rural en territorios de baja densidad demográfica en España. Boletín de la Asociación de Geógrafos Españoles, 71, 247-272. doi:10.21138/bage.2282

Jurado-Almonte, J. M., & Pazos-García, F. J. (2018). La Frontera como Recurso Turístico. Posibilidades del Bajo/Baixo Guadiana. *Revista de Estudios Andaluces*, (35), 1–34. doi:10.12795/rea.2018.i35.01

Kabakchieva, D., & Vasileva, V. (2020). Cross-Border Territories and Tourism. Published in Collection of Scientifical articles from Innovation and Entrepreneurship Conference, Ajax Publishing.

Karababa, E., & Develi Isikli, N. (2005). Pekmez: A traditional concentrated fruit product. *Food Reviews International*, *21*(4), 357–366. doi:10.1080/87559120500222714

Karakurt Tosun, E. (2013). Yaşam kalitesi ekseninde şekillenen alternatif bir kentsel yaşam modeli: Yavaş kentleşme hareketi. *Uludağ Üniversitesi İktisadi ve İdari Bilimler Fakültesi Dergisi, 32*(1), 215–237.

Kartal, M. Y. (2020). *Muğla Alternatif Turizm Sektörü Yatırım Olanakları.* https://geka.gov.tr/tr

Kaya, C., Yıldız, M., Hayoğlu, İ., & Kola, O. (2005). Pekmez üretim teknikleri. *GAP VI. Tarım Kongresi*, 1482–1490.

Keskin, E. B. (2012). Sürdürülebilir kent kavramına farklı bir bakış: Yavaş şehirler (Cittaslow). *Paradoks Ekonomi. Sosyoloji ve Politika Dergisi, 8*(1), 81–99.

Kilic, G., Yucedag, N., & Aytekin, E. (2017). An Evaluation Towards Gastronomy Tourism. *Developments in Social Sciences, 551.*

Kim, Y. G., Eves, A., & Scarles, C. (2009). Building a model of local food consumption on trips and holidays: A grounded theory approach. *International Journal of Hospitality Management, 28*(3), 3. doi:10.1016/j.ijhm.2008.11.005

Kinley, A. (2012). Local food on a global scale: Anexploration of the ınternational slow food movement. *Journal of International Students, 1*(3), 1–14.

Kiráľová, A. (2020). Creativity as a Tool of Tourism Development. In Destination Management and Marketing: Breakthroughs in Research and Practice (pp. 504-530). IGI Global.

Királová, A. (2017). Creativity as a Tool of Tourism Development. In A. Kiráľová (Ed.), *Driving Tourism through Creative Destinations and Activities* (pp. 67–93). IGI Global. doi:10.4018/978-1-5225-2016-0.ch004

Kovács, I. & Nagy, G. (2013). Tourism in peripheric regions – The possibilities of thematic routes. *Seria Geografie, 23*(2), 220-228.

Kovács, Z., Vida, G., Lados, G., & Tóth, B. (2020). The Potentials of Cross-Border Tourism Development in the Lower-Tisa Valley. *GeoJournal of Tourism and Geosites, 28*(1), 360–375. doi:10.30892/gtg.28128-475

Kozak, M. & Buhalis, D. (2019). Cross–border tourism destination marketing: Prerequisites and critical success factors. *Journal of Destination Marketing & Management, 14*, 1-9. doi:10.1016/j.jdmm.2019.100392

Krippendorf, J. (1980). *Marketing im Fremdenverkehr.* Academic Press.

Kumar, N., Singh, G., & Parikh, T. S. (2011). Folk Music Goes Digital in India. *CHI 2011, Session: User Studies/Ethnography in Developing Regions.* 10.1145/1978942.1979151

Lange, E. (2012). De 'Países Subdesenvolvidos' à 'Fronteira do Subdesenvolvimento' – Contributo para um Debate Conceitual do Desenvolvimento. *Anais... 18th APDR Congress - Innovation and Regional Dynamics*, 47-59.

Lange, E. (2011). EGTC – a 'Breath of Fresh Air' for Practical Cross-Border Cooperation. Tracing the Legal Cross-Border Co-operation from the European Level to the Galicia-North of Portugal Cross-Border Region. *Regional Studies Association Annual International Conference.*

Leira, G. (2009). Responsabilidad Social y Desarrollo Turístico. Oportunidades en Ferrol-Ortegal. *Revista de Ocio y Turismo, 2*(1), 195–228. doi:10.17979/rotur.2009.2.1.1241

Leloup, F., & Stoffel, S. (2001). Intégration régionale et frontière dans le Mercosur: Entre théorie, principes et réalités. *Mondes en Developpement, 3*(115-116), 73–80. doi:10.3917/med.115.0073

Lemus, J., Dagostino, R., Olivas, M., & Huízar, A. (2018). *La pesca Deportiva em Bahía de Banderas.* Universidad de Guadalajara.

Lev-Tov, D. (2019). UNESCO Cities of Literature Every Bookworm Should Visit. *AFAR*. Retrieved from https://www.afar.com/magazine/unesco-cities-of-literature-every-book-lover-should-visit

Li, J., Krishnamurthy, S., Roders, A. P., & van Wesemael, P. (2020a). Community participation in cultural heritage management: A systematic literature review comparing Chinese and international practices. *Cities (London, England)*, *96*, 102476. doi:10.1016/j.cities.2019.102476

Li, J., Krishnamurthy, S., Roders, A. P., & van Wesemael, P. (2020b). Informing or consulting? Exploring community participation within urban heritage management in China. *Habitat International*, *105*, 102268. doi:10.1016/j.habitatint.2020.102268

Li, J., Krishnamurthy, S., Roders, A. P., & van Wesemael, P. (2021). Imagine the Old Town of Lijiang: Contextualising community participation for urban heritage management in China. *Habitat International*, *108*, 102321. doi:10.1016/j.habitatint.2021.102321

Lima, F. (2012). As dinâmicas territoriais no espaço de fronteira na fachada atlântica peninsular: a eurocidade Valença/Tuy. *Cadernos curso de doutoramento em Geografia FLUP*. https://ler.letras.up.pt/uploads/ficheiros/9963.pdf

Lindroth, K., Ritalahti, J., & Soisalon-Soininen, T. (2007). Creative tourism in destination development. *Tourism Review*, *62*(3), 58–58.

Lin, L., & Mao, P. (2015). Food for memories and culture – A content analysis study of food specialties and souvenirs. *Journal of Hospitality and Tourism Management*, *22*, 22. doi:10.1016/j.jhtm.2014.12.001

LITESCAPE.PT. (n.d.). *Atlas das Paisagens Literárias de Portugal Continental*. https://litescape.ielt.fcsh.unl.pt/.

Lois, R. C. & Carballo, A. C. (2015). La frontera hispano-lusa en la actualidad: una visión geográfica. *Revista de historiografía*, *23*, 191-21.

Lombard, N., & Labescat, G. (2010). Les îles, laboratoires du tourisme durable. *Espaces Tourisme & Loisirs*, *278*, 8–37.

Long, L. (2004). *Culinary Tourism: Exploring the Other through Food*. The University Press of Kentucky.

López López, P., & Tortosa Gil, F. M. (2002). Los métodos bibliométricos en Psicología. In *Nuevas Tecnologías de la Información y Documentación en Psicología* (pp. 199–225). Ariel Psicología.

Lowenthal, D. (1998). Fabricating Heritage. *History & Memory*, *10*(1), 5–24.

Lowe, P., Ray, C., Ward, N., Wood, D., & Woodward, R. (1999). The Forces of Change Affecting Rural Areas. In *Participation in Rural Development* (pp. 1–13). European Foundation for the Improvement of Living and Working Conditions.

MacCannell, D. (1976). *The tourist: A New Theory of the Leisure*. Class, MacMillan.

Madariaga, C., & García del Hoyo, J. J. (2014). Cultura de la pesca, patrimonio y turismo. In *Cultura, mercados y gestión de la pesca artesanal en el Golfo de Cádiz* (pp. 109–131). Universidad de Huelva. doi:10.13140/2.1.3189.5047

Mahon, M., & Hyyrylainen, T. (2019). Rural Arts Festivals as Contributors to Rural Development and Resilience. Sociologia Ruralis, 59(4), 612-635. doi:10.1111oru.12231

Maksić, M., Cekić, N., Vasov, M., & Igić, M. (2015). Problems of Archaeological Site of Felix Romuliana in Serbia. In *Proceedings of the III International Conference "Modern Technologies in Cultural Heritage"* (pp 22-28). TU Sofia.

Maksić, M. (2019). Governance Capacity of the City of Niš, Serbia, in the Spatial Planning Process. *Journal of Planning Education and Research*. Advance online publication. doi:10.1177/0739456X19844564

Maksić, M., Dobričić, M., & Trkulja, S. (2018). Institutional limitations in the management of UNESCO cultural heritage in Serbia: The case of Gamzigrad-Romuliana archaeological site. *Land Use Policy*, *78*, 195–206. doi:10.1016/j.landusepol.2018.06.055

Malachovský, A. (2020). Creative Tourism in Successful Destination Management as a Solution of Systemic Crisis in Slovak Tourism. In Destination Management and Marketing: Breakthroughs in Research and Practice (pp. 854-875). IGI Global. doi:10.4018/978-1-7998-2469-5.ch047

Mansvelt Beck, J. & Hortelano Mínguez, L. A. (2016). La apertura de la frontera: ¿nuevas identidades transfronterizas? *Boletín de la Asociación de Geógrafos Españoles*, *72*, 271-284. doi:10.21138/bage.2340

Marango, S., Bosworth, G., & Curry, N. (2021). Applying Neo-Endogenous Development Theory to Delivering Sustainable Local Nature Conservation. *Sociologia Ruralis*, *61*(1), 116–139. doi:10.1111oru.12315

Marić, D., Leković, K., & Tomić, S. (2021). Slow Tourism Consumers´s Recommendations. *Teme*, 1429-1440.

Markwell, K., & Fullagar, S. & Wilson, E. (2012), Reflecting Upon Slow Travel and Tourism Experiences. In S. Fullagar, K. Markwell & E. Wilson (Eds.), Slow Tourism Experiences and Mobilities (pp. 227-233). Channel View Publications.

Martinengo, M. C., & Savoja, L. (1999). *Il turismo dell'ambiente*. Guerini Studio.

Martínez Quintana, V. (2020). *Eco-Cultural Tourism: Sustainable Development and Promotion of Natural and Cultural Heritage*. doi:10.5772/intechopen.93897

Martínez-Filgueira, X., David Peón, D., & López-Iglesias, E. (2017). Intra-rural divides and regional planning: An analysis of a traditional emigration region (Galicia, Spain). *European Planning Studies*, *25*(7), 1237–1255. doi:10.1080/09654313.2017.1319465

Martín-Navarro, A., Sancho, M. P. L., & Medina-Garrido, J. A. (2018). BPMS para la gestión: Una revisión sistemática de la literatura. *Revista Espanola la de Documentacion Cientifica*, *41*(3), 213. doi:10.3989/redc.2018.3.1532

Martins, R. C. (2007). Fronteira e função: o caso europeu. In Mare Oceanus. Atlântico: Espaço de Diálogos.

Marujo, N., Borges, M. do R., Serra, J., & Lourenço, P. (2020). Handicraft and Creative Tourism: The Project Mãos de Cá. In S. M. Cabeça, A. R. Gonçalves, J. F. Marques, & M. Tavares (Eds.), Creative Tourism Dynamics:Connecting Travellers, Communities, Cultures and Places (pp. 171–180). Grácio Editor.

Mayer, M.; Zbaraszewski, W.; Pieńkowski, D.; Gach, G. & Gernert, J. (2019). Cross-Border Tourism in Protected Areas Potentials, Pitfalls and Perspectives. In *Geographies of Tourism and Global Change*. Springer Nature.

Mayer, H., & Knox, P. L. (2005). Slow cities: Sustainable places in a fast world. *Journal of Urban Affairs*, *28*(4), 321–334. doi:10.1111/j.1467-9906.2006.00298.x

McElroy, J. L. (2003). Tourism development in small islands across the world. *Geografiska Annaler*, *85*(4), 231–242. doi:10.1111/j.0435-3684.2003.00145.x

McElroy, J. L., & Parry, C. E. (2010). The characteristics of small island tourist economies. *Tourism and Hospitality Research*, *10*(4), 315–328. doi:10.1057/thr.2010.11

McManus, P. A., Hunt, W., Storey, J., & White, J. (2011). *Identifying the health and well-being benefits of recreational fishing*. Retrieved from http://www. tarfish.org/documents/Health and Wellbeing of Rec Fishing 2012.pdf

MêdaC. M. (2018). *A Queijaria*. https://cm-meda.pt/diretorio/a-queijaria

Medeiros, E. (2009). *The Cross-border cooperation in Raia Ibérica Geographical synthesis of INTERREG-A territorial impacts*. CEG.

Medeiros, E. (2010). Old vs Recent Cross-Border Cooperation: Portugal-Spain and Sweden-Norway. *Area*, *42*(4), 434–443. doi:10.1111/j.1475-4762.2010.00940.x

Medeiros, E. (2016). Territorial Cohesion: An EU Concept. *European Journal of Spatial Development*, (60), 1–30.

Medeiros, E. (2018). Should EU cross-border cooperation programmes focus mainly on reducing border obstacles. *Documents d'Analisi Geografica*, *64*(3), 467–49. doi:10.5565/rev/dag.517

Medrado, L., & Jackson, L. A. (2015). (in press). Corporate nonfinancial disclosures: An illuminating look at the corporate social responsibility and sustainability reporting practices of hospitality and tourism firms. *Tourism and Hospitality Research*. Advance online publication. doi:10.1177/1467358415600210

Melo, F., & Farias, S. (2018). Sustainability communication and its effect in consumer intention to visit a tourist destination. *Tourism & Management Studies*, *14*(2), 36–44. doi:10.18089/tms.2018.14204

Menadžment plan arheološkog lokaliteta Caričin grad (Justiniana Prima). (2014). [Management plan for the archaeological site of Caričin grad (Justiniana Prima)]. Academic Press.

Meydan Uygur, S., & Baykan, E. (2007). Kültür Turizmi ve Turizmin Kültürel Varlıklar Üzerindeki Etkileri. *Ticaret ve Turizm Eğitim Fakültesi Dergisi Yıl*, 30-49.

Miles, I., & Green, L. (2010). Innovation in creative services. In F. Gallouj & F. Djellal (Eds.), *The handbook of innovation in services: A multi-disciplinary perspective* (pp. 178–196). Edward Elgar. doi:10.4337/9781849803304.00018

Ministarstvo poljoprivrede, šumarstva i vodoprivrede. (2009). *Plan strategije ruralnog razvoja 2009–2013* [Strategic plan for rural development 2009-2013]. Academic Press.

Ministério do Turismo. (2010). *Turismo de Pesca: orientações básicas. Ministério do Turismo, Secretaria Nacional de Políticas de Turismo, Departamento de Estruturação, Articulação e Ordenamento Turístico, Coordenação Geral de Segmentação. – 2*. Ministério do Turismo.

Miranda, F. (2012). *A Pesca Lúdica e Desportiva no Nordeste Transmontano (Bacia do Douro, Portugal)*. Dissertação de Mestrado em Gestão de Recursos Florestais Escola Superior Agrária de Bragança, IPB.

Mitchell, P., & Fisher, R. (2010). From passenger to driver: Creativity and culture in rural communities. *Tourism, Culture & Communication*, *10*(3), 187–200. doi:10.3727/109830410X12910355180865

Mitchell, R., & Hall, C. M. (2006). Wine tourism research: The state of play. *Tourism Review International*, *9*(4), 307–332. doi:10.3727/154427206776330535

Moneva, J. M., & Hernández, J. C. (2009). Responsabilidad social corporativa e información de sostenibilidad en la Pyme. *Revista Internacional de la Pequeña y Mediana Empresa*, *1*(2), 23–41.

Moral, S., Cañero, P., Jimber, J., & Orgaz, F. (2016). Turismo fronterizo como motor de desarrollo de la frontera. Una revisión de la literatura. *International Journal of Scientific Managment Tourism*, *2*(2), 249–265.

Morar, C., & Pop, A.-C. (2016, November). Water, Tourism and Sport. A Conceptual Approach. *GeoJournal of Tourism and Geosites Year IX*, *18*(2), 249–258.

Moravsick, M. J. (1989). ¿Cómo evaluar a la Ciencia y a los Científicos? *Revista Espanola la de Documentacion Cientifica*, *12*, 313–325.

Mota, I., Pinto, M., Sá, J., Marques, V., & Ribeiro, J. (2005). *Estratégia Nacional para o Desenvolvimento Sustentável ENDS 2005-2015.* Retrieved from https://infoeuropa.eurocid.pt/files/database/000015001-000020000/000019537.pdf

Mubaideen, S., & Al Kurdi, N. (2017). Heritage conservation and urban development: A supporting management model for the effective incorporation of archaeological sites in the planning process. *Journal of Cultural Heritage*, *28*, 117–128. doi:10.1016/j.culher.2017.05.007

MunicipalityA. (2021). http://www.ahlat.gov.tr/kaymakamlik-birimleri

Naldi, L., Nilsson, P., Westlund, H., & Wixe, S. (2015). What is smart rural development? *Journal of Rural Studies*, *40*, 90–101. doi:10.1016/j.jrurstud.2015.06.006

Natário, M., Braga, A., & Fernandes, G. (2018). Determinantes do desempenho empreendedor e inovador nas regiões transfronteiriças. *Innovar: Revista de Ciencias Administrativas y Sociales*, *28*(70), 39–54. doi:10.15446/innovar.v28n70.74445

Ng'ombe, J. L. (2000). *The communication model and folk media in development.* Sustainable Development Dimensions. http://www.fao.org/sd/CDdirect/CDan0029.htm

Nieto Masot, A., Cárdenas Alonso, G., & Engelmo Moriche, A. (2020). Spatial Analysis of the Rural-Urban Structure of the Spanish Municipalities. *International Journal of Geo-Information*, *9*(4), 213. doi:10.3390/ijgi9040213

Nocifora, E., de Salvo, P., & Calzati, V. (2011). *Territori lenti e turismo di qualità, prospettive innovative per lo sviluppo di un turismo sostenibile.* Franco Angeli.

Nunes, N., Söderström, H., & Hipke, S. (2017). Culture In Sustainability Towards A Transdisciplinary Approach, In S. Asikainen, C.Brites, K. Plebańczyk, L.Rogač Mijatović, & K.Soini (Eds.), Understanding Cultural Sustainability (pp. 28-44). Sophi Publishing.

Nunes, P. A. (2010). *Desporto, Turismo e Natureza - O Turismo de Natureza como pólo de atração turística complementar ao produto Sol e Mar na sub-região do Litoral Alentejano.* Universidade Técnica de Lisboa - Faculdade de Motricidade Humana.

Nyseth, T., & Sognnæs, J. (2013). Preservation of old towns in Norway: Heritage discourses, community processes and the new cultural economy. *Cities (London, England)*, *31*, 69–75. doi:10.1016/j.cities.2012.05.014

Observer Science Tourism. (2021). *Montañas de Courel* [Courel Mountains]. Retrieved January 18, 2021, from https://www.observersciencetourism.com/es/destino/monta-as-de-courel

Ohridska-Olson, R. V., Ivanov, S. H., Ohridska-Olson, R., & Ivanov, S. (2010). Creative Tourism Business Model And Its Application In Bulgaria. *Proceedings of the Black Sea Tourism Forum'Cultural Tourism–The Future of Bulgaria.*

Olímpio, M., Baltazar, M., & Grosso, L. (2013). Empreendedorismo em territórios de baixa densidade. In *Futuro do Mundo Rural em Questão: Atas do I Congresso de Estudos Rurais do Norte Alentejano.* Estudo Rurais.

Oliveira, T. (2005). Tipologia das relações fronteiriças: elementos para o debate teórico-práticos. In *Território sem limites* (pp. 377–408). Editora da UFMS.

Ölmez, F. N., & Etikan, S. (2013a). Fethiye Alara Düz Dokumaları. *Süleyman Demirel Üniversitesi Güzel Sanatlar Fakültesi Hakemli Dergisi*, *13*(11), 75–99.

Ölmez, F. N., & Etikan, S. (2013b). Muğla İli Fethiye İlçesi Yantırlı Kilimleri. *Art-e Sanat Dergisi*, *13*(12), 62–82.

Organisation for Economic Cooperation and Development. (2020). *The territorial impact of COVID-19: Managing the crisis across levels of government.* Retrieved December 14, 2020, from https://www.oecd.org/coronavirus/policy-responses/the-territorial-impact-of-covid-19-managing-the-crisis-across-levels-of-government-d3e314e1/

Organización Mundial del Turismo. (2007). *Tourism Development and Climate Change: Understanding, Anticipating, Adapting, Participating in the Common Effort.* https://www.un.org/apps/sg/sgstats.asp?nid=2603

Orús, A. H., & Carcavilla Urquí, L. (2020). Twenty Years of Spanish Geoparks: Analysis and Future Prospects. *Geoheritage, 12*(4), 87. doi:10.100712371-020-00510-2

Osorio, S. (2020). Review of slow tourism, food and cities slow tourism, food and cities: Pace and the search for the "Good Life". *Tourism Geographies, 22*(4-5), 4–5, 874–876. doi:10.1080/14616688.2019.1618907

OTA. (2021). *Satisfação do turista que visita os Açores 2019.* Ponta Delgada: Observatório do Turismo dos Açores.

Oyman, N. R., & Erdal Sideli, E. (2018). Kayaköy (Muğla, Fethiye) Halıları. *The Journal of Academic Social Science Studies, 3*(67), 225–243.

Özdemir, B. & Seyitoğlu, F. (2017). A conceptual study of gastronomical quests of tourists: authenticity or safety and comfort? *Tourism Management Perspective, 23.*

Özdemir, G., & Çelebi, D. (2018). Exploring dimensions of slow tourism motivation. *Anatolia, 29*(4), 540–552. doi:10.1080/13032917.2018.1460854

Özgürel, G., Alan, R., & Bingöl, Z. (2015). Yerel Kalkınmada Bir Araç Olarak Festivaller: Munzur Kültür ve Doğa Festivali. *International Journal of Social and Economic Sciences, 5*(2), 113–121.

Özmen Uysal, Ö. (2010). Business Ethics Research with an Accounting Focus: A Bibliometric Analysis from 1988 to 2007. *Journal of Business Ethics, 93*(1), 137–160. doi:10.100710551-009-0187-9

Paasi, A., & Prokkola, E.-K. (2008). Territorial Dynamics, Cross-border Work and Everyday Life in the Finnish–Swedish Border Area. Space and Polity, 12(1), 13-29. doi:10.1080/13562570801969366

Page, S. (2015). *Tourism management* (5th ed.). Routledge.

Parkins, W., & Craig, G. (2009). Culture and the politics of alternative food networks. *Food, Culture, & Society, 12*(1), 77–103. doi:10.2752/155280109X368679

Paül, V. (2013). Hopes for the countryside's future. An analysis of two endogenous development experiences in South-eastern Galicia. *Journal of Urban and Regional Analysis, 5*(2), 169–192. doi:10.37043/JURA.2013.5.2.5

Pavlidis, G., & Markantonatou, S. (2020). Gastronomic tourism in Greece and beyond: A thorough review. *International Journal of Gastronomy and Food Science, 21*, 21. doi:10.1016/j.ijgfs.2020.100229 PMID:32834883

Pearce, D. G., & Butler, R. W. (1993). *Tourism research: Critiques and challenges (Issues in Tourism).* Routledge.

Pecsek, B. (2016). Revitalizing tourism in small regional towns through folklore-driven slow tourism: The example of Matyó land, Hungary. *Dos Algarves: A Multidisciplinary e-Journal, 27*, 94-119. . doi:10.18089/DAMeJ.2016.27.6

Peña, D., Guevara, A., & Fraiz, J. (2016). La investigación de la responsabilidad social empresarial en el sector hotelero. Análisis y revisión de la literatura científica. *Turismo e Sociedade, XVIII*, 137–158. doi:10.18601/01207555.n18.08

Peón Pose, D., Martínez-Filgueira, X. M., & López-Iglesias, E. (2020). Productive vs. residential economy: Factors behind the recovery of rural areas in socioeconomic decline. *Revista Galega de Economía, 29*(2), 67–44. doi:10.15304/rge.29.2.6744

Pérez Andrés, C., Estrada Lorenzo, J. M., Villar Álvarez, F., & Rebollo Rodríguez, M. (2002). Estudio bibliométrico de los artículos originales de la Revista Española de Salud Pública (1991-2000). Parte Primera: Indicadores generales. *Revista Espanola de Salud Publica, 76*(6), 659–672. doi:10.1590/S1135-57272002000600003 PMID:12602135

Pérez-Serrano, G., Anta Cabreros, C., Miguel Badesa, S., García Llamas, J. L., Pérez de Guzmán, M. V., & Pérez Serrano, M. (2004). Análisis bibliométrico en educación. Incidencia en la calidad universitaria. Proyecto subvencionado por la Subdirección General de Estudios y Análisis; España. *Referência (Coimbra)*, EA2003–EA0007.

Petrić, L., Hell, M., & van der Borg, J. (2020). Process orientation of the world heritage city management system. *Journal of Cultural Heritage*, *46*, 259–267. doi:10.1016/j.culher.2020.07.009

Pike, A., Rodríguez-Pose, A., & Tomaney, J. (2006). *Local and Regional Development*. Routledge. doi:10.4324/9780203003060

Pilati, M., & Hunter, A. (2020). *Research for REGI Committee – EU Lagging Regions: state of play and future challeges*. European Parliament.

Pilgrim, S., Samson, C., & Pretty, J. (2010). Ecocultural revitalization: Replenishing Community Connections to the Land. In S. Pilgrim & J. Pretty (Eds.), *Nature and Culture: Rebuilding Lost Connections* (pp. 235–256). Earthscan Publications. doi:10.4324/9781849776455

Pillet Capdepón, F. (2012). El turismo de interior en la España peninsular: El patrimonio territorial como destino turístico [Inland Tourism in Peninsular Spain: Territorial Heritage as a Tourist Destination]. *Boletín de la Asociación de Geógrafos Españoles*, *59*, 345–365. doi:10.21138/bage.1461

Pilz, D., Mclain, R., Alexander, S., Villarreal-Ruiz, S. B., Wurtz, Parks, C.G., McFarlane, E., Baker, B., Molina, R., & Smith, J. E. (2007). *Ecology and Management of Morels Harvested From the Forests of Western North America.* United States Department of Agriculture Forest Service Pacific Nortwest Research Station. General Technical Report, PNW-GTR-710, March 2007.

Pine, B. J., & Gilmore, J. H. (1998). Welcome to the experience economy. *Harvard Business Review*, *76*(4), 97–105. https://www.ncbi.nlm.nih.gov/pubmed/10181589 PMID:10181589

Pine, I. I. J., & Gilmore, J. (1998). *Welcome to the experience economy*. Harvard Business Review, July-August.

Pine, I. I. J., & Gilmore, J. (1999). *The Experience Economy*. Harvard Business School Press.

Plan detaljne regulacije "prostora zone I stepena zaštite arheološkog nalazišta Romulijana-Gamzigrad sa proširenjem. (2018). *Službeni list grada Zaječara, 14*.

Plan detaljne regulacije arheološkog nalazišta Caričin grad. (2018). *Službeni glasnik grada Leskovca, 14*.

Policarpo, F. & Mogollón, R. (2015). A actividade empreendedora na fronteira portugal-espanha alentejo- extremadura. *Anais... XVII seminário luso-espanhol de economia empresarial, sl3e'15*.

PORDATA. (2011). https://www.pordata

Prado-Lorenzo, J. M., Gallego-Álvarez, I., García-Sánchez, I. M., & Rodríguez-Domínguez, L. (2008). Social responsibility in Spain: Practices and motivations in firms. *Management Decision*, *46*(8), 1247–1271. doi:10.1108/00251740810901417

Prebensen, N. K. (2014). Facilitating for enhanced experience value. In G. A. Alsos, D. Eide, & E. L. Madson (Eds.), *Handbook of research on innovation in tourism industries* (pp. 154–180). Edward Elgar.

Prentice, R., & Andersen, V. (2003). Festival as creative destination. *Annals of Tourism Research*, *30*(1), 7–30. doi:10.1016/S0160-7383(02)00034-8

Pritchard, A. (1969). Statistical bibliography or bibliometrics. *The Journal of Documentation*, *25*(4), 348–349.

Privitera, D., Nedelcu, A., & Nicula, V. (2018). Gastronomic and food tourism as an economic local resource: Cases studies from Romania and Italy. *Geo Journal of Tourism and Geosites, 2018*, 1.

Pró Raia – Associação de Desenvolvimento Integrado da Raia Centro Norte (n.d.). *Guarda*. Pró-Raia / Câmara Municipal da Guarda – Sec. Turismo.

Prokkola, E.-K. (2008). Resources and barriers in tourism development: Cross-border cooperation, regionalization and destination building at the Finnish- Swedish border. *Fennia, Helsinki, 186*(1), 31–46.

Prokkola, E.-K. (2010). Borders in tourism: The transformation of the Swedish–Finnish border landscape. *Current Issues in Tourism, 13*(3), 223–238. doi:10.1080/13683500902990528

Prostorni plan opštine Lebane 2009-2024. (2011). *Sl. glasnik grada Leskovca, 13*.

Prostorni plan područja arheološkog nalazišta Romulijana-Gamzigrad. (2004). *Službeni glasnik RS, 131*.

Prostorni plan područja posebne namene Radan planine. (2014). *Sl. glasnik RS, 110*.

Prostorni plan Republike Srbije od 2021. do 2035. godine (nacrt) [Law on Spatial Plan of the Republic of Serbia from 2021 to 2035 (draft)] (in Serban).

Prostorni plan teritorije grada Zaječara (Sl. list grada Zaječara, 15/12) [Spatial plan of the territory of City of Zaječar) (Official Gazette of city of Zaječar, 15/12)] (in Serban).

Quadri-Felitti, D., & Fiore, A. M. (2012). Experience economy constructs as a framework for understanding wine tourism. *Journal of Vacation Marketing, 18*(1), 3–15. doi:10.1177/1356766711432222

Quan, S., & Wang, N. (2008). Towards a structural model of the tourist experience: An illustration from food experiences in tourism. *Tourism Management, 25*(3), 297–305. doi:10.1016/S0261-5177(03)00130-4

Quartesan, A., & Romis, M. (2011). Oaxaca de Juбrez, Mexico. In E. Rojas & F. Lanzafame (Eds.), *City Development: Experiences in the Preservation of Ten World Heritage Sites* (pp. 13–59). Inter-American Development Bank.

Queiroz, E. (2008). The City and the Mountains. New Directions Books.

Quinteiro, S. & Baleiro, R. (Orgs.). (2014a). *Lit&tour: Ensaios sobre literatura e turismo*. Vila Nova de Famalicão: Húmus.

Quinteiro, S., & Baleiro, R. (2014b). Uma personagem à procura da literatura: A ficção literária e a prática turística. *Dos Algarves. A Multidisciplinary e-Journal, 24*, 9–27.

Quinteiro, S., & Baleiro, R. (2017a). Turismo literario. In *Tendencias actuales del turismo en el mundo* (pp. 137–144). Editorial Síntesis.

Quinteiro, S., & Baleiro, R. (2017b). *Estudos em Literatura e Turismo: conceitos Fundamentais*. Universidade de Lisboa. Faculdade de Letras. Centro de Estudos Comparatistas.

Quinteiro, S., & Baleiro, R. (2018). *Key concepts in literature and tourism studies*. University of Lisbon.

Quinteiro, S., Baleiro, R., & Santos, I. D. (Eds.). (2016). *Literatura e turismo: Turistas, viajantes e lugares literários*. Universidade do Algarve.

Ray, C. (2006). Neo-endogenous rural development in the EU. In P. Cloke, T. Marsden, & P. Mooney (Eds.), *The handbook of rural studies* (pp. 278–292). Sage Publications. doi:10.4135/9781848608016.n19

Raymond, C. (2007). Creative Tourism New Zealand: The Practical Challenges Of Developing Creative Tourism. In Tourism, creativity and development (pp. 145–157). London: Routledge.

Remoaldo, P., & Cadima-Ribeiro, J. (2019). Creative Tourism as a New Challenge to the Development of Destinations: The Portuguese Case Study. In M. Peris-Ortiz, M. R. Cabrera-Flores, & A. Serrano-Santoyo (Eds.), *Cultural and Creative Industries A Path to Entrepreneurship and Innovation* (pp. 81–99). Springer.

Report, B. (1987). *Our Common Future*. United Nations World Commission on Environment and Development. Retrieved from https://en.wikisource.org/wiki/Brundtland_Report

Republic of Turkey Ministry of Culture and Tourism. (2021). *Fethiye, Mugla*. https://mugla.ktb.gov.tr

República Portuguesa. (2017). *Programa Nacional para a Coesão Territorial*. Unidade de Missão para a valorização do Interior.

Republički zavod za statistiku (2014). *Popis stanovništva, domaćinstava i stanova 2011. u Republici Srbiji: Stanovništvo* [2011 Census of Population, Households and Dwellings in the Republic of Serbia: Population]. Academic Press.

Ribeira Sacra Tourism. (2021). *Descubre Ribeira Sacra* [Discovering Sacred Riverside]. Retrieved January 12, 2021, from https://turismo.ribeirasacra.org/

Ribeiro, M. T., & Valente, I. M. (2013). *Portugal Europe: Spaces Border and Crossborder Cooperation «Portugal Europe: Spaces Border and Crossborder Cooperation». In Comunicatin the EU policies beyond the bordes*. Oradea Uniersity Pressa.

Richards, G. (2009). Creative Tourism and Local Development: A global conversation. In Creative Tourism: A global conversation (pp. 78–90). Academic Press.

Richards, G. (2012). Tourism, creativity and creative industries. In Conference Creativity and Creative Industries in Challenging Times. NHTV Breda.

Richards, G. (2019). Creative tourism: opportunities for smaller places? *Tourism & Management Studies, 15*(SI), 7–10.

Richards, G. (Ed.) (1996). Cultural Tourism in Europe. Wallingford: CABI Publishing.

Richards, G. (2002). Creative Tourism as a Factor Destination Development. In S. Flanagan, J. Ruddy, & N. Andrews (Eds.), *Innovation in Tourism Planning* (pp. 118–127). Dublin Institute of Technology.

Richards, G. (2010). Tourism development trajectories: From culture to creativity? *Tourism & Management Studies*, *6*, 9–15.

Richards, G. (2011). Creativity and tourism. The state of the art. *Annals of Tourism Research*, *38*(4), 1225–1253. doi:10.1016/j.annals.2011.07.008

Richards, G. (2013). The Routledge Handbook of Cultural Tourism. In M. Smith & G. Richards (Eds.), *The Routledge Handbook of Cultural Tourism* (Vol. 6). Routledge.

Richards, G. (2014). Creativity and tourism in the city. *Current Issues in Tourism*, *17*(2), 119–144. doi:10.1080/13683500.2013.783794

Richards, G. (2016). The challenge of creative tourism. *Ethnologies (Québec)*, *38*(1–2), 31–42.

Richards, G. (2017). Creative tourism: Opportunities for smaller places? *Tourism & Management Studies*, *15*(SI), 7–10. Advance online publication. doi:10.18089/tms.2019.15SI01

Richards, G. (2020). Designing creative places: The role of creative tourism. *Annals of Tourism Research*, *85*, 102922. doi:10.1016/j.annals.2020.102922

Richards, G., & Marques, L. (2012). Exploring Creative Tourism: Editors Introduction. *Journal of Tourism Consumption and Practice*, *4*(2), 1–11.

Richards, G., & Raymond, C. (2000). Creative tourism. *ATLAS News*, *23*, 16–20.

Richards, G., & Wilson, J. (2006a). Developing creativity in tourist experiences: A solution to the serial reproduction of culture? *Tourism Management*, *27*(6), 1209–1223. doi:10.1016/j.tourman.2005.06.002

Richards, G., & Wilson, J. (2006b). The creative turn in regeneration: Creative spaces, spectacles and tourism in cities. In M. K. Smith (Ed.), *Tourism, Culture and Regeneration* (pp. 12–24). CABI. doi:10.1079/9781845931308.0012

Richards, G., & Wilson, J. (2007). Tourism, creativity and development. In G. Richards & J. Wilson (Eds.), *Tourism, Creativity and Development*. Routledge. doi:10.4324/9780203933695

Rifkin, J. (2000). *L'era dell'accesso. La rivoluzione della new economy*. Oscar Mondatori.

Ritzer, G. (2003). Rethinking globalization: Glocalization/grobalization and something/nothing. *Sociological Theory*, *21*(3), 193–209. doi:10.1111/1467-9558.00185

Robinson, M., & Andersen, H. C. (2002). Beyond and between the pages: Literature and tourism relationships. In M. Robinson & H. C. Andersen (Eds.), *Literature and Tourism: Essays in the Reading and Writing on Tourism* (pp. 39–79). Continuum.

Robinson, M., & Andersen, H. C. (2004). *Literature and tourism: Essays in the reading and writing on tourism*. Thomas Learning.

Rodrigues, J. (2009). Geoturismo – Uma Abordagem Emergente. In C. Neto de Carvalho & J. C. Rodrigues (Eds.), *Geoturismo & Desenvolvimento Local* (pp. 38–60p). Idanha-a-Nova.

Rodríguez-Pose, A. (2000). Economic convergence and regional development strategies in Spain: The case of Galicia and Navarre. *European Investment Bank Papers*, *5*(1), 88–115.

Rogerson, C. M., & Visser, G. (2011). Rethinking South African urban tourism research. *Tourism Review International*, *15*(1-2), 77–90. doi:10.3727/154427211X13139345020336

Rosa, V. (2013). Da literatura ao itinerário turístico: as casas-museu e o turismo literário. *Research Gate*. Retrieved from https://www.researchgate.net/publication/344544005_Da_literatura_ao_itinerario_turistico_as_housemuseums_e_o_turismo_literario

Ruiz Pulpón, A. R., & Cañizares Ruiz, M. C. (2020). Enhancing the Territorial Heritage of Declining Rural Areas in Spain: Towards Integrating Top-Down and Bottom-Up Approaches. *Land (Basel)*, *9*(7), 216. doi:10.3390/land9070216

Sabrin, M. (2002). A ranking of the most productive business ethics scholars: A five years study. *Journal of Business Ethics*, *36*(4), 355–379. doi:10.1023/A:1014432004643

Sabugal, C. M. (2018). *12 em Rede Aldeias em Festa, AHP_Marialva_Castelo Novo_Linhares da Beira*. https://www.cm-sabugal.pt/12-rede-aldeias-festa-ahp_marialva_castelo-novo_linhares-da-beira_/#more-16690

Şahin, G. (2019). Türk Kültür Coğrafyasından Özel Bir Örnek: Türk Sabunları. *Ağrı İbrahim Çeçen Üniversitesi Sosyal Bilimler Enstitüsü Dergisi*, *5*(2), 371–404.

Şahinkaya, S. (2010). *Bir Yerel Kalkınma Modeli: Cittaslow ve Seferihisar Üzerine Değerlendirmeler*. http://www.bagimsizsosyalbilimciler.org/ Yazilar_Uye/Sahin Tem10.pdf

Sahlins, P. (1996). *Frontières et identités nationales, la France et l'Espagne dans les Pyrénées depuis le XVIIe siècle*. Paris: Belin.

Salgado, A. R. (2010). La cooperación transfronteriza y sus consecuencias: Hacia la reestructuración territorial en Europa. *Investigaciones Regionales*, (18), 141–152.

Saliba, C. (2007, Spring). Financial services and small island jurisdictions. *Bank of Valletta Review*, (35), 39–54.

Sánchez-Mesa Martínez, L. J. (2019). Planning Territorial Policies against Inner Areas Depopulation in Spain: Keys for sustainable management of cultural and environmental resources. *Il Capitale Culturale*, *19*, 53–81. doi:10.13138/2039-2362/1980

Santos, L. (2017). Prefácio. In A Viagem que Ainda Ninguém Fez. Por 49 Aldeias – Maravilha de Portugal. Edição 7 Maravilhas.

Saramago, J. (1989). *Viagem a Portugal*. Porto Editora.

Sardo, A. (2008). Literary Tourism: a way of enriching the local heritage and culture. Egitania Sciencia (vol. 2). Guarda: Instituto Politécnico da Guarda.

Sardo, A. (2009). Turismo literário: A importância do património e dos sítios literários para o desenvolvimento turístico regional. In J. M. Simões & C. C. Ferreira (Eds.), Turismos de nichos: Motivações, produtos, territórios (pp. 339-352). Centro de Estudos Geográficos. Lisboa: Universidade de Lisboa.

Sassi, P. (2006). *Strategies For Sustainable Architecture*. Taylor & Francis. doi:10.4324/9780203480106

Saygılı, D., Demirci, H., & Samav, U. (2020). Coğrafi işaretli Türkiye peynirleri. *Aydın Gastronomy*, *4*(1), 11–21. doi:10.17932/IAU.GASTRONOMY.2017.016/2020.401/gas_v04i1002

Scaccheri, A. (2010). *Vino e territorio. Tipicità del vino e gusti del consumatore*. Franco Angeli.

Scott, K., Rowe, F., & Pollock, V. (2018). Creating the good life? A wellbeing perspective on cultural value in rural development. *Journal of Rural Studies*, *59*(April), 173–182. doi:10.1016/j.jrurstud.2016.07.001

Semmens, J., & Freeman, C. (2012). The value of Cittaslow as an approach to local sustainable development: A New Zealand perspective. *International Planning Studies*, *17*(4), 353–375. doi:10.1080/13563475.2012.726851

Seo, U. S. (2020). Urban regeneration governance, community organizing, and artists' commitment: A case study of Seongbuk-dong in Seoul. *City. Cultura e Scuola*, *21*, 100328. doi:10.1016/j.ccs.2019.100328

Serdane, Z., Maccarrone-Eaglen, A., & Sharifi, S. (2020). Conceptualising slow tourism: A perspective from Latvia. *Tourism Recreation Research*, *45*(3), 337–350. doi:10.1080/02508281.2020.1726614

Serra, P., Vera, A., Tulla, A. F., & Salvati, L. (2014). Beyond urban-rural dichotomy: Exploring socioeconomic and land-use processes of change in Spain (1991-2011). *Applied Geography (Sevenoaks, England)*, *55*, 71–81. doi:10.1016/j.apgeog.2014.09.005

Servon, L. J., & Pink, S. (2015). Cittaslow: Going Glocal In Spain. *Journal of Urban Affairs*, *37*(3), 327–340. doi:10.1111/juaf.12169

Seyitoğlu, F., & Ivanov, S. (2020). A conceptual study of the strategic role of gastronomy in tourism destinations. *International Journal of Gastronomy and Food Science*, *21*, 21. doi:10.1016/j.ijgfs.2020.100230

Shang, W., Yuan, Q., & Chen, N. (2020). Examining structural relationships among brand experience, existential authenticity, and place attachment in slow tourism destinations. *Sustainability (Switzerland)*, *12*(7), 1–23. doi:10.3390u12072784

Shao, Y. (2014). *Local Identity Identification & Assessment. University of Sheffield, Department of Landscape*. University of Sheffield.

Sharples, L., & Hall, C. M. (2004). The consumption of experiences or the experience of consumption? An introduction to the tourism of taste. In *Food Tourism Around the World*. Routledge.

Sharpley, R. (2014). Host perceptions of tourism: A review of the research. *Tourism Management, 42*(1), 37–49. doi:10.1016/j.tourman.2013.10.007

Sigala, M. (2016). Learning with the market A market approach and framework for developing social entrepreneurship in tourism and hospitality. *International Journal of Contemporary Hospitality Management, 28*(6), 1245–1286. doi:10.1108/IJCHM-06-2014-0285

Silva, A., Lima, F., & Chamusca, P. (2010). Estratégias de eficiência colectiva em territórios de baixa densidade: reflexões a propósito do Minho-Lima e do Tâmega. XII Colóquio Ibérico de Geografia. Porto: Faculdade de Letras da Universidade do Porto.

Silva, F. (2013). *Turismo na natureza como base do desenvolvimento turístico responsável nos Açores* (Unpublished doctoral dissertation). Universidade de Lisboa, Lisboa.

Silva, C., Abrantes, J., & Kastenholz, E. (2018). A Imagem da Serra da Estrela na per-spetiva dos turistas. In *Turismo no Centro de Portugal – Potencialidades e Tendências* (pp. 279–298). Actual.

Silva, F. (2017). Planeamento turístico nos espaços insulares. In F. Silva & J. Umbelino (Eds.), *Planeamento e desenvolvimento turístico* (pp. 467–482). LIDEL.

Simões, O., & Cristóvão, A. (2003). *Turismo em Espaços Rurais e Naturais*. Edições IPC.

Sims, R. (2009). Food, place and authenticity: Local food and the sustainable tourism experience. *Journal of Sustainability, 3*.

Sineiro-García, F., Vázquez-González, I., & García-Arias, A. I. (2014). Mountain Family Farms in Galicia, Spain: Challenges and Strategies. *Mountain Research and Development, 34*(4), 375–385. doi:10.1659/MRD-JOURNAL-D-14-00015.1

Singsomboon, T. (2014). Tourism promotion and the use of local wisdom through creative tourism process. *IJBTS International Journal of Business Tourism and Applied Sciences, 2*(2), 32–37.

Sırım, V. (2012). Çevreyle Bütünleşmiş Bir Yerel Yönetim Örneği Olarak "Sakin Şehir" Hareketi ve Türkiye'nin Potansiyeli. *Tarih Kültür ve Sanat Araştırmaları, 1*(4), 119–131.

Slee, B. (1993), Endogenous development: a concept in search of a theory. *Options Mediterraneenns, A*(23).

Slowmotionporto. (2018). *Yavaş turlar*. https://slowmotionporto.wordpress.com/about/

Smith, S. L., & Xiao, H. (2008). Culinary tourism supply chains: A preliminary examination. *Journal of Travel Research, 46*(3), 289–299. doi:10.1177/0047287506303981

Sofield, T. (2006). Border Tourism and Border Communities: An Overview. *Tourism Geographies, 8*(2), 102–121. doi:10.1080/14616680600585489

Soini, K., & Birkeland, I. (2014). Exploring the scientific discourse on cultural sustainability. *Geoforum, 51*, 213–223. doi:10.1016/j.geoforum.2013.12.001

Sönmez, S., Efe, R., Cürebal, İ., & Soykan, A. (2015). Fethiye ve Çevresinde Zeytinin Yetişmesinde İklimin Rolü. *1.Teke Yöresi Sempozyum*, 622–630.

Sormaz, U., Akmese, H., Gunes, E., & Aras, S. (2016). Gastronomy in tourism. *Procedia Economics and Finance*, 39.

Sortino, A., & Chang, M. (2008). Pattern endogeni di sviluppo dell'agricoltura dicotomizzata: basi teoriche per una nuova politica agricola comunitaria. In *Acqua, agricoltura e ambiente nei nuovi scenari di politica comunitaria*. Franco Angeli.

Soysaldı, A. (2010). Bolu-Göynük Tokalı Örtmeleri. *Motif Academy Journal of Folklore, 3*(5), 150–163.

Spanish Observatory of Rural Tourism and Ecotourism. (2021). *Tendencias turísticas: Ecoturismo y turismo rural* [Tourist trends: Ecotourism and rural tourism]. Retrieved January 22, 2021, from https://www.escapadarural.com/blog/ecoturismo-y-turismo-rural/

Spiekermann, M. (2011). Aleppo, Syria. In E. Rojas & F. Lanzafame (Eds.), *City Development: Experiences in the preservation of ten World Heritage Sites* (pp. 309–329). Inter-American Development Bank.

Spinak, E. (1996). *Diccionario enciclopédico de bibliometría, cienciometría e informetría*. UNESCO.

SREA. (2019). *Inquérito aos Residentes sobre o Turismo nos Açores 2018*. Angra do Heroísmo: Serviço Regional de Estatística dos Açores.

SREA. (2021). *Statistics Azores*. Retrieved Mai 12, 2021, from Serviço Regional de Estatística dos Açores https://srea.azores.gov.pt/default.aspx?lang_id=2

Stilwell, D. (2011). *Indicadores de sustentabilidade aplicados ao contexto do desenvolvimento turístico português*. Dissertação de Mestrado, Faculty of Sciences, University of Lisbon.

Stipanovic, C., & Rudan, E. (2014). Development concept and strategy for creative tourism of the Kvarner destination. *22nd Biennial International Congress Tourism & Hospitality Industry 2014*, 507–517.

Stobart, H. (1994). Flourishing Horns and Enchanted Tubers: Music and Potatoes in Highland Bolivia. *British Journal of Ethnomusicology*, *3*(1), 35–48. doi:10.1080/09681229408567224

Stone, M. J., & Migacz, S. (2016). *The American culinary traveler: Profiles, behaviors, and attitudes*. World Food Travel Association.

Stueve, A. M., Cooks, S., & Drew, D. (2002). *The Geotourism Study: Phase I. Executive Summary*. Travel Industry Association of America.

Stunpo, S. (2011). Siracusa, Italy. In E. Rojas & F. Lanzafame (Eds.), *City Development: Experiences in the preservation of ten World Heritage Sites* (pp. 255–283). Inter-American Development Bank.

Sustainability Index, G. D. S. (2017). *Sustainable destination management trends and insights: A path to a brighter future*. Retrieved from https://www.100carbonneutral.com/assets/sustainable-destination-management-trends-and-insights--a-path-to-a-brighter-future.pdf

Taneja, S., Taneja, P., & Gupta, R. (2011). Researches in corporate social responsibility: A review of shifting focus, paradigms and methodologies. *Journal of Business Ethics*, *101*(3), 343–364. doi:10.100710551-010-0732-6

Tan, S. K., Tan, S. H., Luh, D. B., & Kung, S. F. (2016). Understanding tourist perspectives in creative tourism. *Current Issues in Tourism*, *19*(10), 981–987. doi:10.1080/13683500.2015.1008427

Tan, S.-K., Luh, D.-B., & Kung, S.-F. (2014). A taxonomy of creative tourists in creative tourism. *Tourism Management*, *42*, 248–259. doi:10.1016/j.tourman.2013.11.008

Tarpai, J. (2010). Development of tourism destination management in the Ukrainian-Hungarian border region. *Jelenkori Társadalmi és Gazdasági Folyamatok*, *5*(1-2), 28-32.

Tasci, A. D., Semrad, K. J., & Yilmaz, S. S. (2013). *Community basted tourism: Finding the equilibrium in the COMCEC context*. Academic Press.

Taşkın, H., & Büyüklaca, S. (2012). Kuzugöbeği mantarı (Morchella spp.)'nın Türkiye'deki bölgesel yayılımı. *Alatarım*, *11*(1), 19–24.

Taylor, W., Bartley, D., Goddard, C., Leonard, N., & Welcomme, R. (Eds.). (2016). Freshwater, fish and the future: Proceedings of the global cross-sectoral conference. Food and Agriculture Organization of the United.

TdP - Turismo De Portugal I.P. (2017). *Estratégia Turismo 2027.* Retrieved from https://estrategia.turismodeportugal.pt/

Templeton, R. (1995). *Freshwater Fisheries Management.* Fishing News Book.

Terrasson, F. (2007). *La peur de la nature: Au plus profond de notre inconscient, les vraies causes de la destruction de la nature* (3rd ed.). Sang de la Terre.

The Travel Foundation. (2016). *Taste of Fethiye.* Retrieved from https://www.thetravelfoundation.org.uk/casestudy/taste-of-fethiye/

Tikkanen, I. (2007). Maslow's hierarchy and food tourism in Finland: Five cases. *British Food Journal, 109*(9), 721–734. doi:10.1108/00070700710780698

Timothy, D. J. (2005). Tourism and Political Boundaries in an Age of Globalization. *Anais... Conference on Border Tourism and Community Development Conference.*

Timothy, D. J. (2006). Relationships between tourism and international boundaries. In Tourism and borders. Ashgate.

Timothy, D. J. (2015). Border and tourism. In Encyclopedia of Tourism online. Springer Publishing House.

Timothy, D. J., & Saarinen, J. (2013). Cross-border co-operation and tourism in Europe. In *Trends in European tourism planning and organisation* (pp. 64–74). Channel View. doi:10.21832/9781845414122-009

Timothy, D. J., & Teye, V. B. (2005). Informal sector business travelers in the developing world: A borderlands perspective. *Journal of Tourism Studies, 16*(1), 82–92.

Trejo Casero, V. (2017). *Análisis de las repercusiones económicas y socio-culturales de un festival: El caso del festival de cultura popular "Guitarvera" en Villanueva de la Vera* (Trabajo Final de Máster, Universitat Oberta de Catalunya). Recuperado de: http://hdl.handle.net/10609/67005

Trigal, L. (2005). La Raya Ibérica como espacio de atracción y cooperación. La Raya Ibérica Centro-Sur nº 21 – Papeles de Economia Española, economia de las comunidades autónomas, fundación de las cajas de ahorros, 11-13.

Trigal, L. (2013). Revisión de la terminología sobre la frontera y lo fronterizo. In *Condições de vida, coesão social e cooperação territorial* (pp. 149–167). Âncora Editora/Centro de Estudos Ibéricos.

Turgay, T., & Erkuş Buyruk, P. (2017). Geleneksel Tarakli Evleri Ve Yapilarda Bozulma Nedenleri. *Journal of Advanced Technology Sciences, 6*(2), 35–46.

Turismo de Portugal. (2017). *Estratégia Turismo 2027. Liderar o turismo do futuro.* Turismo de Portugal I.P.

Turismo de Portugal. (2021). *TravelBI. Taxa de sazonalidade.* Retrieved Mai 10, 2021, from Turismo de Portugal, https://travelbi.turismodeportugal.pt/pt-pt/Paginas/PowerBI/Sustentabilidade/taxa-de-sazonalidade.aspx

Turkish Patent and Trademark Office. (2020). *Patents.* Retrieved from https://www.turkpatent.gov.tr/

Ugarte, I. A. (2017). Asociación Española de Casas-Museo y Fundaciones de Escritores. RdM. Revista de Museología: Publicación científica al servicio de la comunidad museológica, 70, 81-83.

UK Music. (2013). *Wish you were here. Music tourism's contribution to the UK economy.* UK Music. www.ukmusic.org/assets/general/LOWRESFORHOMEPRINTING.pdf

Ünal, Ç. (2016). Turizm Coğrafyasinda Yeni Bir Kavram "Yavaş Şehirler". *Doğu Coğrafya Dergisi, 21*(36), 13–28. doi:10.17295/dcd.09928

UN-DESA. (2010). *Trends in sustainable development small island developing states*. United Nations.

UNESCO WHC. (2016). *Operational Guidelines for the Implementation of the World Heritage Convention*. UNESCO. Retrieved: December 09, 2020 from http://whc.unesco.org/en/list/1253

UNESCO. (2006). *Discussion Report of the Planning Meeting for 2006 International Conference on Creative Tourism*. UNESCO.

UNESCO. (2015). *International Geoscience and Geoparks Programme*. Retrieved December 16, 2020, from http://www.unesco.org/new/en/natural-sciences/environment/earth-sciences/international-geoscience-and-geoparks-programme/

UNESCO. (2015). *Statutes of the International Geoscience and Geoparks Programme and Operational Guidelines for UNESCO Global Geoparks*. UNESCO.

UNESCO. (2018). *Basic Texts of the 2003 Convention for the Safeguarding of the Intangible Cultural Heritage*. https://ich.unesco.org/doc/src/2003_Convention_Basic_Texts-_2018_version-EN.pdf

UNESCO. (2021). *Creative Cities Network*. Retrieved from https://en.unesco.org/creative-cities/

UNESCO. (2021). *Intangible Cultural Heritage*. https://ich.unesco.org/

UNESCO. (n.d.). *Creative Cities Network*. Retrieved from https://en.unesco.org/

UN-OHRLLS. (2011). *Small island developing states - Small islands big(ger) takes*. United Nations - Office of the High Representative for the Least Developed Countries, Landlocked Developing Countries and Small Island Developing States.

UNWTO. (2004). *ST-EP Programme Sustainable Tourism Eliminating Poverty*. UNWTO, OMT and IOHBTO.

Urriolagoitia, L., Murillo, D., & Lozano, J.M. (2009). *CSR and SMEs: speech to the implementation, a European perspective, contains major European perspective, contains major policies to bring CSR to SMEs developed in Germany, Austria, Denmark, France, Greece, Italy, Netherlands and United Kingdom, and analyze the underlying central lines*. Ministry of Industry, Tourism and Trade of Spain.

Urry, J. (1995). *Consuming Places*. Routledge.

Uslu, A. (2009). *"Sürdürülebilir Yeşil Kent Fikirleri, Örnekleri ve Türkiye İçin Dersler", XXI. Uluslararası Yapı ve Yaşam Kongresi Bildiriler Kitabı*. TMMOB Mimarlar Odası Yayını.

Üzümcü, O. O., & Koç, B. (2017). Ekolojik turizm kavramı: Fethiye örneği. *International Journal of Agricultural and Natural Sciences, 10*(1), 14–19.

Van Der Ploeg, J. D., & Renting, H. (2000). Impact and Potential: A Comparative Review of European Rural Development Practices. *Sociologia Ruralis, 40*(4), 529–543. doi:10.1111/1467-9523.00165

Vázquez-Carrasco y López-Pérez, M. (2013). Small & medium-sized enterprises and corporate social responsibility: A systematic review of the literature. *Quality & Quantity, 47*(6), 3205–3218. doi:10.100711135-012-9713-4

Veron, S., Mouchet, M., Govaerts, R., Haevermans, T., & Pellens, R. (2019). Vulnerability to climate change of islands worldwide and its impact on the tree of life. *Scientific Reports, 9*(1), 14471. doi:10.103841598-019-51107-x PMID:31597935

Vestrum, I., Rasmussen, E., & Carter, S. (2017). How nascent community enterprises build legitimacy in internal and external environments. *Regional Studies, 51*(11), 1721–1734. doi:10.1080/00343404.2016.1220675

Visit Portugal. (n.d.). *Roteiros literários*. Retrieved from https://www.visitportugal.com/pt-pt/content/roteiros-literarios

Vivo, D. (2018). *Viver no interior vai aliviar IRS*. https://www.dinheirovivo.pt/economia/viver-no-interior-vai-aliviar-irs-12795759.html

Waitt, G., & Gibson, C. (2014). Tourism and Creative Economies. In A. A. Lew, C. M. Hall, & A. M. Williams (Eds.), *The Wiley Blackwell Companion to Tourism* (pp. 230–239). John Wiley & Sons, Ltd. doi:10.1002/9781118474648.ch18

Walker, B., Holling, S., Carpenter, R., & Kinzig, A. (2004). Resilience, adaptability and transformability in social–ecological systems. *Ecology and Society*, *9*(2), 5. doi:10.5751/ES-00650-090205

Wallace, G., & Russel, A. (2004). Eco-cultural tourism as a means for the sustainable development of culturally marginal and environmentally sensitive regions. *Tourist Studies*, *4*(3), 235–254. doi:10.1177/1468797604057326

Watene, K., & Yap, M. (2015). Culture and sustainable development: Indigenous contributions. *Journal of Global Ethics*, *11*(1), 51–55. doi:10.1080/17449626.2015.1010099

Watson, J. L. (2020, November 4). *Cultural globalization: Additional Information*. Britannica: https://www.britannica.com/science/cultural-globalization/The-persistence-of-local-culture

Wearing, S., Wearing, M., & McDonald, M. (2012). *Slow'n Down the Town to Let Nature Grow: Ecotourism, Social Justice and Sustainability. In Slow Tourism:Experiences and Mobilities*. Channel View Publications.

White, P. (2011). *Equitable Access: Remote and Rural Communities "Transport Needs"*. International Transport Forum. Discussion Paper No. 2011-19.

Williams, S., & Lew, A. (2015). *Tourism Geography. Critical understandings of Place, Space and Experience*. Routledge.

Wimmler, C., Hejazi, G., Fernandes, E., Moreira, C., & Connors, S. (2015). Multi-criteria decision support methods for renewable energy systems on islands. *Journal of Clean Energy Technologies*, *3*(3), 185–195. doi:10.7763/JOCET.2015.V3.193

Winter, T. (2004). Landscape, memory and heritage: New Year celebrations at Angkor, Cambodia. *Current Issues in Tourism*, *7*(4-5), 330–345. doi:10.1080/13683500408667989

Woods, M. (2007). Engaging the global countryside: Globalization, hybridity and the reconstitution of rural place. *Progress in Human Geography*, *31*(4), 485–507. doi:10.1177/0309132507079503

World Economic Forum. (2017). *Travel & Tourism Competitiveness Report 2017*. World Economic Forum. Retrieved from http://reports.weforum.org/travel-and-tourism-competitiveness-report-2017/

World Encyclopedia of Puppetry Arts. (2016). *Bonecos de Santo Aleixo*. https://wepa.unima.org/es/bonecos-de-santo-aleixo/

World Tourism Organization. (2018). *Annual Report 2017*. UNWTO.

World Tourism Organization. (2019*). UNWTO Tourism Definitions | Définitions du tourisme de l'OMT | Definiciones de turismo de la OMT*. Retrieved from www.e-unwto.org/doi/book/10.18111/9789284420858

WTO - World Tourism Organization. (2004). Indicators of sustainable development for tourism destinations. A guidebook. WTO.

WTO – World Tourism Organization. (2006). Parte I – Introdução e Conceitos Básicos. OMT, Introdução à Metodologia da Pesquisa em Turismo.

WTO – World Tourism Organization. (2007). *A Practical Guide to Tourism Destination Management*. WTO.

WTO - World Tourism Organization. (2013). *Sustainable Tourism for Development*. European Commission. Retrieved from https://www.e-unwto.org/doi/pdf/10.18111/9789284415496

Wurzburger, R., Aageson, T., Pattakos, A., & Pratt, S. (2009). *Creative tourism: A global conversation: How to provide unique creative experiences for travelers worldwide*. Sunstone Press.

Xunta de Galicia. (2017). *Resolución de la Dirección General del Patrimonio Cultural* [Resolution of the General Directorate of Cultural Heritage]. Retrieved December 16, 2020, from https://www.xunta.gal/dog/Publicados/2017/20171229/AnuncioG0164-261217-0001_es.html

Yazıcıoğlu, T., & Gökçen, J. (1976). *Kuru Üzümlerden Diffüzyon Yoluyla Pekmez (Konsantre) Elde Edilmesi İçin Geliştirilen Bir Yöntem TÜBİTAK Marmara Bilimsel ve Endüstriyel Araştırma Enstitüsü*. Gebze. Yayın No: 11.

Yepes Temiño, J. (2002). *Geomorfología de un sector comprendido entre las provincias de Lugo y Ourense* [Geomorphology of a sector comprised between the Lugo and Ourense provinces]. A Coruña, Spain: Publications of the Laboratorio Xeolóxico de Laxe, Serie Nova Terra.

Yeşilay, S., & Özsolak, İ. (2017). Sicak Cam Üfleme Tekniğinde Lüster Etkisinin Araştırılması. *TÜBAV Bilim Dergisi, 10*(4), 21–28.

Yetiş, Ardıç, Ş. (2015). Kapadokya yemek kültürü ve Mustafapaşa beldesi (Sinasos) örneği. *Journal of Tourism and Gastronomy Studies*, *3*(12), 12–19.

Yılmaz, İ., & Çokal, Z. (2019). *Alternatif Turizm*. Detay Yayıncılık.

Yuan, J., Cai, L. A., Morrison, A. M., & Linton, S. (2005). An analysis of wine festival attendees' motivations: a synergy of wine, travel and special events? *Journal of Vacation Marketing 11*(1).

Yurtseven, H. R., Kaya, O., & Harman, S. (2010). Yavaş hareketi. Ankara: Detay Yayıncılık.

Zakon o kulturnim dobrima (Službeni glasnik RS, 71/94, 52/2011 - dr. zakoni i 99/2011 - dr. zakon) [Law on Cultural Heritage ("Official Gazette of RS", 71/94, 52/ 2011 – other laws and 99/2011 – other laws)] (in Serban).

Zakon o planiranju i izgradnji (Službeni glasnik RS, 72/2009, 81/2009 - ispr., 64/ 2010 - odluka US, 24/2011, 121/2012, 42/2013 - odluka US, 50/2013 - odluka US, 98/2013 - odluka US, 132/2014, 145/2014, 83/2018, 31/2019, 37/2019 i 9/2020) [Law on Planning and Construction (Official Gazette of RS, 72/2009, 81/2009 - April, 64/2010 - Decision CC, 24/ 2011, 121/2012, 42/2013 - Decision CC, 50/2013 - Decision CC, 98/2013 – Decision CC, 132/2014, 145/2014, 83/2018, 31/2019, 37/2019 and 9/2020)] (in Serban).

Zakon o Prostornom planu Republike Srbije od 2010. do 2020. godine (Službeni glasnik RS, 88/2010) [Law on Spatial Plan of the Republic of Serbia from 2010 to 2020 (Official Gazette of RS, 88/2010)] (in Serban).

Zhao, Y., Ponzini, D., & Zhang, R. (2020). The policy networks of heritage-led development in Chinese historic cities: The case of Xi'an's Big Wild Goose Pagoda area. *Habitat International*, *96*, 102106. doi:10.1016/j.habitatint.2019.102106

Zielińska-Szczepkowska, J. (2020). Slow Tourism – The Fashion for Slowness or a Conscious Choice? *Folia Oeconomica Stetinensia*, *20*(2), 468–483. doi:10.2478/foli-2020-0060

Zoğal, V., & Emekli, G. (2017). Yaratıcı turizme kavramsal ve Coğrafi Bir Yaklaşım. *Eğe Coğrafya Dergisi*, *26*(1), 21–34.

Zouros, N. (2004). The European Geoparks Network. Geological heritage protection and local development. *Episodes*, *7*(3), 165–171. doi:10.18814/epiiugs/2004/v27i3/002

Zubeidat, I., Desvarieux, A. R., Salamanca, Y., & Sierra, J. C. (2004). Análisis bibliométrico de la revista journal of sex research (1980-2003). *Universitas Psychologica*, *3*(1), 47–54.

About the Contributors

Gonçalo Poeta Fernandes, Professor at Polytechnic Institute of Guarda – School of Tourism and Hospitality Management/Dept. Tourism and Hotel Industry. PhD in Geography and Regional Planning, specializing in Tourism and Territorial Management, Faculty of Social Sciences and Humanities, New University of Lisbon. Research Areas: Tourism and Development, Planning and Territorial Management, Public Policy and Border Areas, Rural Entrepreneurship, Territory and Tourist Destinations, Mountains and geotourism. Researcher at Centre for Tourism Research, Development and Innovation – CITUR & Interdisciplinary Centre of Social Sciences - CICS:NOVA. Member of the Scientific Council of the Douro International Natural Park and Strategic Council of the Serra da Estrela Natural Park until 2019. Member of the Scientific Committee Estrela UNESCO Global Geopark and coordinator of the Territory, Tourism and Sustainability Research Area in the Science and Education Network for the Sustainability of Estrela Geopark (RCEGE). Member of the Executive Board of the National Mountain Research Network of FCT and Scientific Council of the Iberian Mountain Research Network (RIIM). Publisher of books in the area of Tourism and Heritage with publications in several international books and journals in the area of tourism, destination and heritage management.

* * *

Meltem Altinay Ozdemir is Assistant Professor in the Department of Tour Guiding in the Faculty of Tourism at Mugla Sitki Kocman University, Turkey. She received her bachelor's (2013) and master's (2016) degrees from Gazi University, Turkey, and her PhD in tourism management from Istanbul University, Turkey in 2020. She currently teaches several courses in tourism marketing, tourist behavior, tour guiding, and recreation management, and is also an official tour guide. After working in different travel agencies and hospitality businesses between 2008 and 2013, she worked as a lecturer at Istanbul Arel University between 2015 and 2018, and at Mugla Sitki Kocman University between 2018 and 2021. She has published papers in several journals and conference proceedings. Her research interests are tourism marketing, destination marketing, tourist behavior, creative tourism, and tour guiding.

Patrícia Azevedo has a degree in Tourism and Leisure by the Escola Superior de Turismo e Hotelaria do Instituto Politécnico da Guarda (2018), and specialized in Nature Sports by the same institution (2015). She has developed unpaid internships at various institutions, namely the Biological Park of Serra da Lousã, Barceló Praga Five, Associação Clube Raia Aventura and in the Hotel Alambique de Ouro. She carried out a Curricular Internship at the Geopark Estrela Association in 2018, having joined the executive team, in the area of Tourism and Community Development, in 2019.

Javier Barra Sanz has a degree in Primary Education, completed with a special mention in music education. Master in Social Anthropology, specialising in Cultural Heritage. Nowadays, PhD student on the management, cataloguing and conservation of musical heritage. Born in Cáceres (Spain), at the age of 18 he started to work for the Catholic Church as organ player and it was there that I came into contact with the world of art, historical heritage and music.

Fatima Carvalho is a researcher and member of the Cinturs (Research Centre for Tourism, Sustainability and Well-being), University of Algarve.

Emanuel Castro has a degree in Geography, Master in Geography and Spatial Planning in 2007 from the University of Coimbra. Professor at the Polytechnic Institute of Guarda, between January 2003 and September 2017. He has participated in different research projects in the areas of Tourism, Landscape Analysis, Climatology and Territorial Development. Since 2016, he holds the position of Executive Coordinator of the Geopark Estrela Association.

Rita Arala Chaves is a Professor of Foreign Language at the Polytechnic of Guarda. She holds a degree in Modern Languages and Literature, a Master's Degree in Translation Studies and a PhD in Education. Her core research interests are Foreign Language Education (with a special focus on English for Tourism and Hospitality, autonomy and collaborative learning) and Eça de Queiroz as a Translator.

Zeynep Çokal is a Dr. in Nevsehir Haci Bektas Veli University Tourism Faculty.

Inês Cravino has a degree in Geography and Regional Planning. Teacher of Secondary and Vocational Education in the area of Geography and Tourism. Coordinator of Professional Tourism courses and supervisor of projects and professional internships. Coordinator of on-the-job training, being the link with the tourism companies in the region, the training entities and the funding programs. Develops cooperation projects in low density territories in terms of training, team coordination and promotion of entrepreneurial projects.

Elena De Uña-Álvarez, PhD in Geography, is full professor of Physical Geography at the University of Vigo (Department of History, Art and Geography-Campus of Ourense, Spain). She is a member of the GEAAT Research Group (Studies on Archaeology, Antiquity and Territory) and a collaborating researcher of the TUDISTAR Research Group (Tourism and Social-Territorial Dynamics in Rural areas, Autonomous University of Barcelona, Spain). Her research interest is related to natural and cultural heritage assets as key resources for territorial development, and she has published articles in scientific journals, chapter books and books about it.

Nilgün Demirel graduated from tourist guiding with a bachelor's degree and a master's degree in tourism management. She has PhD degree in tourism management at Nevsehir Hacı Bektas Veli University (Nevsehir, Turkey) in 2019. She has been working as a lecturer at Igdir University (Iğdır, Turkey) since 2015. Her research areas include tourism marketing, sustainable tourism, customer relations and complaint management, as well as tourist guiding. She continues to work on articles and book chapters and conference presentations in national and international fields.

Emre Ergun is lecturer Dr. in Travel and Tourism Services department at Muğla Sıtkı Koçman University. He focuses on social customer relationship management, tourism management, tourist guiding and current issues in tourism.

Silvia Fernandes is an Assistant Professor at the Faculty of Economics of the University of Algarve. Holds a PhD from this university and is member of the Cinturs (Research Centre for Tourism, Sustainability and Well-being).

Hugo Gomes has a PhD in Geology in Quaternary Materials and Cultures, from the University of Trás-os-Montes and Alto Douro since 2015.Graduated in Geology since 2007, Master in Geosciences since 2010, from the University of Coimbra and Since 2007, he has been researcher at the Center for Geosciences at F.C.T.U.C, in which he has been involved in several multidisciplinary research projects in Portugal, Spain, Italy, Brazil, Angola and Ethiopia. Since 2016, he has been part of the executive team of Geopark Estrela Association, performing duties in the area of Science and Research.

Martín Gómez-Ullate, Ph.D. in Sociology (esp. Social Anthropology), researcher at the University of Extremadura, Spain. Member of the research group "Musical Heritage, Culture and Education" funded by the Government of Extremadura. He has coordinated Project Cultour+, an Erasmus+ Strategic Partnership researching on cultural pilgrimage and religious routes to enhance and empower entrepreneurs with business projects in cultural tourism. In the framework of Cultour+ project, he has done team ethnographic research on pilgrimage routes to Santiago, the Via de la Plata and the Camino Interior de Santiago. He is coordinating another two Erasmus+ Strategic Partnerships (Cultrural+ and Learning Villages) in the field of cultural heritage as a factor for rural development. He is author and co-author of several publications in these fields of knowledge as "Cultural Routes & Heritage Tourism & Rural Development" M Gómez-Ullate, LO Siguencia, J Álvarez-García. Book of Proceedings, INBIE, 2018; "Demographic Challenges in Rural Europe and Cases of Resilience Based on Cultural Heritage Management. A Comparative Analysis in Mediterranean Countries Inner Regions" European Countryside, vol.12, no.3, 2020, pp.408-431, "Pilgrimage Tourism and Cultural Route Team Ethnographies in the Iberian Peninsula: A Collaborative Study" (Routledge, 2018),: Rethinking safety issues within the context of pilgrimage routes (CABI, 2018), "Cultour+: Building Professional Skills on Religious and Thermal Tourism" (Springer, 2017).

María Paula Lechuga Sancho is an Assistant Professor in Business and Organization at the University of Cádiz (Spain). She holds a master's degree in business administration (MBA). To date, she has attended more than 50 training actions and have participated in several teaching innovation projects. Her research focuses on i) entrepreneurship, ii) the impact of BPMs on the management of processes and knowledge, and iii) the management of social responsibility in the context of SMEs. She has made multiple contributions to national and international congresses, conferences and seminars and published in leading journals (JCR or SJR). She has participated as a member of organizing committees of several research conferences (national and international) and has taken part in at least 5 research projects funded by different institutions.

Fábio Loureiro has a degree in Biology and Geology (2012) from the University of Aveiro and Master in Geomaterials and Geological Resources (2014) from the Universities of Aveiro and Porto. He worked as a museology technician between 2015 and 2016. Since 2017, integrates the team of Associação Geopark Estrela, performing functions in the area of Geology and Geoconservation.

Milica Maksić Mulalić holds a PhD in urban planning and is a senior planner and a research associate at the Institute of Urban Planning Niš, in Serbia. She is the author of two scientific monographs and 65 papers published in journals and conference proceedings. She has participated in the development of 30 spatial and urban plans and studies as head of an expert team, and as a member of a synthesis and expert team. She has also taken part in the development of programs in the field of housing and local development planning implemented by UN Habitat and the City of Niš. She is a member of the National Commission of the Ministry in charge of construction for expert control of spatial plans, the Supervisory Board of the Serbian Spatial Planners Association, and the Serbian Chamber of Engineers. Her research interests are primarily focused on the institutional framework of planning and spatial development and diverse governance mechanisms.

Ana Mafalda Marques is a Product Manager at General Sales Agent (GSA) in Portugal. Degree in Tourism and Sea and Master in Planning and Management of Nature and Adventure Tourism, researcher and passionate about the concept small is beautiful, who found in slow tourism a way to contribute to a better world, traveling and having meaningful experiences.

Alicia Martín-Navarro is an Assistant Professor in Business Management Department, at Cádiz University, Spain. She holds an MBA master degree and earned her PhD in Social Science. She has written some books about business decision making. Her research interests are in the area of entrepreneurship, Information System for management and Knowledge Management. She has published in JCR journals and presented her research work at various international and national conferences, serving as a reviewer for some of them.

Mario Ossorio is Assistant professor in Management at Università degli Studi della Campania Luigi Vanvitelli. He earned his Masters in Economics and Finance from University of Naples "Federico II" and his PhD in Entrepreneurship and Innovation from University of Naples II. His current research interests include family businesses, enogastronomic tourism and corporate governance. Recently, he published in Management Decision, Journal of Management and Governance, EuroMed Journal of Business, International Journal of Managerial and Financial Accounting.

Elsa Ventura Ramos is a Specialist Professor in Tourism and Leisure at Polytechnic of Guarda, since 2018. Master's Degree in Management and Sustainability in Tourism from the Polytechnic of Guarda. Member of the Pedagogical Council at School of Tourism and Hospitality Management/IPG and Coordinator of under graduated course for Health and Wellness Tourism. Member of Centre for Tourism Research, Development and Innovation CITUR. Lectures under graduated and graduated courses of Tourism and leisure, organization management, destination management and methodology of research in tourism recreation, Extensive professional experience in the area of tourist entertainment and leisure.

João Reis has a PhD in Geography (Urban and Regional Planning at the Institute of Geography and Spatial Planning of the University of Lisbon). Degree in Geography (Geography Teaching at Faculty of Arts, University of Lisbon). Lecturer, coordination of the Department of Tourism and Leisure and Direction of the degree in Tourism Management at Higher Institute for Tourism and Hotel Studies. Researcher at the Center for Tourism Research, Development and Innovation (CiTUR). Collaborates with other academic institutions, provides tourism advisory services and participates in seminars and conferences (tourism planning, territorial development and GIS).

Manuel Rodríguez Palacios (Cáceres, Spain) graduated in Primary Education specialising in music from the University of Extremadura (UEx) and holds a master's degree in Social Anthropology from the University of Extremadura (UEx). On the other hand, he has taken training courses in Bilingual Methodology; Physical Activity and healthy habits as educational resources for children and young people; and Monitor of free time for children and young people. Currently enrolled in the Doctorate programme in Heritage (Interuniversity programme of the Universities of Cordoba, Extremadura, Huelva and Jaen focused on the study of Cultural and Natural Heritage), whose thesis focuses on the role of choirs in the didactics of musical expression. Choral groups as non-formal music schools.

Anabela Sardo currently works at the Research Unit for Inland Development (UDI), Polytechnic Institute of Guarda, in Portugal. Her present projects are "The literary work of the Portuguese writer Ana Teresa Perreira" and "Azorean Writers". She collaborates on the project "Tourism Studies". She is also a researcher collaborator in the CI&DEI, an interdisciplinary network research centre of the Polytechnic Institute of Viseu and in the Topus Research Group.

Francisco Silva completed his PhD in Geography, specializing in Urban and Regional Planning (IGOT-UL), and holds a specialist degree in Travel and Leisure (ESHTE), a Master's degree in Science and Geographic Information Systems (ISEGI-UNL) and a Degree in Geography (FL-UL). He's a professor at the Estoril Higher Institute for Tourism and Hotel Studies (ESHTE), where he coordinates the Master Program in Tourism. Researcher at the Centre for Geographical Studies (CEG, IGOT-UL). His professional activity has been quite varied. Has been photojournalist, works as a consultant in tourism, and as an instructor and coach in several adventure activities including mountain climbing, canyoning and rope work courses. He has been working in tourism planning, namely as a consultant in some tourism strategic plans and has developed projects applied to nature tourism. Author of several articles and books on tourism, adventure sports and geography.

Montserrat Villarino-Pérez, PhD in Geography, has been full professor of Human Geography until 2017 at the University of Santiago de Compostela (Spain). She is a collaborator researcher of the TUDISTAR Group (Autonomous University of Barcelona, Spain). Her research interest is mainly related to the revitalization and processes of change that affect inland and rural territories.

Index

F

G

H

I

L

M

N

O

P

R

S

T

U

W

www.ingramcontent.com/pod-product-compliance
Lightning Source LLC
LaVergne TN
LVHW060424300926
849162LV00006B/1522